Causal
Inference

T0315134

The
Mixtape

Scott
Cunningham

Causal Inference

The Mixtape

Yale
UNIVERSITY
PRESS

NEW HAVEN
& LONDON

Copyright © 2021 by Scott Cunningham.

All rights reserved.

This book may not be reproduced, in whole or in part, including illustrations, in any form (beyond that copying permitted by Sections 107 and 108 of the U.S. Copyright Law and except by reviewers for the public press), without written permission from the publishers.

Yale University Press books may be purchased in quantity for educational, business, or promotional use. For information, please e-mail sales.press@yale.edu (U.S. office) or sales@yaleup.co.uk (U.K. office).

Set in Roboto type by Newgen.

Title-page illustration: iStock.com/2p2play.

Printed in Great Britain by Clays Ltd, Elcograf S.p.A.

Library of Congress Control Number: 2020939011

ISBN 978-0-300-25168-5 (pbk. : alk. paper).

A catalogue record for this book is available from the British Library.

10 9 8 7 6 5 4

MIX
Paper | Supporting
responsible forestry
FSC
www.fsc.org   FSC® C018072

To my son, Miles, one of my favorite people.
I love you. You've tagged my head and heart.

Contents

Acknowledgments

Just as it takes a village to raise a child, it takes many people to help me write a book like this. The people to whom I am indebted range from the scholars whose work has inspired me—Alberto Abadie, Josh Angrist, Susan Athey, David Card, Esther Duflo, Guido Imbens, Alan Krueger, Robert LaLonde, Steven Levitt, Alex Tabarrok, John Snow, and many more—to friends, mentors, and colleagues.

I am most indebted first of all to my former advisor, mentor, coauthor, and friend Christopher Cornwell. I probably owe Chris my entire career. He invested in me and taught me econometrics as well as empirical designs more generally when I was a grad student at the University of Georgia. I was brimming with a million ideas and he somehow managed to keep me focused. Always patient, always holding me to high standards, always believing I could achieve them, always trying to help me correct fallacious reasoning and poor knowledge of econometrics. I would also like to thank Alvin Roth, who has encouraged me over the last decade in my research. That encouragement has buoyed me throughout my career repeatedly. Finally, I'd like to thank Judea Pearl for inviting me to UCLA for a day of discussions around an earlier draft of the Mixtape and helping improve it.

But a book like this is also due to countless conversations with friends over the years, as well as reading carefully their own work and learning from them. People like Mark Hoekstra, Rebecca Thornton, Paul Goldsmith-Pinkham, Mark Anderson, Greg DeAngelo, Manisha Shah, Christine Durrance, Melanie Guldi, Caitlyn Myers, Bernie Black, Keith Finlay, Jason Lindo, Andrew Goodman-Bacon, Pedro Sant'anna, Andrew Baker, Rachael Meager, Nick Papageorge, Grant McDermott, Salvador Lozano, Daniel Millimet, David Jaeger, Berk Ozler, Erin Hengel, Alex Bartik, Megan Stevenson, Nick Huntington-Klein, Peter Hull, as well as many many more on #EconTwitter, a vibrant community of social scientists on Twitter.

I would also like to thank my two students Hugo Rodrigues and Terry Tsai. Hugo and Terry worked tirelessly to adapt all of my blue collar Stata code into R programs. Without them, I would have been lost. I would also like to thank another student, Brice Green, for early trials of the code to confirm it worked by non-authors. Blagoj Gegov helped create many of the figures in Tikz. I would like to thank Ben Chidmi for adapting a simulation from R into Stata, and Yuki Yanai for allowing me to use his R code for a simulation. Thank you to Zeljko Hrcek for helping make amendments to the formatting of the LaTeX when I was running against deadline. And thank you to my friend Seth Hahne for creating several beautiful illustrations in the book. I would also like to thank Seth Ditchik for believing in this project, my agent Lindsay Edgecombe for her encouragement and work on my behalf, and Yale University Press. And to my other editor, Charlie Clark, who must have personally read this book fifty times and worked so hard to improve it. Thank you, Charlie. And to the musicians who have sung the soundtrack to my life, thanks to Chance, Drake, Dr. Dre, Eminem, Lauryn Hill, House of Pain, Jay-Z, Mos Def, Notorious B.I.G., Pharcyde, Tupac Shakur, Tribe, Kanye West, Young MC, and many others.

I would like to thank the health economist Marcelo Perraillon, whose insightful approach to using simulated variables and figures to illustrate estimation challenges with regression discontinuity shaped my own pedagogy. I also acknowledge him for introducing me to the "cmogram" command in Stata.

Finally, I'd like to thank my close friends, Baylor colleagues, students, and family for tolerating my eccentric enthusiasm for causal inference and economics for years. I have benefited tremendously from many opportunities and resources, and for that and other things I am very grateful.

This book, and the class it was based on, is a distillation of countless journal articles, books, as well as classes I have taken in person and studied from afar. It is also a product of numerous conversations I've had with colleagues, students and teachers for many years. I have attempted to give credit where credit is due. All errors in this book were caused entirely by me, not the people listed above.

Causal
Inference

Introduction

My path to economics was not linear. I didn't major in economics, for instance. I didn't even take an economics course in college. I majored in English, for Pete's sake. My ambition was to become a poet. But then I became intrigued with the idea that humans can form plausible beliefs about causal effects even without a randomized experiment. Twenty-five years ago, I wouldn't have had a clue what that sentence even meant, let alone how to do such an experiment. So how did I get here? Maybe you would like to know how I got to the point where I felt I needed to write this book. The TL;DR version is that I followed a windy path from English to causal inference.[1] First, I fell in love with economics. Then I fell in love with empirical research. Then I noticed that a growing interest in causal inference had been happening in me the entire time. But let me tell the longer version.

I majored in English at the University of Tennessee at Knoxville and graduated with a serious ambition to become a professional poet. But, while I had been successful writing poetry in college, I quickly realized that finding the road to success beyond that point was probably not realistic. I was newly married, with a baby on the way, and working as a qualitative research analyst doing market research. Slowly, I had stopped writing poetry altogether.[2]

My job as a qualitative research analyst was eye opening, in part because it was my first exposure to empiricism. My job was to do "grounded theory"—a kind of inductive approach to generating explanations of human behavior based on observations. I did this by running focus groups and conducting in-depth interviews, as well as

---

1 "Too long; didn't read."

2 Rilke said you should quit writing poetry when you can imagine yourself living without it [Rilke, 2012]. I could imagine living without poetry, so I took his advice and quit. Interestingly, when I later found economics, I went back to Rilke and asked myself if I could live without it. This time, I decided I couldn't, or wouldn't—I wasn't sure which. So I stuck with it and got a PhD.

through other ethnographic methods. I approached each project as an opportunity to understand why people did the things they did (even if what they did was buy detergent or pick a cable provider). While the job inspired me to develop my own theories about human behavior, it didn't provide me a way of falsifying those theories.

I lacked a background in the social sciences, so I would spend my evenings downloading and reading articles from the Internet. I don't remember how I ended up there, but one night I was on the University of Chicago Law and Economics working paper series website when a speech by Gary Becker caught my eye. It was his Nobel Prize acceptance speech on how economics applies to all of human behavior [Becker, 1993], and reading it changed my life. I thought economics was about stock markets and banks until I read that speech. I didn't know economics was an engine that one could use to analyze all of human behavior. This was overwhelmingly exciting, and a seed had been planted.

But it wasn't until I read an article on crime by Lott and Mustard [1997] that I became truly enamored of economics. I had no idea that there was an empirical component where economists sought to estimate causal effects with quantitative data. A coauthor of that paper was David Mustard, then an associate professor of economics at the University of Georgia, and one of Gary Becker's former students. I decided that I wanted to study with Mustard, and so I applied to the University of Georgia's doctoral program in economics. I moved to Athens, Georgia, with my wife, Paige, and our infant son, Miles, and started classes in the fall of 2002.

After passing my first-year comprehensive exams, I took Mustard's labor economics field class and learned about a variety of topics that would shape my interests for years. These topics included the returns to education, inequality, racial discrimination, crime, and many other fascinating topics in labor. We read many, many empirical papers in that class, and afterwards I knew that I would need a strong background in econometrics to do the kind of research I cared about. In fact, I decided to make econometrics my main field of study. This led me to work with Christopher Cornwell, an econometrician and labor economist at Georgia. I learned a lot from Chris, both about econometrics and about research itself. He became a mentor, coauthor, and close friend.

Econometrics was difficult. I won't even pretend I was good at it. I took all the econometrics courses offered at the University of Georgia, some more than once. They included classes covering topics like probability and statistics, cross-sections, panel data, time series, and qualitative dependent variables. But while I passed my field exam in econometrics, I struggled to understand econometrics at a deep level. As the saying goes, I could not see the forest for the trees. Something just wasn't clicking.

I noticed something, though, while I was writing the third chapter of my dissertation that I hadn't noticed before. My third chapter was an investigation of the effect of abortion legalization on the cohort's future sexual behavior [Cunningham and Cornwell, 2013]. It was a revisiting of Donohue and Levitt [2001]. One of the books I read in preparation for my study was Levine [2004], which in addition to reviewing the theory of and empirical studies on abortion had a little table explaining the difference-in-differences identification strategy. The University of Georgia had a traditional econometrics pedagogy, and most of my field courses were theoretical (e.g., public economics, industrial organization), so I never really had heard the phrase "identification strategy," let alone "causal inference." Levine's simple difference-in-differences table for some reason opened my eyes. I saw how econometric modeling could be used to isolate the causal effects of some treatment, and that led to a change in how I approach empirical problems.

What Is Causal Inference?

My first job out of graduate school was as an assistant professor at Baylor University in Waco, Texas, where I still work and live today. I was restless the second I got there. I could feel that econometrics was indispensable, and yet I was missing something. But what? It was a theory of causality. I had been orbiting that theory ever since seeing that difference-in-differences table in Levine [2004]. But I needed more. So, desperate, I did what I always do when I want to learn something new—I developed a course on causality to force myself to learn all the things I didn't know.

I named the course Causal Inference and Research Design and taught it for the first time to Baylor master's students in 2010. At the time, I couldn't really find an example of the sort of class I was looking for, so I cobbled together a patchwork of ideas from several disciplines and authors, like labor economics, public economics, sociology, political science, epidemiology, and statistics. You name it. My class wasn't a pure econometrics course; rather, it was an applied empirical class that taught a variety of contemporary research designs, such as difference-in-differences, and it was filled with empirical replications and readings, all of which were built on the robust theory of causality found in Donald Rubin's work as well as the work of Judea Pearl. This book and that class are in fact very similar to one another.[3]

So how would I define causal inference? Causal inference is the leveraging of theory and deep knowledge of institutional details to estimate the impact of events and choices on a given outcome of interest. It is not a new field; humans have been obsessing over causality since antiquity. But what is new is the progress we believe we've made in estimating causal effects both inside and outside the laboratory. Some date the beginning of this new, modern causal inference to Fisher [1935], Haavelmo [1943], or Rubin [1974]. Some connect it to the work of early pioneers like John Snow. We should give a lot of credit to numerous highly creative labor economists from the late 1970s to late 1990s whose ambitious research agendas created a revolution in economics that continues to this day. You could even make an argument that we owe it to the Cowles Commission, Philip and Sewall Wright, and the computer scientist Judea Pearl.

But however you date its emergence, causal inference has now matured into a distinct field, and not surprisingly, you're starting to see more and more treatments of it as such. It's sometimes reviewed in a lengthy chapter on "program evaluation" in econometrics textbooks

---

3 I decided to write this book for one simple reason: I didn't feel that the market had provided the book that I needed for my students. So I wrote this book for my students and me so that we'd all be on the same page. This book is my best effort to explain causal inference to myself. I felt that if I could explain causal inference to myself, then I would be able to explain it to others too. Not thinking the book would have much value outside of my class, I posted it to my website and told people about it on Twitter. I was surprised to learn that so many people found the book helpful.

[Wooldridge, 2010], or even given entire book-length treatments. To name just a few textbooks in the growing area, there's Angrist and Pischke [2009], Morgan and Winship [2014], Imbens and Rubin [2015], and probably a half dozen others, not to mention numerous, lengthy treatments of specific strategies, such as those found in Angrist and Krueger [2001] and Imbens and Lemieux [2008]. The market is quietly adding books and articles about identifying causal effects with data all the time.

So why does *Causal Inference: The Mixtape* exist? Well, to put it bluntly, a readable introductory book with programming examples, data, and detailed exposition didn't exist until this one. My book is an effort to fill that hole, because I believe what researchers really need is a guide that takes them from knowing almost nothing about causal inference to a place of competency. Competency in the sense that they are conversant and literate about what designs can and cannot do. Competency in the sense that they can take data, write code and, using theoretical and contextual knowledge, implement a reasonable design in one of their own projects. If this book helps someone do that, then this book will have had value, and that is all I can and should hope for.

But what books out there do I like? Which ones have inspired this book? And why don't I just keep using them? For my classes, I mainly relied on Morgan and Winship [2014], Angrist and Pischke [2009], as well as a library of theoretical and empirical articles. These books are in my opinion definitive classics. But they didn't satisfy my needs, and as a result, I was constantly jumping between material. Other books were awesome but not quite right for me either. Imbens and Rubin [2015] cover the potential outcomes model, experimental design, and matching and instrumental variables, but not directed acyclic graphical models (DAGs), regression discontinuity, panel data, or synthetic control. Morgan and Winship [2014] cover DAGs, the potential outcomes model, and instrumental variables, but have too light a touch on regression discontinuity and panel data for my tastes. They also don't cover synthetic control, which has been called the most important innovation in causal inference of the last 15 years by Athey and Imbens [2017b]. Angrist and Pischke [2009] is very close to what I need but does not include anything on synthetic control or on the graphical models that I find so critically useful. But maybe most importantly, Imbens and Rubin

[2015], Angrist and Pischke [2009], and Morgan and Winship [2014] do not provide *any* practical programming guidance, and I believe it is in replication and coding that we gain knowledge in these areas.[4]

This book was written with a few different people in mind. It was written first and foremost for *practitioners*, which is why it includes easy-to-download data sets and programs. It's why I have made several efforts to review papers as well as replicate the models as much as possible. I want readers to understand this field, but as important, I want them to feel empowered so that they can use these tools to answer their own research questions.

Another person I have in mind is the experienced social scientist who wants to retool. Maybe these are people with more of a theoretical bent or background, or maybe they're people who simply have some holes in their human capital. This book, I hope, can help guide them through the modern theories of causality so common in the social sciences, as well as provide a calculus in directed acyclic graphical models that can help connect their knowledge of theory with estimation. The DAGs in particular are valuable for this group, I think.

A third group that I'm focusing on is the nonacademic person in industry, media, think tanks, and the like. Increasingly, knowledge about causal inference is expected throughout the professional world. It is no longer simply something that academics sit around and debate. It is crucial knowledge for making business decisions as well as for interpreting policy.

Finally, this book is written for people very early in their careers, be they undergraduates, graduate students, or newly minted PhDs. My hope is that this book can give them a jump start so that they don't have to meander, like many of us did, through a somewhat labyrinthine path to these methods.

Do Not Confuse Correlation with Causality

It is very common these days to hear someone say "correlation does not mean causality." Part of the purpose of this book is to help

4 Although Angrist and Pischke [2009] provides an online data warehouse from dozens of papers, I find that students need more pedagogical walk-throughs and replications for these ideas to become concrete and familiar.

readers be able to understand exactly why correlations, particularly in observational data, are unlikely to be reflective of a causal relationship. When the rooster crows, the sun soon after rises, but we know the rooster didn't cause the sun to rise. Had the rooster been eaten by the farmer's cat, the sun still would have risen. Yet so often people make this kind of mistake when naively interpreting simple correlations.

But weirdly enough, sometimes there are causal relationships between two things and yet *no observable correlation*. Now that is definitely strange. How can one thing cause another thing without any discernible correlation between the two things? Consider this example, which is illustrated in Figure 1. A sailor is sailing her boat across the lake on a windy day. As the wind blows, she counters by turning the rudder in such a way so as to exactly offset the force of the wind. Back and forth she moves the rudder, yet the boat follows a straight line across the lake. A kindhearted yet naive person with no knowledge of wind or boats might look at this woman and say, "Someone get this sailor a new rudder! Hers is broken!" He thinks this because he cannot see any relationship between the movement of the rudder and the direction of the boat.

Figure 1. No correlation doesn't mean no causality. Artwork by Seth Hahne © 2020.

But does the fact that he cannot see the relationship mean there isn't one? Just because there is no observable relationship does not mean there is no causal one. Imagine that instead of perfectly countering the wind by turning the rudder, she had instead flipped a coin— heads she turns the rudder left, tails she turns the rudder right. What do you think this man would have seen if she was sailing her boat according to coin flips? If she *randomly* moved the rudder on a windy day, then he would see a sailor zigzagging across the lake. Why would he see the relationship if the movement were randomized but not be able to see it otherwise? Because the sailor is *endogenously* moving the rudder in response to the unobserved wind. And as such, the relationship between the rudder and the boat's direction is canceled—even though there is a causal relationship between the two.

This sounds like a silly example, but in fact there are more serious versions of it. Consider a central bank reading tea leaves to discern when a recessionary wave is forming. Seeing evidence that a recession is emerging, the bank enters into open-market operations, buying bonds and pumping liquidity into the economy. Insofar as these actions are done optimally, these open-market operations will show no relationship whatsoever with actual output. In fact, in the ideal, banks may engage in aggressive trading in order to stop a recession, and we would be unable to see any evidence that it was working *even though it was*!

Human beings engaging in optimal behavior are the main reason correlations almost never reveal causal relationships, because rarely are human beings acting randomly. And as we will see, it is the presence of randomness that is crucial for identifying causal effect.

Optimization Makes Everything Endogenous

Certain presentations of causal inference methodologies have sometimes been described as atheoretical, but in my opinion, while some practitioners seem comfortable flying blind, the actual methods employed in causal designs are always deeply dependent on theory and local institutional knowledge. It is my firm belief, which I will emphasize over and over in this book, that without prior knowledge, estimated causal effects are rarely, if ever, believable. Prior knowledge is *required* in order to justify any claim of a causal finding. And economic theory also highlights why causal inference is necessarily a thorny task. Let me explain.

There's broadly thought to be two types of data. There's experimental data and non-experimental data. The latter is also sometimes called *observational* data. Experimental data is collected in something akin to a laboratory environment. In a traditional experiment, the researcher participates actively in the process being recorded. It's more difficult to obtain data like this in the social sciences due to feasibility, financial cost, or moral objections, although it is more common now than was once the case. Examples include the Oregon Medicaid Experiment, the RAND health insurance experiment, the field experiment movement inspired by Esther Duflo, Michael Kremer, Abhijit Banerjee, and John List, and many others.

Observational data is usually collected through surveys in a retrospective manner, or as the by-product of some other business activity ("big data"). In many observational studies, you collect data about what happened previously, as opposed to collecting data as it happens, though with the increased use of web scraping, it may be possible to get observational data closer to the exact moment in which some action occurred. But regardless of the timing, the researcher is a passive actor in the processes creating the data itself. She observes actions and results but is not in a position to interfere with the environment in which the units under consideration exist. This is the most common form of data that many of us will ever work with.

Economic theory tells us we should be suspicious of correlations found in observational data. In observational data, correlations are almost certainly not reflecting a causal relationship because the variables were endogenously chosen by people who were making decisions they thought were best. In pursuing some goal while facing constraints, they chose certain things that created a spurious correlation with other things. And we see this problem reflected in the potential outcomes model itself: a correlation, in order to be a measure of a causal effect, must be based on a choice that was made independent of the potential outcomes under consideration. Yet if the person is making some choice *based* on what she thinks is best, then it necessarily is based on potential outcomes, and the correlation does not remotely satisfy the conditions we need in order to say it is causal. To put it as bluntly as I can, economic theory says choices are endogenous, and therefore since they are, the correlations between those choices and outcomes in the aggregate will rarely, if ever, represent a causal effect.

Now we are veering into the realm of epistemology. Identifying causal effects involves assumptions, but it also requires a particular kind of belief about the work of scientists. Credible and valuable research requires that we believe that it is more important to do our work *correctly* than to try and achieve a certain outcome (e.g., confirmation bias, statistical significance, asterisks). The foundations of scientific knowledge are scientific methodologies. True scientists do not collect evidence in order to prove what they want to be true or what others want to believe. That is a form of deception and manipulation called *propaganda*, and propaganda is not science. Rather, scientific methodologies are devices for forming a particular kind of belief. Scientific methodologies allow us to accept unexpected, and sometimes undesirable, answers. They are process oriented, not outcome oriented. And without these values, causal methodologies are also not believable.

Example: Identifying Price Elasticity of Demand

One of the cornerstones of scientific methodologies is empirical analysis.[5] By empirical analysis, I mean the use of data to test a theory or to estimate a relationship between variables. The first step in conducting an empirical economic analysis is the careful formulation of the question we would like to answer. In some cases, we would like to develop and test a formal economic model that describes mathematically a certain relationship, behavior, or process of interest. Those models are valuable insofar as they both describe the phenomena of

---

5 It is not the only cornerstone, or even necessarily the most important cornerstone, but empirical analysis has always played an important role in scientific work.

interest and make falsifiable (testable) predictions. A prediction is fal-sifiable insofar as we can evaluate, and potentially reject, the prediction with data.[6] A model is the framework with which we describe the rela-tionships we are interested in, the intuition for our results, and the hypotheses we would like to test.[7]

After we have specified a model, we turn it into what is called an econometric model, which can be estimated directly with data. One clear issue we immediately face is regarding the functional form of the model, or how to describe the relationships of the variables we are interested in through an equation. Another important issue is how we will deal with variables that cannot be directly or reasonably observed by the researcher, or that cannot be measured very well, but which play an important role in our model.

A generically important contribution to our understanding of causal inference is the notion of comparative statics. Comparative statics are theoretical descriptions of causal effects contained within the model. These kinds of comparative statics are always based on the idea of *ceteris paribus*—or "all else constant." When we are trying to describe the causal effect of some intervention, for instance, we are always assuming that the other relevant variables in the model are not changing. If they were changing, then they would be correlated with the variable of interest and it would confound our estimation.[8]

To illustrate this idea, let's begin with a basic economic model: supply and demand equilibrium and the problems it creates for esti-mating the price elasticity of demand. Policy-makers and business managers have a natural interest in learning the price elasticity of

6 You can also obtain a starting point for empirical analysis through an intuitive and less formal reasoning process. But economics favors formalism and deductive methods.

7 Scientific models, be they economic ones or otherwise, are abstract, not realis-tic, representations of the world. That is a strength, not a weakness. George Box, the statistician, once quipped that "all models are wrong, but some are useful." A model's usefulness is its ability to unveil hidden secrets about the world. No more and no less.

8 One of the things implied by *ceteris paribus* that comes up repeatedly in this book is the idea of covariate balance. If we say that everything is the same except for the movement of one variable, then everything is the same on both sides of that vari-able's changing value. Thus, when we invoke *ceteris paribus*, we are implicitly invoking covariate balance—both the observable and the unobservable covariates.

demand because knowing it enables firms to maximize profits and governments to choose optimal taxes, and whether to restrict quantity altogether [Becker et al., 2006]. But the problem is that we do not observe demand curves, because demand curves are theoretical objects. More specifically, a demand curve is a collection of paired potential outcomes of price and quantity. We observe *price and quantity equilibrium values*, not the potential price and potential quantities along the entire demand curve. Only by tracing out the potential outcomes along a demand curve can we calculate the elasticity.

To see this, consider this graphic from Philip Wright's Appendix B [Wright, 1928], which we'll discuss in greater detail later (Figure 2). The price elasticity of demand is the ratio of percentage changes in quantity to price *for a single demand curve*. Yet, when there are shifts in supply and demand, a sequence of quantity and price pairs emerges in history that reflect neither the demand curve nor the supply curve. In fact, connecting the points does not reflect any meaningful or useful object.

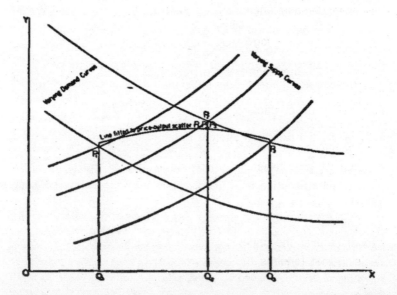

Figure 2. Wright's graphical demonstration of the identification problem. Figure from Wright, P. G. (1928). *The Tariff on Animal and Vegetable Oils*. The Macmillan Company.

The price elasticity of demand is the solution to the following equation:

$$\epsilon = \frac{\partial \log Q}{\partial \log P}$$

But in this example, the change in P is *exogenous*. For instance, it holds supply fixed, the prices of other goods fixed, income fixed, preferences fixed, input costs fixed, and so on. In order to estimate the price elasticity of demand, we need changes in P that are completely and utterly independent of the otherwise normal determinants of supply and the other determinants of demand. Otherwise we get shifts in either supply or demand, which creates new pairs of data for which any correlation between P and Q will not be a measure of the elasticity of demand.

The problem is that the elasticity is an important object, and we need to know it, and therefore we need to solve this problem. So given this theoretical object, we must write out an econometric model as a starting point. One possible example of an econometric model would be a linear demand function:

$$\log Q_d = \alpha + \delta \log P + \gamma X + u$$

where α is the intercept, δ is the elasticity of demand, X is a matrix of factors that determine demand like the prices of other goods or income, γ is the coefficient on the relationship between X and Q_d, and u is the error term.[9]

Foreshadowing the content of this mixtape, we need two things to estimate price elasticity of demand. First, we need numerous rows of data on price and quantity. Second, we need for the variation in price in our imaginary data set to be independent of u. We call this kind of independence *exogeneity*. Without both, we cannot recover the price elasticity of demand, and therefore any decision that requires that information will be based on stabs in the dark.

---

9 More on the error term later.

Conclusion

This book is an introduction to research designs that can recover causal effects. But just as importantly, it provides you with hands-on practice to implement these designs. Implementing these designs means writing code in some type of software. I have chosen to illustrate these designs using two popular software languages: Stata (most commonly used by economists) and R (most commonly used by everyone else).

The book contains numerous empirical exercises illustrated in the Stata and R programs. These exercises are either simulations (which don't need external data) or exercises requiring external data. The data needed for the latter have been made available to you at Github. The Stata examples will download files usually at the start of the program using the following command: use https://github.com/scunning1975/mixtape/raw/master/DATAFILENAME.DTA, where DATAFILENAME.DTA is the name of a particular data set.

For R users, it is a somewhat different process to load data into memory. In an effort to organize and clean the code, my students Hugo Sant'Anna and Terry Tsai created a function to simplify the data download process. This is partly based on a library called haven, which is a package for reading data files. It is secondly based on a set of commands that create a function that will then download the data directly from Github.[10]

Some readers may not be familiar with either Stata or R but nonetheless wish to follow along. I encourage you to use this opportunity to invest in learning one or both of these languages. It is beyond the scope of this book to provide an introduction to these languages, but fortunately, there are numerous resources online. For instance, Christopher Baum has written an excellent introduction to Stata at https://fmwww.bc.edu/GStat/docs/StataIntro.pdf. Stata is popular among microeconomists, and given the amount of coauthoring involved in modern economic research, an argument could be

---

10 This was done solely for aesthetic reasons. Often the URL was simply too long for the margins of the book otherwise.

made for investing in it solely for its ability to solve basic coordination problems between you and potential coauthors. But a downside to Stata is that it is proprietary and must be purchased. And for some people, that may simply be too big of a barrier—especially for anyone simply wanting to follow along with the book. R on the other hand is open-source and free. Tutorials on Basic R can be found at https://cran.r-project.org/doc/contrib/Paradis-rdebuts_en.pdf, and an introduction to Tidyverse (which is used throughout the R programming) can be found at https://r4ds.had.co.nz. Using this time to learn R would likely be well worth your time.

Perhaps you already know R and want to learn Stata. Or perhaps you know Stata and want to learn R. Then this book may be helpful because of the way in which both sets of code are put in sequence to accomplish the same basic tasks. But, with that said, in many situations, although I have tried my best to reconcile results from Stata and R, I was not always able to do so. Ultimately, Stata and R are different programming languages that sometimes yield different results because of different optimization procedures or simply because the programs are built slightly differently. This has been discussed occasionally in articles in which authors attempt to better understand what accounts for the differing results. I was not always able to fully reconcile different results, and so I offer the two programs as simply alternative approaches. You are ultimately responsible for anything you do on your own using either language for your research. I leave it to you ultimately to understand the method and estimating procedure contained within a given software and package.

In conclusion, simply finding an association between two variables might be suggestive of a causal effect, but it also might not. Correlation doesn't mean causation unless key assumptions hold. Before we start digging into the causal methodologies themselves, though, I need to lay down a foundation in statistics and regression modeling. Buckle up! This is going to be fun.

Probability and Regression Review

Numbers is hardly real and they never have feelings. But you push too hard, even numbers got limits.

Mos Def

Basic probability theory. In practice, causal inference is based on statistical models that range from the very simple to extremely advanced. And building such models requires some rudimentary knowledge of probability theory, so let's begin with some definitions. A random process is a process that can be repeated many times with different outcomes each time. The sample space is the set of all the possible outcomes of a random process. We distinguish between discrete and continuous random processes Table 1 below. Discrete processes produce, integers, whereas continuous processes produce fractions as well.

We define independent events two ways. The first refers to logical independence. For instance, two events occur but there is no reason to believe that the two events affect each other. When it is assumed that they *do* affect each other, this is a logical fallacy called *post hoc ergo propter hoc*, which is Latin for "after this, therefore because of this." This fallacy recognizes that the temporal ordering of events is not sufficient to be able to say that the first thing caused the second.

Table 1. Examples of discrete and continuous random processes.

Description	Type	Potential outcomes
12-sided die	Discrete	1, 2, 3, 4, 5, 6, 7, 8, 9, 10, 11, 12
Coin	Discrete	Heads, Tails
Deck of cards	Discrete	$2 \diamond, 3 \diamond, \ldots$ King $\heartsuit$, Ace $\heartsuit$
Gas prices	Continuous	$P \geq 0$

The second definition of an independent event is statistical independence. We'll illustrate the latter with an example from the idea of sampling with and without replacement. Let's use a randomly shuffled deck of cards for an example. For a deck of 52 cards, what is the probability that the first card will be an ace?

$$\text{Pr(Ace)} = \frac{\text{Count Aces}}{\text{Sample Space}} = \frac{4}{52} = \frac{1}{13} = 0.077$$

There are 52 possible outcomes in the sample space, or the set of all possible outcomes of the random process. Of those 52 possible outcomes, we are concerned with the frequency of an ace occurring. There are four aces in the deck, so $\frac{4}{52} = 0.077$.

Assume that the first card was an ace. Now we ask the question again. If we shuffle the deck, what is the probability the next card drawn is also an ace? It is no longer $\frac{1}{13}$ because we did not sample with replacement. We sampled *without* replacement. Thus the new probability is

$$\text{Pr}\left(\text{Ace} \mid \text{Card 1} = \text{Ace}\right) = \frac{3}{51} = 0.059$$

Under sampling without replacement, the two events—ace on Card 1 and an ace on Card 2 if Card 1 was an ace—aren't independent events. To make the two events independent, you would have to put the ace back and shuffle the deck. So two events, A and B, are independent if and only if:

$$\text{Pr}(A \mid B) = \text{Pr}(A)$$

An example of two independent events would be rolling a 5 with one die after having rolled a 3 with another die. The two events are independent, so the probability of rolling a 5 is always 0.17 regardless of what we rolled on the first die.[1]

But what if we want to know the probability of some event occurring that requires that multiple events first to occur? For instance, let's say we're talking about the Cleveland Cavaliers winning the NBA championship. In 2016, the Golden State Warriors were 3–1 in a best-of-seven playoff. What had to happen for the Warriors to lose the playoff? The Cavaliers had to win three in a row. In this instance, to find the probability, we have to take the product of all marginal probabilities, or $\Pr(\cdot)^n$, where $\Pr(\cdot)$ is the marginal probability of one event occurring, and n is the number of repetitions of that one event. If the unconditional probability of a Cleveland win is 0.5, and each game is independent, then the probability that Cleveland could come back from a 3–1 deficit is the *product* of each game's probability of winning:

$$\text{Win probability} = \Pr(W, W, W) = (0.5)^3 = 0.125$$

Another example may be helpful. In Texas Hold'em poker, each player is dealt two cards facedown. When you are holding two of a kind, you say you have two "in the pocket." So, what is the probability of being dealt pocket aces? It's $\dfrac{4}{52} \times \dfrac{3}{51} = 0.0045$. That's right: it's 0.45%.

Let's formalize what we've been saying for a more generalized case. For independent events, to calculate *joint probabilities*, we multiply the marginal probabilities:

$$\Pr(A, B) = \Pr(A)\Pr(B)$$

where $\Pr(A, B)$ is the joint probability of both A and B occurring, and $\Pr(A)$ is the marginal probability of A event occurring.

Now, for a slightly more difficult application. What is the probability of rolling a 7 using two six-sided dice, and is it the same as the probability of rolling a 3? To answer this, let's compare the two probabilities. We'll use a table to help explain the intuition. First, let's look at all the ways to get a 7 using two six-sided dice. There are 36 total possible outcomes ($6^2 = 36$) when rolling two dice. In Table 2 we see that there are six different ways to roll a 7 using only two dice. So the probability

---

[1] The probability rolling a 5 using one six-sided die is $\dfrac{1}{6} = 0.167$.

Table 2. Total number of ways to get a 7 with two six-sided dice.

Die 1	Die 2	Outcome
1	6	7
2	5	7
3	4	7
4	3	7
5	2	7
6	1	7

Table 3. Total number of ways to get a 3 using two six-sided dice.

Die 1	Die 2	Outcome
1	2	3
2	1	3

of rolling a 7 is $6/36 = 16.67\%$. Next, let's look at all the ways to roll a 3 using two six-sided dice. Table 3 shows that there are only two ways to get a 3 rolling two six-sided dice. So the probability of rolling a 3 is $2/36 = 5.56\%$. So, no, the probabilities of rolling a 7 and rolling a 3 are different.

Events and conditional probability. First, before we talk about the three ways of representing a probability, I'd like to introduce some new terminology and concepts: events and conditional probabilities. Let *A* be some event. And let *B* be some other event. For two events, there are four possibilities.

1. A and B: Both A and B occur.
2. ∼ A and B: A does not occur, but B occurs.
3. A and ∼ B: A occurs, but B does not occur.
4. ∼ A and ∼ B: Neither A nor B occurs.

I'll use a couple of different examples to illustrate how to represent a probability.

Probability tree. Let's think about a situation in which you are trying to get your driver's license. Suppose that in order to get a driver's license, you have to pass the written exam and the driving exam. However, if you fail the written exam, you're not allowed to take the driving exam. We can represent these two events in a probability tree.

Probability trees are intuitive and easy to interpret.[2] First, we see that the probability of passing the written exam is 0.75 and the probability of failing the exam is 0.25. Second, at every branching off from a node, we can further see that the probabilities associated with a given branch are summing to 1.0. The joint probabilities are also all summing to 1.0. This is called the law of total probability and it is equal to the sum of all joint probability of A and B_n events occurring:

$$\Pr(A) = \sum_n \Pr(A \cup B_n)$$

We also see the concept of a conditional probability in the driver's license tree. For instance, the probability of failing the driving exam, conditional on having passed the written exam, is represented as $\Pr(\text{Fail} \mid \text{Pass}) = 0.45$.

Venn diagrams and sets. A second way to represent multiple events occurring is with a Venn diagram. Venn diagrams were first conceived by John Venn in 1880. They are used to teach elementary set theory, as well as to express set relationships in probability and statistics. This example will involve two sets, *A* and *B*.

---

2 The set notation ∪ means "union" and refers to two events occurring together.

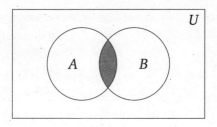

The University of Texas's football coach has been on the razor's edge with the athletic director and regents all season. After several mediocre seasons, his future with the school is in jeopardy. If the Longhorns don't make it to a great bowl game, he likely won't be rehired. But if they do, then he likely will be rehired. Let's discuss elementary set theory using this coach's situation as our guiding example. But before we do, let's remind ourselves of our terms. A and B are events, and U is the universal set of which A and B are subsets. Let A be the probability that the Longhorns get invited to a great bowl game and B be the probability that their coach is rehired. Let $\Pr(A) = 0.6$ and let $\Pr(B) = 0.8$. Let the probability that both A and B occur be $\Pr(A, B) = 0.5$.

Note, that $A + \sim A = U$, where $\sim A$ is the complement of A. The complement means that it is everything in the universal set that is not A. The same is said of B. The sum of B and $\sim B = U$. Therefore:

$$A + \sim A = B + \sim B$$

We can rewrite out the following definitions:

$$A = B + \sim B - \sim A$$
$$B = A + \sim A - \sim B$$

Whenever we want to describe a set of events in which either A or B could occur, it is: $A \cup B$. And this is pronounced "A union B," which means it is the new set that contains every element from A and every element from B. Any element that is in either set A or set B, then, is also in the new union set. And whenever we want to describe a set of events that occurred together—the joint set—it's $A \cap B$, which is pronounced "A intersect B." This new set contains every element that is in both the A and B sets. That is, only things inside both A and B get added to the new set.

Now let's look closely at a relationship involving the set A.

$$A = A \cup B + A \cup \sim B$$

Notice what this is saying: there are two ways to identify the A set. First, you can look at all the instances where A occurs with B. But then what about the rest of A that is not in B? Well, that's the $A \cup B$ situation, which covers the rest of the A set.

A similar style of reasoning can help you understand the following expression.

$$A \cap B = A \cup \sim B + \sim A \cup B + A \cup B$$

To get the A intersect B, we need three objects: the set of A units outside of B, the set of B units outside A, and their joint set. You get all those, and you have $A \cap B$.

Now it is just simple addition to find all missing values. Recall that A is your team making playoffs and $\Pr(A) = 0.6$. And B is the probability that the coach is rehired, $\Pr(B) = 0.8$. Also, $\Pr(A,B) = 0.5$, which is the probability of both A and B occurring. Then we have:

$$A = A \cup B + A \cup \sim B$$

$$A \cup \sim B = A - A \cup B$$

$$\Pr(A, \sim B) = \Pr(A) - \Pr(A,B)$$

$$\Pr(A, \sim B) = 0.6 - 0.5$$

$$\Pr(A, \sim B) = 0.1$$

When working with sets, it is important to understand that probability is calculated by considering the share of the set (for example A) made up by the subset (for example $A \cup B$). When we write down that the probability that $A \cup B$ occurs at all, it is with regards to U. But what if we were to ask the question "What share of A is due to $A \cup B$?" Notice, then, that we would need to do this:

$$? = A \cup B \div A$$

$$? = 0.5 \div 0.6$$

$$? = 0.83$$

Table 4. Twoway contingency table.

Event labels	Coach is rehired ($\sim B$)	Coach is not rehired (B)	**Total**
(A) Bowl game	$\Pr(A,\sim B) = 0.1$	$\Pr(A,B) = 0.5$	$\Pr(A) = 0.6$
($\sim A$) No bowl game	$\Pr(\sim A,\sim B) = 0.1$	$\Pr(\sim A,B) = 0.3$	$\Pr(B) = 0.4$
Total	$\Pr(\sim B) = 0.2$	$\Pr(B) = 0.8$	1.0

I left this intentionally undefined on the left side so as to focus on the calculation itself. But now let's define what we are wanting to calculate: In a world where A has occurred, what is the probability that B will also occur? This is:

$$\text{Prob}(B \mid A) = \frac{\Pr(A,B)}{\Pr(A)} = \frac{0.5}{0.6} = 0.83$$

$$\text{Prob}(A \mid B) = \frac{\Pr(A,B)}{\Pr(B)} = \frac{0.5}{0.8} = 0.63$$

Notice, these conditional probabilities are not as easy to see in the Venn diagram. We are essentially asking what percentage of a subset— e.g., $\Pr(A)$—is due to the joint set, for example, $\Pr(A,B)$. This reasoning is the very same reasoning used to define the concept of a conditional probability.

Contingency tables. Another way that we can represent events is with a contingency table. Contingency tables are also sometimes called twoway tables. Table 4 is an example of a contingency table. We continue with our example about the worried Texas coach.

Recall that $\Pr(A) = 0.6$, $\Pr(B) = 0.8$, and $Pr(A,B) = 0.5$. Note that to calculate conditional probabilities, we must know the frequency of the element in question (e.g., $\Pr(A,B)$) relative to some other larger event (e.g., $\Pr(A)$). So if we want to know what the conditional probability of B is given A, then it's:

$$\Pr(B \mid A) = \frac{\Pr(A,B)}{\Pr(A)} = \frac{0.5}{0.6} = 0.83$$

But note that knowing the frequency of $A \cup B$ in a world where B occurs is to ask the following:

$$\Pr(A \mid B) = \frac{\Pr(A,B)}{\Pr(B)} = \frac{0.5}{0.8} = 0.63$$

So, we can use what we have done so far to write out a definition of joint probability. Let's start with a definition of conditional probability first. Given two events, A and B:

$$\Pr(A \mid B) = \frac{\Pr(A,B)}{\Pr(B)} \tag{2.1}$$

$$\Pr(B \mid A) = \frac{\Pr(B,A)}{\Pr(A)} \tag{2.2}$$

$$\Pr(A,B) = \Pr(B,A) \tag{2.3}$$

$$\Pr(A) = \Pr(A, \sim B) + \Pr(A,B) \tag{2.4}$$

$$\Pr(B) = \Pr(A,B) + \Pr(\sim A,B) \tag{2.5}$$

Using equations 2.1 and 2.2, I can simply write down a definition of joint probabilities.

$$\Pr(A,B) = \Pr(A \mid B)\Pr(B) \tag{2.6}$$

$$\Pr(B,A) = \Pr(B \mid A)\Pr(A) \tag{2.7}$$

And this is the formula for joint probability. Given equation 2.3, and using the definitions of ($\Pr(A,B$ and $\Pr(B,A)$), I can also rearrange terms, make a substitution, and rewrite it as:

$$\Pr(A \mid B)\Pr(B) = \Pr(B \mid A)\Pr(A)$$
$$\Pr(A \mid B) = \frac{\Pr(B \mid A)\Pr(A)}{\Pr(B)} \tag{2.8}$$

Equation 2.8 is sometimes called the naive version of Bayes's rule. We will now decompose this equation more fully, though, by substituting equation 2.5 into equation 2.8.

$$\Pr(A \mid B) = \frac{\Pr(B \mid A)\Pr(A)}{\Pr(A,B) + \Pr(\sim A,B)} \tag{2.9}$$

Substituting equation 2.6 into the denominator for equation 2.9 yields:

$$\Pr(A \mid B) = \frac{\Pr(B \mid A)\Pr(A)}{\Pr(B \mid A)\Pr(A) + \Pr(\sim A,B)} \tag{2.10}$$

Finally, we note that using the definition of joint probability, that $\Pr(B, \sim A) = \Pr(B \mid \sim A)\Pr(\sim A)$, which we substitute into the denominator of equation 2.10 to get:

$$\Pr(A \mid B) = \frac{\Pr(B \mid A) \Pr(A)}{\Pr(B \mid A) \Pr(A) + \Pr(B \mid \sim A) \Pr(\sim A)} \qquad (2.11)$$

That's a mouthful of substitutions, so what does equation 2.11 mean? This is the Bayesian decomposition version of Bayes's rule. Let's use our example again of Texas making a great bowl game. A is Texas making a great bowl game, and B is the coach getting rehired. And $A \cap B$ is the joint probability that both events occur. We can make each calculation using the contingency tables. The questions here is this: If the Texas coach is rehired, what's the probability that the Longhorns made a great bowl game? Or formally, $\Pr(A \mid B)$. We can use the Bayesian decomposition to find this probability.

$$
\begin{aligned}
\Pr(A \mid B) &= \frac{\Pr(B \mid A) \Pr(A)}{\Pr(B \mid A) \Pr(A) + \Pr(B \mid \sim A) \Pr(\sim A)} \\
&= \frac{0.83 \cdot 0.6}{0.83 \cdot 0.6 + 0.75 \cdot 0.4} \\
&= \frac{0.498}{0.498 + 0.3} \\
&= \frac{0.498}{0.798} \\
\Pr(A \mid B) &= 0.624
\end{aligned}
$$

Check this against the contingency table using the definition of joint probability:

$$\Pr(A \mid B) = \frac{\Pr(A, B)}{\Pr(B)} = \frac{0.5}{0.8} = 0.625$$

So, if the coach is rehired, there is a 63 percent chance we made a great bowl game.[3]

Monty Hall example. Let's use a different example, the Monty Hall example. This is a fun one, because most people find it counterintu-

---

3 Why are they different? Because 0.83 is an approximation of $\Pr(B \mid A)$, which was technically 0.833... trailing.

itive. It even is used to stump mathematicians and statisticians.[4] But Bayes's rule makes the answer very clear—so clear, in fact, that it's somewhat surprising that Bayes's rule was actually once controversial [McGrayne, 2012].

Let's assume three closed doors: door 1 (D_1), door 2 (D_2), and door 3 (D_3). Behind one of the doors is a million dollars. Behind each of the other two doors is a goat. Monty Hall, the game-show host in this example, asks the contestants to pick a door. After they pick the door, but before he opens the door they picked, he opens one of the other doors to reveal a goat. He then asks the contestant, "Would you like to switch doors?"

A common response to Monty Hall's offer is to say it makes no sense to change doors, because there's an equal chance that the million dollars is behind either door. Therefore, why switch? There's a 50–50 chance it's behind the door picked and there's a 50–50 chance it's behind the remaining door, so it makes no rational sense to switch. Right? Yet, a little intuition should tell you that's not the right answer, because it would seem that when Monty Hall opened that third door, he made a statement. But what exactly did he say?

Let's formalize the problem using our probability notation. Assume that you chose door 1, D_1. The probability that D_1 had a million dollars when you made that choice is $\Pr(D_1 = 1 \text{ million}) = \frac{1}{3}$. We will call that event A_1. And the probability that D_1 has a million dollars at the start of the game is $\frac{1}{3}$ because the sample space is 3 doors, of which one has a million dollars behind it. Thus, $\Pr(A_1) = \frac{1}{3}$. Also, by the law of total probability, $\Pr(\sim A_1) = \frac{2}{3}$. Let's say that Monty Hall had opened door 2, D_2, to reveal a goat. Then he asked, "Would you like to change to door number 3?"

We need to know the probability that door 3 has the million dollars and compare that to Door 1's probability. We will call the opening of

---

4 There's an ironic story in which someone posed the Monty Hall question to the columnist, Marilyn vos Savant. Vos Savant had an extremely high IQ and so people would send in puzzles to stump her. Without the Bayesian decomposition, using only logic, she got the answer right. Her column enraged people, though. Critics wrote in to mansplain how wrong she was, but in fact it was they who were wrong.

door 2 event B. We will call the probability that the million dollars is behind door i, A_i. We now write out the question just asked formally and decompose it using the Bayesian decomposition. We are ultimately interested in knowing what the probability is that door 1 has a million dollars (event A_1) given that Monty Hall opened door 2 (event B), which is a conditional probability question. Let's write out that conditional probability using the Bayesian decomposition from equation 2.11.

$$Pr(A_1 \mid B) = \frac{Pr(B \mid A_1)\,Pr(A_1)}{Pr(B \mid A_1)\,Pr(A_1) + Pr(B \mid A_2)\,Pr(A_2) + Pr(B \mid A_3)\,Pr(A_3)}$$

(2.12)

There are basically two kinds of probabilities on the right side of the equation. There's the marginal probability that the million dollars is behind a given door, $Pr(A_i)$. And there's the conditional probability that Monty Hall would open door 2 given that the million dollars is behind door A_i, $Pr(B \mid A_i)$.

The marginal probability that door i has the million dollars behind it without our having any additional information is $\frac{1}{3}$. We call this the *prior probability*, or *prior belief*. It may also be called the *unconditional probability*.

The conditional probability, $Pr(B|A_i)$, requires a little more careful thinking. Take the first conditional probability, $Pr(B \mid A_1)$. If door 1 has the million dollars behind it, what's the probability that Monty Hall would open door 2?

Let's think about the second conditional probability: $Pr(B \mid A_2)$. If the money is behind door 2, what's the probability that Monty Hall would open door 2?

And then the last conditional probability, $Pr(B \mid A_3)$. In a world where the money is behind door 3, what's the probability Monty Hall will open door 2?

Each of these conditional probabilities requires thinking carefully about the feasibility of the events in question. Let's examine the easiest question: $Pr(B \mid A_2)$. If the money is behind door 2, how likely is it for Monty Hall to open that same door, door 2? Keep in mind: this is a game show. So that gives you some idea about how the game-show host will behave. Do you think Monty Hall would open a door that had the million dollars behind it? It makes no sense to think he'd ever open a door that actually had the money behind it—he will always open a door with a goat. So don't you think he's only opening doors with goats? Let's see what happens if take that intuition to its logical extreme and conclude that Monty Hall *never* opens a door if it has a million dollars. He *only* opens a door if the door has a goat. Under that assumption, we can proceed to estimate $\Pr(A_1 \mid B)$ by substituting values for $\Pr(B \mid A_i)$ and $\Pr(A_i)$ into the right side of equation 2.12.

What then is $\Pr(B \mid A_1)$? That is, in a world where *you* have chosen door 1, and the money is behind door 1, what is the probability that he would open door 2? There are two doors he could open if the money is behind door 1—he could open either door 2 or door 3, as both have a goat behind them. So $\Pr(B \mid A_1) = 0.5$.

What about the second conditional probability, $\Pr(B \mid A_2)$? If the money is behind door 2, what's the probability he will open it? Under our assumption that he never opens the door if it has a million dollars, we know this probability is 0.0. And finally, what about the third probability, $\Pr(B \mid A_3)$? What is the probability he opens door 2 given that the money is behind door 3? Now consider this one carefully— the contestant has already chosen door 1, so he can't open that one. And he can't open door 3, because that has the money behind it. The only door, therefore, he could open is door 2. Thus, this probability is 1.0. Furthermore, all marginal probabilities, $\Pr(A_i)$, equal 1/3, allowing us to solve for the conditional probability on the left side through substitution, multiplication, and division.

$$\Pr(A_1 \mid B) = \frac{\frac{1}{2} \cdot \frac{1}{3}}{\frac{1}{2} \cdot \frac{1}{3} + 0 \cdot \frac{1}{3} + 1.0 \cdot \frac{1}{3}}$$

$$= \frac{\frac{1}{6}}{\frac{1}{6} + \frac{2}{6}}$$

$$= \frac{1}{3}$$

Aha. Now isn't that just a little bit surprising? The probability that the contestant chose the correct door is $\frac{1}{3}$, just as it was before Monty Hall opened door 2.

But what about the probability that door 3, the door you're holding, has the million dollars? Have your beliefs about that likelihood changed now that door 2 has been removed from the equation? Let's crank through our Bayesian decomposition and see whether we learned anything.

$$\Pr(A_3 \mid B) = \frac{\Pr(B \mid A_3)\Pr(A_3)}{\Pr(B \mid A_3)\Pr(A_3) + \Pr(B \mid A_2)\Pr(A_2) + \Pr(B \mid A_1)\Pr(A_1)}$$

$$= \frac{1.0 \cdot \frac{1}{3}}{1.0 \cdot \frac{1}{3} + 0 \cdot \frac{1}{3} + \frac{1}{2} \cdot \frac{1}{3}}$$

$$= \frac{2}{3}$$

Interestingly, while your beliefs about the door you originally chose haven't changed, your beliefs about the other door have changed. The prior probability, $\Pr(A_3) = \frac{1}{3}$, increased through a process called *updating* to a new probability of $\Pr(A_3 \mid B) = \frac{2}{3}$. This new conditional probability is called the *posterior probability*, or *posterior belief*. And it simply means that having witnessed B, you learned information that allowed you to form a new belief about which door the money might be behind.

As was mentioned in footnote 14 regarding the controversy around vos Sant's correct reasoning about the need to switch doors, deductions based on Bayes's rule are often surprising even to smart people—probably because we lack coherent ways to correctly incorporate information into probabilities. Bayes's rule shows us how to do that in a way that is logical and accurate. But besides being insightful, Bayes's rule also opens the door for a different kind of reasoning about cause and effect. Whereas most of this book has to do with estimating effects from known causes, Bayes's rule reminds us that we can form reasonable beliefs about causes from known effects.

Summation operator. The tools we use to reason about causality rest atop a bedrock of probabilities. We are often working with mathematical tools and concepts from statistics such as expectations and probabilities. One of the most common tools we will use in this book is the linear regression model, but before we can dive into that, we have to build out some simple notation.[5] We'll begin with the summation operator. The Greek letter Σ (the capital Sigma) denotes the summation operator. Let $x_1, x_2, \ldots, x_n$ be a sequence of numbers. We can compactly write a sum of numbers using the summation operator as:

$$\sum_{i=1}^{n} x_i \equiv x_1 + x_2 + \ldots + x_n$$

The letter i is called the index of summation. Other letters, such as j or k, are sometimes used as indices of summation. The subscript variable simply represents a specific value of a random variable, x. The numbers 1 and n are the lower limit and the upper limit, respectively, of the summation. The expression $\sum_{i=1}^{n} x_i$ can be stated in words as "sum the numbers x_i for all values of i from 1 to n." An example can help clarify:

$$\sum_{i=6}^{9} x_i = x_6 + x_7 + x_8 + x_9$$

The summation operator has three properties. The first property is called the constant rule. Formally, it is:

$$\text{For any constant } c: \quad \sum_{i=1}^{n} c = nc \qquad (2.13)$$

Let's consider an example. Say that we are given:

$$\sum_{i=1}^{3} 5 = (5 + 5 + 5) = 3 \cdot 5 = 15$$

A second property of the summation operator is:

$$\sum_{i=1}^{n} c x_i = c \sum_{i=1}^{n} x_i \qquad (2.14)$$

---

5 For a more complete review of regression, see Wooldridge [2010] and Wooldridge [2015]. I stand on the shoulders of giants.

Again let's use an example. Say we are given:

$$\sum_{i=1}^{3} 5x_i = 5x_1 + 5x_2 + 5x_3$$

$$= 5(x_1 + x_2 + x_3)$$

$$= 5\sum_{i=1}^{3} x_i$$

We can apply both of these properties to get the following third property:

$$\text{For any constant } a \text{ and } b: \quad \sum_{i=1}^{n}(ax_i + by_i) = a\sum_{i=1}^{n} x_i + b\sum_{j=1}^{n}$$

Before leaving the summation operator, it is useful to also note things which are not properties of this operator. First, the summation of a ratio is not the ratio of the summations themselves.

$$\sum_{i}^{n} \frac{x_i}{y_i} \neq \frac{\sum_{i=1}^{n} x_i}{\sum_{i=1}^{n} y_i}$$

Second, the summation of some squared variable is not equal to the squaring of its summation.

$$\sum_{i=1}^{n} x_i^2 \neq \left(\sum_{i=1}^{n} x_i\right)^2$$

We can use the summation indicator to make a number of calculations, some of which we will do repeatedly over the course of this book. For instance, we can use the summation operator to calculate the average:

$$\bar{x} = \frac{1}{n}\sum_{i=1}^{n} x_i$$

$$= \frac{x_1 + x_2 + \cdots + x_n}{n}$$

(2.15)

where $\bar{x}$ is the average (mean) of the random variable x_i. Another calculation we can make is a random variable's deviations from its own

Table 5. Sum of deviations equaling 0.

x	$x - \bar{x}$
10	2
4	-4
13	5
5	-3
Mean=8	Sum=0

mean. The sum of the deviations from the mean is always equal to 0:

$$\sum_{i=1}^{n}(x_i - \bar{x}) = 0 \tag{2.16}$$

You can see this in Table 5.

Consider a sequence of two numbers $\{y_1, y_2, \ldots, y_n\}$ and $\{x_1, x_2, \ldots, x_n\}$. Now we can consider double summations over possible values of x's and y's. For example, consider the case where $n = m = 2$. Then, $\sum_{i=1}^{2}\sum_{j=1}^{2}x_i y_j$ is equal to $x_1 y_1 + x_1 y_2 + x_2 y_1 + x_2 y_2$. This is because

$$x_1 y_1 + x_1 y_2 + x_2 y_1 + x_2 y_2 = x_1(y_1 + y_2) + x_2(y_1 + y_2)$$

$$= \sum_{i=1}^{2}x_i(y_1 + y_2)$$

$$= \sum_{i=1}^{2}x_i\left(\sum_{j=1}^{2}y_j\right)$$

$$= \sum_{i=1}^{2}\left(\sum_{j=1}^{2}x_i y_j\right)$$

$$= \sum_{i=1}^{2}\sum_{j=1}^{2}x_i y_j$$

One result that will be very useful throughout the book is:

$$\sum_{i=1}^{n}(x_i - \bar{x})^2 = \sum_{i=1}^{n}x_i^2 - n(\bar{x})^2 \tag{2.17}$$

An overly long, step-by-step proof is below. Note that the summation index is suppressed after the first line for easier reading.

$$\sum_{i=1}^{n}(x_i - \bar{x})^2 = \sum_{i=1}^{n}(x_i^2 - 2x_i\bar{x} + \bar{x}^2)$$

$$= \sum x_i^2 - 2\bar{x}\sum x_i + n\bar{x}^2$$

$$= \sum x_i^2 - 2\frac{1}{n}\sum x_i\sum x_i + n\bar{x}^2$$

$$= \sum x_i^2 + n\bar{x}^2 - \frac{2}{n}\left(\sum x_i\right)^2$$

$$= \sum x_i^2 + n\left(\frac{1}{n}\sum x_i\right)^2 - 2n\left(\frac{1}{n}\sum x_i\right)^2$$

$$= \sum x_i^2 - n\left(\frac{1}{n}\sum x_i\right)^2$$

$$= \sum x_i^2 - n\bar{x}^2$$

A more general version of this result is:

$$\sum_{i=1}^{n}(x_i - \bar{x})(y_i - \bar{y}) = \sum_{i=1}^{n}x_i(y_i - \bar{y})$$

$$= \sum_{i=1}^{n}(x_i - \bar{x})y_i \qquad (2.18)$$

$$= \sum_{i=1}^{n}x_iy_i - n(\overline{xy})$$

Or:

$$\sum_{i=1}^{n}(x_i - \bar{x})(y_i - \bar{y}) = \sum_{i=1}^{n}x_i(y_i - \bar{y}) = \sum_{i=1}^{n}(x_i - \bar{x})y_i = \sum_{i=1}^{n}x_iy_i - n(\overline{xy}) \quad (2.19)$$

Expected value. The expected value of a random variable, also called the expectation and sometimes the population mean, is simply the weighted average of the possible values that the variable can take, with the weights being given by the probability of each value occurring in the population. Suppose that the variable X can take on values $x_1, x_2, \ldots, x_k$, each with probability $f(x_1), f(x_2), \ldots, f(x_k)$, respectively. Then we define the expected value of X as:

$$E(X) = x_1 f(x_1) + x_2 f(x_2) + \cdots + x_k f(x_k)$$

$$= \sum_{j=1}^{k} x_j f(x_j) \qquad (2.20)$$

Let's look at a numerical example. If X takes on values of -1, 0, and 2, with probabilities 0.3, 0.3, and 0.4, respectively.[6] Then the expected value of X equals:

$$E(X) = (-1)(0.3) + (0)(0.3) + (2)(0.4)$$

$$= 0.5$$

In fact, you could take the expectation of a function of that variable, too, such as X^2. Note that X^2 takes only the values 1, 0, and 4, with probabilities 0.3, 0.3, and 0.4. Calculating the expected value of X^2 therefore is:

$$E(X^2) = (-1)^2(0.3) + (0)^2(0.3) + (2)^2(0.4)$$

$$= 1.9$$

The first property of expected value is that for any constant c, $E(c) = c$. The second property is that for any two constants a and b, then $E(aX + b) = E(aX) + E(b) = aE(X) + b$. And the third property is that if we have numerous constants, $a_1, \ldots, a_n$ and many random variables, $X_1, \ldots, X_n$, then the following is true:

$$E(a_1 X_1 + \cdots + a_n X_n) = a_1 E(X_1) + \cdots + a_n E(X_n)$$

We can also express this using the expectation operator:

$$E\left(\sum_{i=1}^{n} a_i X_i \right) = \sum_{i=1}^{n} a_i E(X_i)$$

And in the special case where $a_i = 1$, then

$$E\left(\sum_{i=1}^{n} X_i \right) = \sum_{i=1}^{n} E(X_i)$$

---

6 The law of total probability requires that all marginal probabilities sum to unity.

Variance. The expectation operator, $E(\cdot)$, is a population concept. It refers to the whole group of interest, not just to the sample available to us. Its meaning is somewhat similar to that of the average of a random variable in the population. Some additional properties for the expectation operator can be explained assuming two random variables, W and H.

$$E(aW + b) = aE(W) + b \text{ for any constants } a, b$$

$$E(W + H) = E(W) + E(H)$$

$$E\left(W - E(W)\right) = 0$$

Consider the variance of a random variable, W:

$$V(W) = \sigma^2 = E\left[(W - E(W))^2\right] \text{ in the population}$$

We can show

$$V(W) = E(W^2) - E(W)^2 \qquad (2.21)$$

In a given sample of data, we can estimate the variance by the following calculation:

$$\widehat{S}^2 = (n-1)^{-1} \sum_{i=1}^{n} (x_i - \overline{x})^2$$

where we divide by $n - 1$ because we are making a degree-of-freedom adjustment from estimating the mean. But in large samples, this degree-of-freedom adjustment has no practical effect on the value of S^2 where S^2 is the average (after a degree of freedom correction) over the sum of all squared deviations from the mean.[7]

A few more properties of variance. First, the variance of a line is:

$$V(aX + b) = a^2 V(X)$$

And the variance of a constant is 0 (i.e., $V(c) = 0$ for any constant, c). The variance of the sum of two random variables is equal to:

$$V(X + Y) = V(X) + V(Y) + 2\left(E(XY) - E(X)E(Y)\right) \qquad (2.22)$$

---

7 Whenever possible, I try to use the "hat" to represent an estimated statistic. Hence $\widehat{S}^2$ instead of just S^2. But it is probably more common to see the sample variance represented as S^2.

If the two variables are independent, then $E(XY) = E(X)E(Y)$ and $V(X + Y)$ is equal to the sum of $V(X) + V(Y)$.

Covariance. The last part of equation 2.22 is called the covariance. The covariance measures the amount of linear dependence between two random variables. We represent it with the $C(X,Y)$ operator. The expression $C(X,Y) > 0$ indicates that two variables move in the same direction, whereas $C(X,Y) < 0$ indicates that they move in opposite directions. Thus we can rewrite equation 2.22 as:

$$V(X + Y) = V(X) + V(Y) + 2C(X,Y)$$

While it's tempting to say that a zero covariance means that two random variables are unrelated, that is incorrect. They could have a nonlinear relationship. The definition of covariance is

$$C(X,Y) = E(XY) - E(X)E(Y) \qquad (2.23)$$

As we said, if X and Y are independent, then $C(X,Y) = 0$ in the population. The covariance between two linear functions is:

$$C(a_1 + b_1X, a_2 + b_2Y) = b_1b_2C(X,Y)$$

The two constants, a_1 and a_2, zero out because their mean is themselves and so the difference equals 0.

Interpreting the magnitude of the covariance can be tricky. For that, we are better served by looking at correlation. We define correlation as follows. Let $W = \dfrac{X - E(X)}{\sqrt{V(X)}}$ and $Z = \dfrac{Y - E(Y)}{\sqrt{V(Y)}}$. Then:

$$\text{Corr}(W,Z) = \frac{C(X,Y)}{\sqrt{V(X)V(Y)}} \qquad (2.24)$$

The correlation coefficient is bounded by -1 and 1. A positive (negative) correlation indicates that the variables move in the same (opposite) ways. The closer the coefficient is to 1 or -1, the stronger the linear relationship is.

Population model. We begin with cross-sectional analysis. We will assume that we can collect a random sample from the population of interest. Assume that there are two variables, x and y, and we want to see how y varies with changes in x.[8]

There are three questions that immediately come up. One, what if y is affected by factors other than x? How will we handle that? Two, what is the functional form connecting these two variables? Three, if we are interested in the causal effect of x on y, then how can we distinguish that from mere correlation? Let's start with a specific model.

$$y = \beta_0 + \beta_1 x + u \qquad (2.25)$$

This model is assumed to hold in the population. Equation 2.25 defines a linear bivariate regression model. For models concerned with capturing causal effects, the terms on the left side are usually thought of as the effect, and the terms on the right side are thought of as the causes.

Equation 2.25 explicitly allows for other factors to affect y by including a random variable called the error term, u. This equation also explicitly models the functional form by assuming that y is linearly dependent on x. We call the β_0 coefficient the intercept parameter, and we call the β_1 coefficient the slope parameter. These describe a population, and our goal in empirical work is to estimate their values. We never directly observe these parameters, because they are not data (I will emphasize this throughout the book). What we can do, though, is estimate these parameters using *data* and *assumptions*. To do this, we need credible assumptions to *accurately* estimate these parameters with data. We will return to this point later. In this simple regression framework, all unobserved variables that determine y are subsumed by the error term u.

First, we make a simplifying assumption without loss of generality. Let the expected value of u be zero in the population. Formally:

$$E(u) = 0 \qquad (2.26)$$

where $E(\cdot)$ is the expected value operator discussed earlier. If we normalize the u random variable to be 0, it is of no consequence. Why? Because the presence of β_0 (the intercept term) always allows us this flexibility. If the average of u is different from 0—for instance, say that

---

8 This is not necessarily causal language. We are speaking first and generally in terms of two random variables systematically moving together in some measurable way.

it's α_0—then we adjust the intercept. Adjusting the intercept has no effect on the β_1 slope parameter, though. For instance:

$$y = (\beta_0 + \alpha_0) + \beta_1 x + (u - \alpha_0)$$

where $\alpha_0 = E(u)$. The new error term is $u - \alpha_0$, and the new intercept term is $\beta_0 + \alpha_0$. But while those two terms changed, notice what did *not* change: the slope, β_1.

Mean independence. An assumption that meshes well with our elementary treatment of statistics involves the mean of the error term for each "slice" of the population determined by values of x:

$$E(u \mid x) = E(u) \text{ for all values } x \qquad (2.27)$$

where $E(u \mid x)$ means the "expected value of u given x." If equation 2.27 holds, then we say that u is mean independent of x.

An example might help here. Let's say we are estimating the effect of schooling on wages, and u is unobserved ability. Mean independence requires that $E(\text{ability} \mid x = 8) = E(\text{ability} \mid x = 12) = E(\text{ability} \mid x = 16)$ so that the average ability is the same in the different portions of the population with an eighth-grade education, a twelfth-grade education, and a college education. Because people choose how much schooling to invest in based on their own unobserved skills and attributes, equation 2.27 is likely violated—at least in our example.

But let's say we are willing to make this assumption. Then combining this new assumption, $E(u \mid x) = E(u)$ (the nontrivial assumption to make), with $E(u) = 0$ (the normalization and trivial assumption), and you get the following new assumption:

$$E(u \mid x) = 0, \text{ for all values } x \qquad (2.28)$$

Equation 2.28 is called the zero conditional mean assumption and is a key identifying assumption in regression models. Because the conditional expected value is a linear operator, $E(u \mid x) = 0$ implies that

$$E(y \mid x) = \beta_0 + \beta_1 x$$

which shows the population regression function is a linear function of x, or what Angrist and Pischke [2009] call the conditional expectation function.[9] This relationship is crucial for the intuition of the parameter, β_1, as a *causal parameter*.

Ordinary least squares. Given data on x and y, how can we estimate the population parameters, β_0 and β_1? Let the pairs of $\{(x_i, \text{ and } y_i) : i = 1, 2, \ldots, n\}$ be random samples of size from the population. Plug any observation into the population equation:

$$y_i = \beta_0 + \beta_1 x_i + u_i$$

where i indicates a particular observation. We observe y_i and x_i but not u_i. We just know that u_i is there. We then use the two population restrictions that we discussed earlier:

$$E(u) = 0$$

$$E(u \mid x) = 0$$

to obtain estimating equations for β_0 and β_1. We talked about the first condition already. The second one, though, means that the mean value of x does not change with different slices of the error term. This independence assumption implies $E(xu) = 0$, we get $E(u) = 0$, and $C(x,u) = 0$. Notice that if $C(x,u) = 0$, then that implies x and u are independent.[10] Next we plug in for u, which is equal to $y - \beta_0 - \beta_1 x$:

$$E(y - \beta_0 - \beta_1 x) = 0$$

$$\left(x[y - \beta_0 - \beta_1 x] \right) = 0$$

These are the two conditions in the population that effectively determine β_0 and β_1. And again, note that the notation here is population concepts. We don't have access to populations, though we do have their sample counterparts:

---

9 Notice that the conditional expectation passed through the linear function leaving a constant, because of the first property of the expectation operator, and a constant times x. This is because the conditional expectation of $E[X \mid X] = X$. This leaves us with $E[u \mid X]$ which under zero conditional mean is equal to 0.

10 See equation 2.23.

$$\frac{1}{n}\sum_{i=1}^{n}\left(y_i - \widehat{\beta}_0 - \widehat{\beta}_1 x_i\right) = 0 \tag{2.29}$$

$$\frac{1}{n}\sum_{i=1}^{n}\left(x_i\left[y_i - \widehat{\beta}_0 - \widehat{\beta}_1 x_i\right]\right) = 0 \tag{2.30}$$

where $\widehat{\beta}_0$ and $\widehat{\beta}_1$ are the estimates from the data.[11] These are two linear equations in the two unknowns $\widehat{\beta}_0$ and $\widehat{\beta}_1$. Recall the properties of the summation operator as we work through the following sample properties of these two equations. We begin with equation 2.29 and pass the summation operator through.

$$\frac{1}{n}\sum_{i=1}^{n}\left(y_i - \widehat{\beta}_0 - \widehat{\beta}_1 x_i\right) = \frac{1}{n}\sum_{i=1}^{n}(y_i) - \frac{1}{n}\sum_{i=1}^{n}\widehat{\beta}_0 - \frac{1}{n}\sum_{i=1}^{n}\widehat{\beta}_1 x_i$$

$$= \frac{1}{n}\sum_{i=1}^{n}y_i - \widehat{\beta}_0 - \widehat{\beta}_1\left(\frac{1}{n}\sum_{i=1}^{n}x_i\right)$$

$$= \bar{y} - \widehat{\beta}_0 - \widehat{\beta}_1\bar{x}$$

where $\bar{y} = \frac{1}{n}\sum_{i=1}^{n}y_i$ which is the average of the n numbers $\{y_i : 1,\ldots,n\}$. For emphasis we will call $\bar{y}$ the sample average. We have already shown that the first equation equals zero (equation 2.29), so this implies $\bar{y} = \widehat{\beta}_0 + \widehat{\beta}_1\bar{x}$. So we now use this equation to write the intercept in terms of the slope:

$$\widehat{\beta}_0 = \bar{y} - \widehat{\beta}_1\bar{x}$$

We now plug $\widehat{\beta}_0$ into the second equation, $\sum_{i=1}^{n}x_i(y_i - \widehat{\beta}_0 - \widehat{\beta}_1 x_i) = 0$. This gives us the following (with some simple algebraic manipulation):

$$\sum_{i=1}^{n}x_i\left[y_i - (\bar{y} - \widehat{\beta}_1\bar{x}) - \widehat{\beta}_1 x_i\right] = 0$$

$$\sum_{i=1}^{n}x_i(y_i - \bar{y}) = \widehat{\beta}_1\left[\sum_{i=1}^{n}x_i(x_i - \bar{x})\right]$$

---

11 Notice that we are dividing by n, not $n - 1$. There is no degrees-of-freedom correction, in other words, when using samples to calculate means. There is a degrees-of-freedom correction when we start calculating higher moments.

So the equation to solve is[12]

$$\sum_{i=1}^{n}(x_i - \overline{x})(y_i - \overline{y}) = \widehat{\beta}_1 \left[\sum_{i=1}^{n}(x_i - \overline{x})^2 \right]$$

If $\sum_{i=1}^{n}(x_i - \overline{x})^2 \neq 0$, we can write:

$$\widehat{\beta}_1 = \frac{\sum_{i=1}^{n}(x_i - \overline{x})(y_i - \overline{y})}{\sum_{i=1}^{n}(x_i - \overline{x})^2}$$

$$= \frac{\text{Sample covariance}(x_i, y_i)}{\text{Sample variance}(x_i)} \tag{2.31}$$

The previous formula for $\widehat{\beta}_1$ is important because it shows us how to take data that we have and compute the slope estimate. The estimate, $\widehat{\beta}_1$, is commonly referred to as the ordinary least squares (OLS) slope estimate. It can be computed whenever the sample variance of x_i isn't 0. In other words, it can be computed if x_i is not constant across all values of i. The intuition is that the variation in x is what permits us to identify its impact in y. This also means, though, that we cannot determine the slope in a relationship if we observe a sample in which everyone has the same years of schooling, or whatever causal variable we are interested in.

Once we have calculated $\widehat{\beta}_1$, we can compute the intercept value, $\widehat{\beta}_0$, as $\widehat{\beta}_0 = \overline{y} - \widehat{\beta}_1\overline{x}$. This is the OLS intercept estimate because it is calculated using sample averages. Notice that it is straightforward because $\widehat{\beta}_0$ is linear in $\widehat{\beta}_1$. With computers and statistical programming languages and software, we let our computers do these calculations because even when n is small, these calculations are quite tedious.

---

12 Recall from much earlier that:

$$\sum_{i=1}^{n}(x_i - \overline{x})(y_i - \overline{y}) = \sum_{i=1}^{n}x_i(y_i - \overline{y})$$

$$= \sum_{i=1}^{n}(x_i - \overline{x})y_i$$

$$= \sum_{i=1}^{n}x_i y_i - n(\overline{xy})$$

For any candidate estimates, $\widehat{\beta}_0, \widehat{\beta}_1$, we define a fitted value for each i as:

$$\widehat{y}_i = \widehat{\beta}_0 + \widehat{\beta}_1 x_i$$

Recall that $i = \{1,\ldots,n\}$, so we have n of these equations. This is the value we predict for y_i given that $x = x_i$. But there is prediction error because $y \neq y_i$. We call that mistake the residual, and here use the $\widehat{u}_i$ notation for it. So the residual equals:

$$\widehat{u}_i = y_i - \widehat{y}_i$$
$$\widehat{u}_i = y_i - \widehat{\beta}_0 - \widehat{\beta}_1 x_i$$

While both the residual and the error term are represented with a u, it is important that you know the differences. The residual is the prediction error based on our fitted $\widehat{y}$ and the actual y. The residual is therefore easily calculated with any sample of data. But u without the hat is the *error term*, and it is by definition unobserved by the researcher. Whereas the residual will appear in the data set once generated from a few steps of regression and manipulation, the error term will never appear in the data set. It is all of the determinants of our outcome not captured by our model. This is a crucial distinction, and strangely enough it is so subtle that even some seasoned researchers struggle to express it.

Suppose we measure the size of the mistake, for each i, by squaring it. Squaring it will, after all, eliminate all negative values of the mistake so that everything is a positive value. This becomes useful when summing the mistakes if we don't want positive and negative values to cancel one another out. So let's do that: square the mistake and add them all up to get $\sum_{i=1}^{n} \widehat{u}_i^2$:

$$\sum_{i=1}^{n} \widehat{u}_i^2 = \sum_{i=1}^{n} (y_i - \widehat{y}_i)^2$$
$$= \sum_{i=1}^{n} \left(y_i - \widehat{\beta}_0 - \widehat{\beta}_1 x_i \right)^2$$

This equation is called the sum of squared residuals because the residual is $\widehat{u}_i = y_i - \widehat{y}$. But the residual is based on estimates of the slope and the intercept. We can imagine any number of estimates of those values. But what if our goal is to *minimize* the sum of squared residuals by choosing $\widehat{\beta}_0$ and $\widehat{\beta}_1$? Using calculus, it can be shown that the solutions to that problem yield parameter estimates that are the same as what we obtained before.

Once we have the numbers $\widehat{\beta}_0$ and $\widehat{\beta}_1$ for a given data set, we write the OLS regression line:

$$\widehat{y} = \widehat{\beta}_0 + \widehat{\beta}_1 x \tag{2.32}$$

Let's consider a short simulation.

STATA
ols.do

```
1   set seed 1
2   clear
3   set obs 10000
4   gen x = rnormal()
5   gen u  = rnormal()
6   gen y  = 5.5*x + 12*u
7   reg y x
8   predict yhat1
9   gen yhat2 = -0.0750109  + 5.598296*x // Compare yhat1 and yhat2
10  sum yhat*
11  predict uhat1, residual
12  gen uhat2=y-yhat2
13  sum uhat*
14  twoway (lfit y x, lcolor(black) lwidth(medium)) (scatter y x, mcolor(black) ///
15  msize(tiny) msymbol(point)), title(OLS Regression Line)
16  rvfplot, yline(0)
```

R
ols.R

```
1   library(tidyverse)
2
3   set.seed(1)
4   tb <- tibble(
5     x = rnorm(10000),
6     u = rnorm(10000),
```

(continued)

R (continued)

```
7    y = 5.5*x + 12*u
8  )
9
10   reg_tb <- tb %>%
11     lm(y ~ x, .) %>%
12     print()
13
14   reg_tb$coefficients
15
16   tb <- tb %>%
17     mutate(
18       yhat1 = predict(lm(y ~ x, .)),
19       yhat2 = 0.0732608 + 5.685033*x,
20       uhat1 = residuals(lm(y ~ x, .)),
21       uhat2 = y - yhat2
22     )
23
24   summary(tb[-1:-3])
25
26   tb %>%
27     lm(y ~ x, .) %>%
28     ggplot(aes(x=x, y=y)) +
29     ggtitle("OLS Regression Line") +
30     geom_point(size = 0.05, color = "black", alpha = 0.5) +
31     geom_smooth(method = lm, color = "black") +
32     annotate("text", x = -1.5, y = 30, color = "red",
33         label = paste("Intercept = ", -0.0732608)) +
34     annotate("text", x = 1.5, y = -30, color = "blue",
35         label = paste("Slope =", 5.685033))
```

Let's look at the output from this. First, if you summarize the data, you'll see that the fitted values are produced both using Stata's Predict command and manually using the Generate command. I wanted the reader to have a chance to better understand this, so did it both ways. But second, let's look at the data and paste on top of it the estimated coefficients, the y-intercept and slope on *x* in Figure 3. The estimated coefficients in both are close to the hard coded values built into the data-generating process.

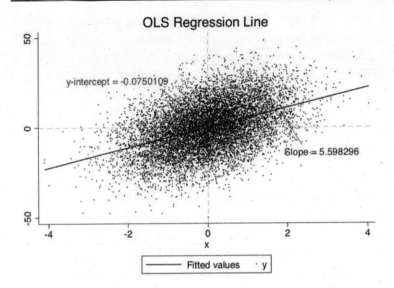

Figure 3. Graphical representation of bivariate regression from y on x.

Once we have the estimated coefficients and we have the OLS regression line, we can predict y (outcome) for any (sensible) value of x. So plug in certain values of x, and we can immediately calculate what y will probably be with some error. The value of OLS here lies in how large that error is: OLS minimizes the error for a linear function. In fact, it is the best such guess at y for all linear estimators because it minimizes the prediction error. There's always prediction error, in other words, with any estimator, but OLS is the least worst.

Notice that the intercept is the predicted value of y if and when $x = 0$. In this sample, that value is -0.0750109.[13] The slope allows us to predict changes in y for any reasonable change in x according to:

$$\Delta \widehat{y} = \widehat{\beta}_1 \Delta x$$

---

13 It isn't exactly 0 even though u and x are independent. Think of it as u and x are independent in the population, but not in the sample. This is because sample characteristics tend to be slightly different from population properties due to sampling error.

And if $\Delta x = 1$, then x increases by one unit, and so $\Delta \widehat{y} = 5.598296$ in our numerical example because $\widehat{\beta}_1 = 5.598296$.

Now that we have calculated $\widehat{\beta}_0$ and $\widehat{\beta}_1$, we get the OLS fitted values by plugging x_i into the following equation for $i = 1,\ldots,n$:

$$\widehat{y}_i = \widehat{\beta}_0 + \widehat{\beta}_1 x_i$$

The OLS residuals are also calculated by:

$$\widehat{u}_i = y_i - \widehat{\beta}_0 - \widehat{\beta}_1 x_i$$

Most residuals will be different from 0 (i.e., they do not lie on the regression line). You can see this in Figure 3. Some are positive, and some are negative. A positive residual indicates that the regression line (and hence, the predicted values) underestimates the true value of y_i. And if the residual is negative, then the regression line overestimates the true value.

Recall that we defined the fitted value as $\widehat{y}_i$ and the residual, $\widehat{u}_i$, as $y_i - \widehat{y}_i$. Notice that the scatter-plot relationship between the residuals and the fitted values created a spherical pattern, suggesting that they are not correlated (Figure 4). This is mechanical—least squares produces residuals which are uncorrelated with fitted values. There's no magic here, just least squares.

Algebraic Properties of OLS. Remember how we obtained $\widehat{\beta}_0$ and $\widehat{\beta}_1$? When an intercept is included, we have:

$$\sum_{i=1}^{n} \left(y_i - \widehat{\beta}_0 - \widehat{\beta}_1 x_i \right) = 0$$

The OLS residual *always* adds up to zero, by *construction*.

$$\sum_{i=1}^{n} \widehat{u}_i = 0 \qquad (2.33)$$

Sometimes seeing is believing, so let's look at this together. Type the following into Stata verbatim.

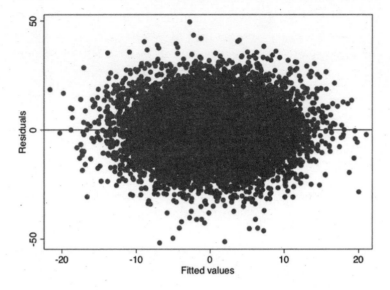

Figure 4. Distribution of residuals around regression line.

```
                          STATA
                         ols2.do
 1  clear
 2  set seed 1234
 3  set obs 10
 4  gen x = 9*rnormal()
 5  gen u  = 36*rnormal()
 6  gen y  = 3 + 2*x + u
 7  reg y x
 8  predict yhat
 9  predict residuals, residual
10  su residuals
11  list
12  collapse (sum) x u y yhat residuals
13  list
```

<table>
<tr><td colspan="2" align="center">R
ols2.R</td></tr>
</table>

```
1   library(tidyverse)
2
3   set.seed(1)
4
5   tb <- tibble(
6     x = 9*rnorm(10),
7     u = 36*rnorm(10),
8     y = 3 + 2*x + u,
9     yhat = predict(lm(y ~ x)),
10    uhat = residuals(lm(y ~ x))
11  )
12
13  summary(tb)
14  colSums(tb)
```

Output from this can be summarized as in the following table (Table 6).

Notice the difference between the u, $\widehat{y}$, and $\widehat{u}$ columns. When we sum these ten lines, neither the error term nor the fitted values of y sum to zero. But the residuals *do* sum to zero. This is, as we said, one of the algebraic properties of OLS—coefficients were optimally chosen to ensure that the residuals sum to zero.

Because $y_i = \widehat{y}_i + \widehat{u}_i$ by definition (which we can also see in Table 6), we can take the sample average of both sides:

$$\frac{1}{n}\sum_{i=1}^{n}y_i = \frac{1}{n}\sum_{i=1}^{n}\widehat{y}_i + \frac{1}{n}\sum_{i=1}^{n}\widehat{u}_i$$

and so $\bar{y} = \bar{\widehat{y}}$ because the residuals sum to zero. Similarly, the way that we obtained our estimates yields

$$\sum_{i=1}^{n}x_i\left(y_i - \widehat{\beta}_0 - \widehat{\beta}_1 x_i\right) = 0$$

The sample covariance (and therefore the sample correlation) between the explanatory variables and the residuals is always zero (see Table 6).

Table 6. Simulated data showing the sum of residuals equals zero.

no.	x	u	y	$\hat{y}$	$\hat{u}$	$x\hat{u}$	$\hat{y}\hat{u}$
1.	−4.381653	−32.95803	−38.72134	−3.256034	−35.46531	155.3967	115.4762
2.	−13.28403	−8.028061	−31.59613	−26.30994	−5.28619	70.22192	139.0793
3.	−.0982034	17.80379	20.60738	7.836532	12.77085	−1.254141	100.0792
4.	−.1238423	−9.443188	−6.690872	7.770137	−14.46101	1.790884	−112.364
5.	4.640209	13.18046	25.46088	20.10728	5.353592	24.84179	107.6462
6.	−1.252096	−34.64874	−34.15294	4.848374	−39.00131	48.83337	−189.0929
7.	11.58586	9.118524	35.29023	38.09396	−2.80373	−32.48362	−106.8052
8.	−5.289957	82.23296	74.65305	−5.608207	80.26126	−424.5786	−450.1217
9.	−.2754041	11.60571	14.0549	7.377647	6.677258	−1.838944	49.26245
10.	−19.77159	−14.61257	−51.15575	−43.11034	−8.045414	159.0706	346.8405
Sum	−28.25072	34.25085	7.749418	7.749418	1.91e − 06	−6.56e − 06	.0000305

$$\sum_{i=1}^{n} x_i \widehat{u}_i = 0$$

Because the $\widehat{y}_i$ are linear functions of the x_i, the fitted values and residuals are uncorrelated too (see Table 6):

$$\sum_{i=1}^{n} \widehat{y}_i \widehat{u}_i = 0$$

Both properties hold by construction. In other words, $\widehat{\beta}_0$ and $\widehat{\beta}_1$ were selected *to make them true*.[14]

A third property is that if we plug in the average for x, we predict the sample average for y. That is, the point $(\overline{x}, \overline{y})$ is on the OLS regression line, or:

$$\overline{y} = \widehat{\beta}_0 + \widehat{\beta}_1 \overline{x}$$

Goodness-of-fit. For each observation, we write

$$y_i = \widehat{y}_i + \widehat{u}_i$$

Define the total sum of squares (SST), explained sum of squares (SSE), and residual sum of squares (SSR) as

$$SST = \sum_{i=1}^{n} (y_i - \overline{y})^2 \qquad (2.34)$$

$$SSE = \sum_{i=1}^{n} (\widehat{y}_i - \overline{y})^2 \qquad (2.35)$$

$$SSR = \sum_{i=1}^{n} \widehat{u}_i^2 \qquad (2.36)$$

---

14 Using the Stata code from Table 6, you can show these algebraic properties yourself. I encourage you to do so by creating new variables equaling the product of these terms and collapsing as we did with the other variables. That sort of exercise may help convince you that the aforementioned algebraic properties always hold.

These are sample variances when divided by $n-1$.[15] $\dfrac{SST}{n-1}$ is the sample variance of y_i, $\dfrac{SSE}{n-1}$ is the sample variance of $\widehat{y}_i$, and $\dfrac{SSR}{n-1}$ is the sample variance of $\widehat{u}_i$. With some simple manipulation rewrite equation 2.34:

$$SST = \sum_{i=1}^{n}(y_i - \overline{y})^2$$

$$= \sum_{i=1}^{n}\left[(y_i - \widehat{y}_i) - (\widehat{y}_i - \overline{y})\right]^2$$

$$= \sum_{i=1}^{n}\left[\widehat{u}_i - (\widehat{y}_i - \overline{y})\right]^2$$

Since equation 2.34 shows that the fitted values are uncorrelated with the residuals, we can write the following equation:

$$SST = SSE + SSR$$

Assuming $SST > 0$, we can define the fraction of the total variation in y_i that is explained by x_i (or the OLS regression line) as

$$R^2 = \frac{SSE}{SST} = 1 - \frac{SSR}{SST}$$

which is called the R-squared of the regression. It can be shown to be equal to the *square* of the correlation between y_i and $\widehat{y}_i$. Therefore $0 \le R^2 \le 1$. An R-squared of zero means no linear relationship between y_i and x_i, and an R-squared of one means a perfect linear relationship (e.g., $y_i = x_i + 2$). As R^2 increases, the y_i are closer and closer to falling on the OLS regression line.

I would encourage you not to fixate on R-squared in research projects where the aim is to estimate some causal effect, though. It's a useful summary measure, but it does not tell us about causality. Remember, you aren't trying to explain variation in y if you are trying to estimate some causal effect. The R^2 tells us how much of the variation in y_i is explained by the explanatory variables. But if we are interested

---

15 Recall the earlier discussion about degrees-of-freedom correction.

in the causal effect of a single variable, R^2 is irrelevant. For causal inference, we need equation 2.28.

Expected value of OLS. Up until now, we motivated simple regression using a population model. But our analysis has been purely algebraic, based on a sample of data. So residuals always average to zero when we apply OLS to a sample, regardless of any underlying model. But our job gets tougher. Now we have to study the statistical properties of the OLS estimator, referring to a population model and assuming random sampling.[16]

The field of mathematical statistics is concerned with questions. How do estimators behave across different samples of data? On average, for instance, will we get the right answer if we repeatedly sample? We need to find the expected value of the OLS estimators—in effect, the average outcome across all possible random samples— and determine whether we are right, on average. This leads naturally to a characteristic called unbiasedness, which is desirable of all estimators.

$$E(\widehat{\beta}) = \beta \qquad (2.37)$$

Remember, our objective is to estimate β_1, which is the slope *population* parameter that describes the relationship between y and x. Our estimate, $\widehat{\beta}_1$, is an estimator of that parameter obtained for a specific sample. Different samples will generate different estimates $(\widehat{\beta}_1)$ for the "true" (and unobserved) β_1. Unbiasedness means that if we could take as many random samples on Y as we want from the population and compute an estimate each time, the average of the estimates would be equal to β_1.

There are several assumptions required for OLS to be unbiased. The first assumption is called linear in the parameters. Assume a population model

$$y = \beta_0 + \beta_1 x + u$$

---

16 This section is a review of traditional econometrics pedagogy. We cover it for the sake of completeness. Traditionally, econometricians motivated their discuss of causality through ideas like unbiasedness and consistency.

where β_0 and β_1 are the unknown population parameters. We view x and u as outcomes of random variables generated by some data-generating process. Thus, since y is a function of x and u, both of which are random, then y is also random. Stating this assumption formally shows that our goal is to estimate β_0 and β_1.

Our second assumption is random sampling. We have a random sample of size n, $\{(x_i, y_i): i = 1, \ldots, n\}$, following the population model. We know how to use this data to estimate β_0 and β_1 by OLS. Because each i is a draw from the population, we can write, for each i:

$$y_i = \beta_0 + \beta_1 x_i + u_i$$

Notice that u_i here is the unobserved error for observation i. It is *not* the residual that we compute from the data.

The third assumption is called sample variation in the explanatory variable. That is, the sample outcomes on x_i are not all the same value. This is the same as saying that the sample variance of x is not zero. In practice, this is no assumption at all. If the x_i all have the same value (i.e., are constant), we cannot learn how x affects y in the population. Recall that OLS is the covariance of y and x divided by the variance in x, and so if x is constant, then we are dividing by zero, and the OLS estimator is undefined.

With the fourth assumption our assumptions start to have real teeth. It is called the zero conditional mean assumption and is probably the most critical assumption in causal inference. In the population, the error term has zero mean given any value of the explanatory variable:

$$E(u \mid x) = E(u) = 0$$

This is the key assumption for showing that OLS is unbiased, with the zero value being of no importance once we assume that $E(u \mid x)$ does not change with x. Note that we can compute OLS estimates whether or not this assumption holds, even if there is an underlying population model.

So, how do we show that $\widehat{\beta_1}$ is an unbiased estimate of β_1 (equation 2.37)? We need to show that under the four assumptions we just outlined, the expected value of $\widehat{\beta}_1$, when averaged across random samples, will center on the true value of β_1. This is a subtle yet critical concept. Unbiasedness in this context means that if we repeatedly sample data from a population and run a regression on each new sample, the average over all those estimated coefficients will equal the true value of β_1. We will discuss the answer as a series of steps.

Step 1: Write down a formula for $\widehat{\beta}_1$. It is convenient to use the $\dfrac{C(x,y)}{V(x)}$ form:

$$\widehat{\beta}_1 = \frac{\sum_{i=1}^{n}(x_i - \overline{x})y_i}{\sum_{i=1}^{n}(x_i - \overline{x})^2}$$

Let's get rid of some of this notational clutter by defining $\sum_{i=1}^{n}(x_i - \overline{x})^2 = SST_x$ (i.e., total variation in the x_i) and rewrite this as:

$$\widehat{\beta}_1 = \frac{\sum_{i=1}^{n}(x_i - \overline{x})y_i}{SST_x}$$

Step 2: Replace each y_i with $y_i = \beta_0 + \beta_1 x_i + u_i$, which uses the first linear assumption and the fact that we have sampled data (our second assumption). The numerator becomes:

$$\sum_{i=1}^{n}(x_i - \overline{x})y_i = \sum_{i=1}^{n}(x_i - \overline{x})(\beta_0 + \beta_1 x_i + u_i)$$

$$= \beta_0 \sum_{i=1}^{n}(x_i - \overline{x}) + \beta_1 \sum_{i=1}^{n}(x_i - \overline{x})x_i + \sum_{i=1}^{n}(x_i + \overline{x})u_i$$

$$= 0 + \beta_1 \sum_{i=1}^{n}(x_i - \overline{x})^2 + \sum_{i=1}^{n}(x_i - \overline{x})u_i$$

$$= \beta_1 SST_x + \sum_{i=1}^{n}(x_i - \overline{x})u_i$$

Note, we used $\sum_{i=1}^{n}(x_i - \overline{x}) = 0$ and $\sum_{i=1}^{n}(x_i - \overline{x})x_i = \sum_{i=1}^{n}(x_i - \overline{x})^2$ to do this.[17]

---

17 Told you we would use this result a lot.

We have shown that:

$$\widehat{\beta}_1 = \frac{\beta_1 SST_x + \sum_{i=1}^{n}(x_i - \overline{x})u_i}{SST_x}$$

$$= \beta_1 + \frac{\sum_{i=1}^{n}(x_i - \overline{x})u_i}{SST_x}$$

Note that the last piece is the slope coefficient from the OLS regression of u_i on x_i, i: $1, \ldots, n$.[18] We cannot do this regression because the u_i are not observed. Now define $w_i = \dfrac{(x_i - \overline{x})}{SST_x}$ so that we have the following:

$$\widehat{\beta}_1 = \beta_1 + \sum_{i=1}^{n} w_i u_i$$

This has showed us the following: First, $\widehat{\beta}_1$ is a linear function of the unobserved errors, u_i. The w_i are all functions of $\{x_1, \ldots, x_n\}$. Second, the random difference between β_1 and the estimate of it, $\widehat{\beta}_1$, is due to this linear function of the unobservables.

Step 3: Find $E(\widehat{\beta}_1)$. Under the random sampling assumption and the zero conditional mean assumption, $E(u_i \mid x_1, \ldots, x_n) = 0$, that means conditional on each of the x variables:

$$E\big(w_i u_i \mid x_1, \ldots, x_n\big) = w_i E\big(u_i \mid x_1, \ldots, x_n\big) = 0$$

because w_i is a function of $\{x_1, \ldots, x_n\}$. This would be true if in the population u and x are correlated.

Now we can complete the proof: conditional on $\{x_1, \ldots, x_n\}$,

$$E(\widehat{\beta}_1) = E\left(\beta_1 + \sum_{i=1}^{n} w_i u_i\right)$$

$$= \beta_1 + \sum_{i=1}^{n} E(w_i u_i)$$

$$= \beta_1 + \sum_{i=1}^{n} w_i E(u_i)$$

---

18 I find it interesting that we see so many $\dfrac{cov}{var}$ terms when working with regression. They show up constantly. Keep your eyes peeled.

$$= \beta_1 + 0$$

$$= \beta_1$$

Remember, β_1 is the fixed constant in the population. The estimator, $\widehat{\beta}_1$, varies across samples and is the random outcome: before we collect our data, we do not know what $\widehat{\beta}_1$ will be. Under the four aforementioned assumptions, $E(\widehat{\beta}_0) = \beta_0$ and $E(\widehat{\beta}_1) = \beta_1$.

I find it helpful to be concrete when we work through exercises like this. So let's visualize this. Let's create a Monte Carlo simulation. We have the following population model:

$$y = 3 + 2x + u \qquad (2.38)$$

where $x \sim Normal(0,9)$, $u \sim Normal(0,36)$. Also, x and u are independent. The following Monte Carlo simulation will estimate OLS on a sample of data 1,000 times. The true β parameter equals 2. But what will the average $\widehat{\beta}$ equal when we use repeated sampling?

STATA
ols3.do

```
1   clear all
2   program define ols, rclass
3   version 14.2
4   syntax [, obs(integer 1) mu(real 0) sigma(real 1) ]
5
6       clear
7       drop _all
8       set obs 10000
9       gen x = 9*rnormal()
10      gen u  = 36*rnormal()
11      gen y  = 3 + 2*x + u
12      reg y x
13      end
14
15  simulate beta=_b[x], reps(1000): ols
16  su
17  hist beta
```

R
ols3.R

```
1   library(tidyverse)
2
3   lm <- lapply(
4     1:1000,
5     function(x) tibble(
6       x = 9*rnorm(10000),
7       u = 36*rnorm(10000),
8       y = 3 + 2*x + u
9     ) %>%
10      lm(y ~ x, .)
11  )
12
13  as_tibble(t(sapply(lm, coef))) %>%
14    summary(x)
15
16  as_tibble(t(sapply(lm, coef))) %>%
17    ggplot()+
18    geom_histogram(aes(x), binwidth = 0.01)
```

Table 7 gives us the mean value of $\widehat{\beta}_1$ over the 1,000 repetitions (repeated sampling). Your results will differ from mine here only in the randomness involved in the simulation. But your results should be similar to what is shown here. While each sample had a different estimated slope, the average for $\widehat{\beta}_1$ over all the samples was 1.998317, which is close to the true value of 2 (see equation 2.38). The standard deviation in this estimator was 0.0398413, which is close to the standard error recorded in the regression itself.[19] Thus, we see that the estimate is the mean value of the coefficient from repeated sampling, and the standard error is the standard deviation from that repeated estimation. We can see the distribution of these coefficient estimates in Figure 5.

The problem is, we don't know which kind of sample we have. Do we have one of the "almost exactly 2" samples, or do we have one of the "pretty different from 2" samples? We can never know whether we are close to the population value. We hope that our sample is "typical" and

19 The standard error I found from running this on one sample of data was 0.0403616.

Table 7. Monte Carlo simulation of OLS.

Variable	Obs.	Mean	*SD*
beta	1,000	1.998317	0.0398413

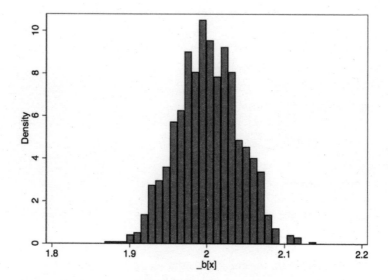

Figure 5. Distribution of coefficients from Monte Carlo simulation.

produces a slope estimate close to $\widehat{\beta}_1$, but we can't know. Unbiasedness is a property of the procedure of the rule. It is not a property of the estimate itself. For example, say we estimated an that 8.2% return on schooling. It is tempting to say that 8.2% is an unbiased estimate of the return to schooling, but that's technically incorrect. The rule used to get $\widehat{\beta}_1 = 0.082$ is unbiased (if we believe that u is unrelated to schooling), not the actual estimate itself.

Law of iterated expectations. The conditional expectation function (CEF) is the mean of some outcome y with some covariate x held fixed. Let's focus more intently on this function.[20] Let's get the notation and

20 I highly encourage the interested reader to study Angrist and Pischke [2009], who have an excellent discussion of LIE there.

some of the syntax out of the way. As noted earlier, we write the CEF as $E(y_i \mid x_i)$. Note that the CEF is explicitly a function of x_i. And because x_i is random, the CEF is random—although sometimes we work with particular values for x_i, like $E(y_i \mid x_i = 8$ years schooling) or $E(y_i \mid x_i = $ Female). When there are treatment variables, then the CEF takes on two values: $E(y_i \mid d_i = 0)$ and $E(y_i \mid d_i = 1)$. But these are special cases only.

An important complement to the CEF is the law of iterated expectations (LIE). This law says that an unconditional expectation can be written as the unconditional average of the CEF. In other words, $E(y_i) = E\{E(y_i \mid x_i)\}$. This is a fairly simple idea: if you want to know the unconditional expectation of some random variable y, you can simply calculate the weighted sum of all conditional expectations with respect to some covariate x. Let's look at an example. Let's say that average grade-point for females is 3.5, average GPA for males is a 3.2, half the population is female, and half is male. Then:

$$E[GPA] = E\{E(GPA_i \mid Gender_i)\}$$
$$= (0.5 \times 3.5) + (3.2 \times 0.5)$$
$$= 3.35$$

You probably use LIE all the time and didn't even know it. The proof is not complicated. Let x_i and y_i each be continuously distributed. The joint density is defined as $f_{xy}(u,t)$. The conditional distribution of y given $x = u$ is defined as $f_y(t \mid x_i = u)$. The marginal densities are $g_y(t)$ and $g_x(u)$.

$$E\{E(y \mid x)\} = \int E(y \mid x = u) g_x(u) du$$
$$= \int \left[\int t f_{y|x}(t \mid x = u) dt \right] g_x(u) du$$
$$= \int \int t f_{y|x}(t \mid x = u) g_x(u) du dt$$
$$= \int t \left[\int f_{y|x}(t \mid x = u) g_x(u) du \right] dt$$
$$= \int t [f_{x,y} du] dt$$

$$= \int t g_y(t) dt$$

$$= E(y)$$

Check out how easy this proof is. The first line uses the definition of expectation. The second line uses the definition of conditional expectation. The third line switches the integration order. The fourth line uses the definition of joint density. The fifth line replaces the prior line with the subsequent expression. The sixth line integrates joint density over the support of x which is equal to the marginal density of y. So restating the law of iterated expectations: $E(y_i) = E\{E(y \mid x_i)\}$.

CEF decomposition property. The first property of the CEF we will discuss is the CEF decomposition property. The power of LIE comes from the way it breaks a random variable into two pieces—the CEF and a residual with special properties. The CEF decomposition property states that

$$y_i = E(y_i \mid x_i) + \varepsilon_i$$

where (i) ε_i is mean independent of x_i, That is,

$$E(\varepsilon_i \mid x_i) = 0$$

and (ii) ε_i is not correlated with any function of x_i.

The theorem says that any random variable y_i can be decomposed into a piece that is explained by x_i (the CEF) and a piece that is left over and orthogonal to any function of x_i. I'll prove the (i) part first. Recall that $\varepsilon_i = y_i - E(y_i \mid x_i)$ as we will make a substitution in the second line below.

$$E(\varepsilon_i \mid x_i) = E\left(y_i - E(y_i \mid x_i) \mid x_i\right)$$

$$= E(y_i \mid x_i) - E(y_i \mid x_i)$$

$$= 0$$

The second part of the theorem states that ε_i is uncorrelated with any function of x_i. Let $h(x_i)$ be any function of x_i. Then $E(h(x_i)\varepsilon_i) =$

$E\{h(x_i)E(\varepsilon_i \mid x_i)\}$ The second term in the interior product is equal to zero by mean independence.[21]

CEF prediction property. The second property is the CEF prediction property. This states that $E(y_i \mid x_i) = \arg\min_{m(x_i)} E[(y - m(x_i))^2]$, where $m(x_i)$ is any function of x_i. In words, this states that the CEF is the minimum mean squared error of y_i given x_i. By adding $E(y_i \mid x_i) - E(y_i \mid x_i) = 0$ to the right side we get

$$\left[y_i - m(x_i)\right]^2 = \left[(y_i - E[y_i \mid x_i]) + (E(y_i \mid x_i) - m(x_i))\right]^2$$

I personally find this easier to follow with simpler notation. So replace this expression with the following terms:

$$(a - b + b - c)^2$$

Distribute the terms, rearrange them, and replace the terms with their original values until you get the following:

$$\arg\min \left(y_i - E(y_i \mid x_i)\right)^2 + 2\left(E(y_i \mid x_i) - m(x_i)\right) \times \left(y_i - E(y_i \mid x_i)\right)$$
$$+ \left(E(y_i \mid x_i) + m(x_i)\right)^2$$

Now minimize the function with respect to $m(x_i)$. When minimizing this function with respect to $m(x_i)$, note that the first term $(y_i - E(y_i \mid x_i))^2$ doesn't matter because it does not depend on $m(x_i)$. So it will zero out. The second and third terms, though, do depend on $m(x_i)$. So rewrite $2(E(y_i \mid x_i) - m(x_i))$ as $h(x_i)$. Also set ε_i equal to $[y_i - E(y_i \mid x_i)]$ and substitute

$$\arg\min \varepsilon_i^2 + h(x_i)\varepsilon_i + \left[E(y_i \mid x_i) + m(x_i)\right]^2$$

Now minimizing this function and setting it equal to zero we get

$$h'(x_i)\varepsilon_i$$

which equals zero by the decomposition property.

21 Let's take a concrete example of this proof. Let $h(x_i) = \alpha + \gamma x_i$. Then take the joint expectation $E(h(x_i)\varepsilon_i) = E([\alpha + \gamma x_i]\varepsilon_i)$. Then take conditional expectations $E(\alpha \mid x_i) + E(\gamma \mid x_i)E(x_i \mid x_i)E(\varepsilon \mid x_i)\} = \alpha + x_i E(\varepsilon_i \mid x_i) = 0$ after we pass the conditional expectation through.

ANOVA theory. The final property of the CEF that we will discuss is the analysis of variance theorem, or ANOVA. According to this theorem, the unconditional variance in some random variable is equal to the variance in the conditional expectation plus the expectation of the conditional variance, or

$$V(y_i) = V\Big[E(y_i \mid x_i)\Big] + E\Big[V(y_i \mid x_i)\Big]$$

where V is the variance and $V(y_i \mid x_i)$ is the conditional variance.

Linear CEF theorem. As you probably know by now, the use of least squares in applied work is extremely common. That's because regression has several justifications. We discussed one—unbiasedness under certain assumptions about the error term. But I'd like to present some slightly different arguments. Angrist and Pischke [2009] argue that linear regression may be useful even if the underlying CEF itself is not linear, because regression is a good approximation of the CEF. So keep an open mind as I break this down a little bit more.

Angrist and Pischke [2009] give several arguments for using regression, and the linear CEF theorem is probably the easiest. Let's assume that we are sure that the CEF itself is linear. So what? Well, if the CEF is linear, then the linear CEF theorem states that the population regression is equal to that linear CEF. And if the CEF is linear, and if the population regression equals it, then of course you should use the population regression to estimate CEF. If you need a proof for what could just as easily be considered common sense, I provide one. If $E(y_i \mid x_i)$ is linear, then $E(y_i \mid x_i) = x'\widehat{\beta}$ for some vector $\widehat{\beta}$. By the decomposition property, you get:

$$E\Big(x(y - E(y \mid x))\Big) = E\Big(x(y - x'\widehat{\beta})\Big) = 0$$

And then when you solve this, you get $\widehat{\beta} = \beta$. Hence $E(y \mid x) = x'\beta$.

Best linear predictor theorem. There are a few other linear theorems that are worth bringing up in this context. For instance, recall that the CEF is the minimum mean squared error predictor of y given x in the class of all functions, according to the CEF prediction property. Given this, the population regression function is the best that we can do in the class of all linear functions.[22]

Regression CEF theorem. I would now like to cover one more attribute of regression. The function $X\beta$ provides the minimum mean squared error linear approximation to the CEF. That is,

$$\beta = \arg\min_{b} E\left\{[E(y_i \mid x_i) - x_i'b]^2\right\}$$

So? Let's try and back up for a second, though, and get the big picture, as all these linear theorems can leave the reader asking, "So what?" I'm telling you all of this because I want to present to you an argument that regression is appealing; even though it's linear, it can still be justified when the CEF itself isn't. And since we don't know with certainty that the CEF is linear, this is actually a nice argument to at least consider. Regression is ultimately nothing more than a crank turning data into estimates, and what I'm saying here is that crank produces something desirable even under bad situations. Let's look a little bit more at this crank, though, by reviewing another theorem which has become popularly known as the regression anatomy theorem.

Regression anatomy theorem. In addition to our discussion of the CEF and regression theorems, we now dissect the regression itself. Here we discuss the regression anatomy theorem. The regression anatomy theorem is based on earlier work by Frisch and Waugh [1933] and Lovell [1963].[23] I find the theorem more intuitive when I think through a specific example and offer up some data visualization. In my opinion, the theorem helps us interpret the individual coefficients of a multiple linear regression model. Say that we are interested in the causal effect of family size on labor supply. We want to regress labor supply on family size:

$$Y_i = \beta_0 + \beta_1 X_i + u_i$$

where Y is labor supply, and X is family size.

22 See Angrist and Pischke [2009] for a proof.

23 A helpful proof of the Frisch-Waugh-Lovell theorem can be found in Lovell [2008].

If family size is truly random, then the number of kids in a family is uncorrelated with the unobserved error term.[24] This implies that when we regress labor supply on family size, our estimate, $\widehat{\beta_1}$, can be interpreted as the causal effect of family size on labor supply. We could just plot the regression coefficient in a scatter plot showing all i pairs of data; the slope coefficient would be the best linear fit of the data for this data cloud. Furthermore, under randomized number of children, the slope would also tell us the average causal effect of family size on labor supply.

But most likely, family size isn't random, because so many people choose the number of children to have in their family—instead of, say, flipping a coin. So how do we interpret $\widehat{\beta_1}$ if the family size is *not* random? Often, people choose their family size according to something akin to an optimal stopping rule. People pick both how many kids to have, when to have them, and when to stop having them. In some instances, they may even attempt to pick the gender. All of these choices are based on a variety of unobserved and observed economic factors that may themselves be associated with one's decision to enter the labor market. In other words, using the language we've been using up until now, it's unlikely that $E(u \mid X) = E(u) = 0$.

But let's say that we have reason to think that the number of kids in a family is *conditionally* random. To make this tractable for the sake of pedagogy, let's say that a particular person's family size is as good as randomly chosen once we condition on race and age.[25] While unrealistic, I include it to illustrate an important point regarding multivariate regressions. If this assumption were to be true, then we could write the following equation:

$$Y_i = \beta_0 + \beta_1 X_i + \gamma_1 R_i + \gamma_2 A_i + u_i$$

where Y is labor supply, X is number of kids, R is race, A is age, and u is the population error term.

---

24 While randomly having kids may sound fun, I encourage you to have kids when you want to have them. Contact your local high school health teacher to learn more about a number of methods that can reasonably minimize the number of random children you create.

25 Almost certainly a ridiculous assumption, but stick with me.

If we want to estimate the average causal effect of family size on labor supply, then we need two things. First, we need a sample of *data* containing all four of those variables. Without all four of the variables, we cannot estimate this regression model. Second, we need for the number of kids, X, to be randomly assigned for a given set of race and age.

Now, how do we interpret $\widehat{\beta}_1$? And how might we visualize this coefficient given that there are six dimensions to the data? The regression anatomy theorem both tells us what this coefficient estimate actually means and also lets us visualize the data in only two dimensions.

To explain the intuition of the regression anatomy theorem, let's write down a population model with multiple variables. Assume that your main multiple regression model of interest has K covariates. We can then write it as:

$$y_i = \beta_0 + \beta_1 x_{1i} + \cdots + \beta_k x_{ki} + \cdots + \beta_K x_{Ki} + e_i \qquad (2.39)$$

Now assume an *auxiliary* regression in which the variable x_{1i} is regressed on all the remaining independent variables:

$$x_{1i} = \gamma_0 + \gamma_{k-1} x_{k-1i} + \gamma_{k+1} x_{k+1i} + \cdots + \gamma_K x_{Ki} + f_i \qquad (2.40)$$

and $\tilde{x}_{1i} = x_{1i} - \widehat{x}_{1i}$ is the residual from that auxiliary regression. Then the parameter β_1 can be rewritten as:

$$\beta_1 = \frac{C(y_i, \tilde{x}_i)}{V(\tilde{x}_i)} \qquad (2.41)$$

Notice that again we see that the coefficient estimate is a scaled covariance, only here, the covariance is with respect to the outcome and residual from the auxiliary regression, and the scale is the variance of that same residual.

To prove the theorem, note that $E[\tilde{x}_{ki}] = E[x_{ki}] - E[\widehat{x}_{ki}] = E[f_i]$, and plug y_i and residual $\tilde{x}_{ki}$ from x_{ki} auxiliary regression into the covariance $cov(y_i, x_{ki})$:

$$\beta_k = \frac{cov(\beta_0 + \beta_1 x_{1i} + \cdots + \beta_k x_{ki} + \cdots + \beta_K x_{Ki} + e_i, \tilde{x}_{ki})}{var(\tilde{x}_{ki})}$$

$$= \frac{cov(\beta_0 + \beta_1 x_{1i} + \cdots + \beta_k x_{ki} + \cdots + \beta_K x_{Ki} + e_i, f_i)}{var(f_i)}$$

Since by construction $E[f_i] = 0$, it follows that the term $\beta_0 E[f_i] = 0$. Since f_i is a linear combination of all the independent variables with the exception of x_{ki}, it must be that

$$\beta_1 E[f_i x_{1i}] = \cdots = \beta_{k-1} E[f_i x_{k-1i}] = \beta_{k+1} E[f_i x_{k+1i}] = \cdots = \beta_K E[f_i x_{Ki}] = 0$$

Consider now the term $E[e_i f_i]$. This can be written as

$$E[e_i f_i] = E[e_i f_i]$$
$$= E[e_i \tilde{x}_{ki}]$$
$$= E\left[e_i(x_{ki} - \widehat{x}_{ki})\right]$$
$$= E[e_i x_{ki}] - E[e_i \tilde{x}_{ki}]$$

Since e_i is uncorrelated with any independent variable, it is also uncorrelated with x_{ki}. Accordingly, we have $E[e_i x_{ki}] = 0$. With regard to the second term of the subtraction, substituting the predicted value from the x_{ki} auxiliary regression, we get

$$E[e_i \tilde{x}_{ki}] = E\left[e_i(\widehat{\gamma}_0 + \widehat{\gamma}_1 x_{1i} + \cdots + \widehat{\gamma}_{k-1i} + \widehat{\gamma}_{k+1} x_{k+1i} + \cdots + \widehat{x}_K x_{Ki})\right]$$

Once again, since e_i is uncorrelated with any independent variable, the expected value of the terms is equal to zero. It follows that $E[e_i f_i] = 0$.

The only remaining term, then, is $[\beta_k x_{ki} f_i]$, which equals $E[\beta_k x_{ki} \tilde{x}_{ki}]$, since $f_i = \tilde{x}_{ki}$. The term x_{ki} can be substituted by rewriting the auxiliary regression model, x_{ki}, such that

$$x_{ki} = E[x_{ki} \mid X_{-k}] + \tilde{x}_{ki}$$

This gives

$$E[\beta_k x_{ki} \tilde{x}_{ki}] = \beta_k E\left[\tilde{x}_{ki}(E[x_{ki} \mid X_{-k}] + \tilde{x}_{ki})\right]$$
$$= \beta_k \left\{ E[\tilde{x}_{ki}^2] + E[(E[x_{ki} \mid X_{-k}]\tilde{x}_{ki})]\right\}$$
$$= \beta_k \, \text{var}(\tilde{x}_{ki})$$

which follows directly from the orthogonality between $E[x_{ki} \mid X_{-k}]$ and $\tilde{x}_{ki}$. From previous derivations we finally get

$$\text{cov}(y_i, \tilde{x}_{ki}) = \beta_k \text{var}(\tilde{x}_{ki})$$

which completes the proof.

I find it helpful to visualize things. Let's look at an example in Stata using its popular automobile data set. I'll show you:

STATA
reganat.do

```
1  ssc install reganat, replace
2  sysuse auto.dta, replace
3  regress price length
4  regress price length weight headroom mpg
5  reganat price length weight headroom mpg, dis(length) biline
```

R
reganat.R

```
1   library(tidyverse)
2   library(haven)
3
4   read_data <- function(df)
5   {
6     full_path <- paste("https://raw.github.com/scunning1975/mixtape/master/",
7                 df, sep = "")
8     df <- read_dta(full_path)
9     return(df)
10  }
11
12
13  auto <- read_data("auto.dta") %>%
14    mutate(length = length - mean(length))
15
16  lm1 <- lm(price ~ length, auto)
17  lm2 <- lm(price ~ length + weight + headroom + mpg, auto)
18
19
```

(continued)

R *(continued)*
20   coef_lm1 <- lm1$coefficients
21   coef_lm2 <- lm2$coefficients
22   resid_lm2 <- lm2$residuals
23
24   y_single <- tibble(price = coef_lm1[1] + coef_lm1[2]*auto$length,
25   length = auto$length)
26
27   y_multi <- tibble(price = coef_lm1[1] + coef_lm2[2]*auto$length,
28   length = auto$length)
29
30
31   ggplot(auto) +
32   geom_point(aes(x = length, y = price)) +
33   geom_smooth(aes(x = length, y = price), data = y_multi, color = "blue") +
34   geom_smooth(aes(x = length, y = price), data = y_single, color="red")

Let's walk through both the regression output that I've reproduced in Table 8 as well as a nice visualization of the slope parameters in what I'll call the short bivariate regression and the longer multivariate regression. The short regression of price on car length yields a coefficient of 57.20 on length. For every additional inch, a car is $57 more expensive, which is shown by the upward-sloping, dashed line in Figure 6. The slope of that line is 57.20.

It will eventually become second nature for you to talk about including more variables on the right side of a regression as "controlling for"

Table 8. Regression estimates of automobile price on length and other characteristics.

Covariates	**Short Regression**	**Long Regression**
Length	57.20	−94.50
	(14.08)	(40.40)
Weight		4.34
		(1.16)
Headroom		−490.97
		(388.49)
Miles per gallon		−87.96
		(83.59)

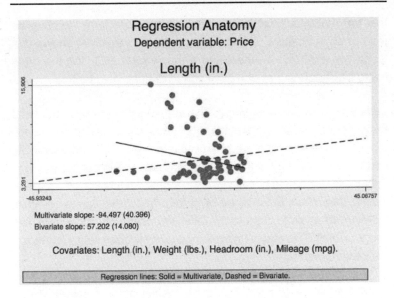

Figure 6. Regression anatomy display.

those variables. But in this regression anatomy exercise, I hope to give a different interpretation of what you're doing when you in fact "control for" variables in a regression. First, notice how the coefficient on length changed signs and increased in magnitude once we controlled for the other variables. Now, the effect on length is -94.5. It appears that the length was confounded by several other variables, and once we conditioned on them, longer cars actually were cheaper. You can see a visual representation of this in Figure 6, where the multivariate slope is negative.

So what exactly going on in this visualization? Well, for one, it has condensed the number of dimensions (variables) from four to only two. It did this through the regression anatomy process that we described earlier. Basically, we ran the auxiliary regression, used the residuals from it, and then calculated the slope coefficient as $\dfrac{\mathrm{cov}(y_i, \tilde{x}_i)}{\mathrm{var}(\tilde{x}_i)}$. This allowed us to show scatter plots of the auxiliary residuals paired with their outcome observations and to slice the slope through them (Figure 6). Notice that this is a useful way to preview the multidimensional correlation between two variables from a multivariate regression. Notice that the solid black line is negative and the slope from the bivariate regression is positive. The regression anatomy theorem shows that these two estimators—one being a multivariate OLS and the other being a bivariate regression price and a residual—are identical.

Variance of the OLS estimators. That more or less summarizes what we want to discuss regarding the linear regression. Under a zero conditional mean assumption, we could epistemologically infer that the rule used to produce the coefficient from a regression in our sample was unbiased. That's nice because it tells us that we have good reason to believe that result. But now we need to build out this epistemological justification so as to capture the inherent uncertainty in the sampling process itself. This added layer of uncertainty is often called inference. Let's turn to it now.

Remember the simulation we ran earlier in which we resampled a population and estimated regression coefficients a thousand times? We produced a histogram of those 1,000 estimates in Figure 5. The mean of the coefficients was around 1.998, which was very close to the true effect of 2 (hard-coded into the data-generating process). But the standard deviation was around 0.04. This means that, basically, in repeated sampling of some population, we got different estimates. But the average of those estimates was close to the true effect, and their spread had a standard deviation of 0.04. This concept of spread in repeated sampling is probably the most useful thing to keep in mind as we move through this section.

Under the four assumptions we discussed earlier, the OLS estimators are unbiased. But these assumptions are not sufficient to tell us anything about the variance in the estimator itself. The assumptions help inform our beliefs that the estimated coefficients, on average, equal the parameter values themselves. But to speak intelligently about the variance of the estimator, we need a measure of dispersion in the sampling distribution of the estimators. As we've been saying, this leads us to the variance and ultimately to the standard deviation. We could characterize the variance of the OLS estimators under the four assumptions. But for now, it's easiest to introduce an assumption that simplifies the calculations. We'll keep the assumption ordering we've been using and call this the fifth assumption.

The fifth assumption is the homoskedasticity or constant variance assumption. This assumption stipulates that our population error term, u, has the same variance given any value of the explanatory variable, x. Formally, this is:

$$V(u \mid x) = \sigma^2 \tag{2.42}$$

When I was first learning this material, I always had an unusually hard time wrapping my head around σ^2. Part of it was because of my humanities background; I didn't really have an appreciation for random variables that were dispersed. I wasn't used to taking a lot of numbers and trying to measure distances between them, so things were slow to click. So if you're like me, try this. Think of σ^2 as just a positive number like 2 or 8. That number is measuring the spreading out of underlying errors themselves. In other words, the variance of the errors conditional on the explanatory variable is simply some finite, positive number. And that number is measuring the variance of the stuff other than x that influence the value of y itself. And because we assume the zero conditional mean assumption, whenever we assume homoskedasticity, we can also write:

$$E(u^2 \mid x) = \sigma^2 = E(u^2) \tag{2.43}$$

Now, under the first, fourth, and fifth assumptions, we can write:

$$E(y \mid x) = \beta_0 + \beta_1 x$$
$$V(y \mid x) = \sigma^2 \tag{2.44}$$

So the average, or expected, value of y is allowed to change with x, but if the errors are homoskedastic, then the variance does not change with x. The constant variance assumption may not be realistic; it must be determined on a case-by-case basis.

Theorem: Sampling variance of OLS. Under assumptions 1 and 2, we get:

$$V(\hat{\beta}_1 \mid x) = \frac{\sigma^2}{\sum_{i=1}^{n}(x_i - \bar{x})^2}$$
$$= \frac{\sigma^2}{SST_x} \tag{2.45}$$

$$V(\widehat{\beta}_0 \mid x) = \frac{\sigma^2 \left(\frac{1}{n} \sum_{i=1}^{n} x_i^2 \right)}{SST_x} \tag{2.46}$$

To show this, write, as before,

$$\widehat{\beta}_1 = \beta_1 + \sum_{i=1}^{n} w_i u_i \tag{2.47}$$

where $w_i = \dfrac{(x_i - \bar{x})}{SST_x}$. We are treating this as nonrandom in the derivation. Because β_1 is a constant, it does not affect $V(\widehat{\beta}_1)$. Now, we need to use the fact that, for uncorrelated random variables, the variance of the sum is the sum of the variances. The $\{u_i : i = 1, \ldots, n\}$ are actually independent across i and uncorrelated. Remember: if we know x, we know w. So:

$$V(\widehat{\beta}_1 \mid x) = \text{Var}\left(\beta_1 + \sum_{i=1}^{n} w_i u_i \mid x \right) \tag{2.48}$$

$$= \text{Var}\left(\sum_{i=1}^{n} w_i u_i \mid x \right) \tag{2.49}$$

$$= \sum_{i=1}^{n} \text{Var}\left(w_i u_i \mid x \right) \tag{2.50}$$

$$= \sum_{i=1}^{n} w_i^2 \text{Var}\left(u_i \mid x \right) \tag{2.51}$$

$$= \sum_{i=1}^{n} w_i^2 \sigma^2 \tag{2.52}$$

$$= \sigma^2 \sum_{i=1}^{n} w_i^2 \tag{2.53}$$

where the penultimate equality condition used the fifth assumption so that the variance of u_i does not depend on x_i. Now we have:

$$\sum_{i=1}^{n} w_i^2 = \sum_{i=1}^{n} \frac{(x_i - \bar{x})^2}{SST_x^2} \tag{2.54}$$

$$= \frac{\sum_{i=1}^{n} (x_i - \bar{x})^2}{SST_x^2} \tag{2.55}$$

$$= \frac{SST_x}{SST_x^2} \tag{2.56}$$

$$= \frac{1}{SST_x} \tag{2.57}$$

We have shown:

$$V(\widehat{\beta_1}) = \frac{\sigma^2}{SST_x} \tag{2.58}$$

A couple of points. First, this is the "standard" formula for the variance of the OLS slope estimator. It is *not* valid if the fifth assumption, of homoskedastic errors, doesn't hold. The homoskedasticity assumption is needed, in other words, to derive this standard formula. But the homoskedasticity assumption is *not* used to show unbiasedness of the OLS estimators. That requires only the first four assumptions.

Usually, we are interested in β_1. We can easily study the two factors that affect its variance: the numerator and the denominator.

$$V(\widehat{\beta_1}) = \frac{\sigma^2}{SST_x} \tag{2.59}$$

As the error variance increases—that is, as σ^2 increases—so does the variance in our estimator. The more "noise" in the relationship between y and x (i.e., the larger the variability in u), the harder it is to learn something about β_1. In contrast, more variation in $\{x_i\}$ is a *good* thing. As SST_x rises, $V(\widehat{\beta_1}) \downarrow$.

Notice that $\frac{SST_x}{n}$ is the sample variance in x. We can think of this as getting close to the population variance of x, σ_x^2, as n gets large. This means:

$$SST_x \approx n\sigma_x^2 \tag{2.60}$$

which means that as n grows, $V(\widehat{\beta_1})$ shrinks at the rate of $\frac{1}{n}$. This is why more data is a good thing: it shrinks the sampling variance of our estimators.

The standard deviation of $\widehat{\beta_1}$ is the square root of the variance. So:

$$sd(\widehat{\beta_1}) = \frac{\sigma}{\sqrt{SST_x}} \tag{2.61}$$

This turns out to be the measure of variation that appears in confidence intervals and test statistics.

Next we look at estimating the error variance. In the formula, $V(\widehat{\beta_1}) = \dfrac{\sigma^2}{SST_x}$, we can compute SST_x from $\{x_i : i = 1,\ldots,n\}$. But we need to estimate σ^2. Recall that $\sigma^2 = E(u^2)$. Therefore, if we could observe a sample on the errors, $\{u_i : i = 1,\ldots,n\}$, an unbiased estimator of σ^2 would be the sample average:

$$\frac{1}{n}\sum_{i=1}^{n} u_i^2 \tag{2.62}$$

But this isn't an estimator that we can compute from the data we observe, because u_i are unobserved. How about replacing each u_i with its "estimate," the OLS residual $\widehat{u_i}$?

$$u_i = y_i - \beta_0 - \beta_1 x_i \tag{2.63}$$

$$\widehat{u_i} = y_i - \widehat{\beta_0} - \widehat{\beta_1} x_i \tag{2.64}$$

Whereas u_i *cannot* be computed, $\widehat{u_i}$ can be computed from the data because it depends on the estimators, $\widehat{\beta_0}$ and $\widehat{\beta_1}$. But, except by sheer coincidence, $u_i \neq \widehat{u_i}$ for any i.

$$\widehat{u_i} = y_i - \widehat{\beta_0} - \widehat{\beta_1} x_i \tag{2.65}$$

$$= (\beta_0 + \beta_1 x_i + u_i) - \widehat{\beta_0} - \widehat{\beta_1} x_i \tag{2.66}$$

$$= u_i - (\widehat{\beta_0} - \beta_0) - (\widehat{\beta_1} - \beta_1)x_i \tag{2.67}$$

Note that $E(\widehat{\beta_0}) = \beta_0$ and $E(\widehat{\beta_1}) = \beta_1$, but the estimators almost always differ from the population values in a sample. So what about this as an estimator of σ^2?

$$\frac{1}{n}\sum_{i=1}^{n} \widehat{u_i}^2 = \frac{1}{n}SSR \tag{2.68}$$

It is a true estimator and easily computed from the data after OLS. As it turns out, this estimator is slightly biased: its expected value is a little less than σ^2. The estimator does not account for the two restrictions on the residuals used to obtain $\widehat{\beta_0}$ and $\widehat{\beta_1}$:

$$\sum_{i=1}^{n} \widehat{u}_i = 0 \qquad (2.69)$$

$$\sum_{i=1}^{n} x_i \widehat{u}_i = 0 \qquad (2.70)$$

There is no such restriction on the unobserved errors. The unbiased estimator, therefore, of σ^2 uses a degrees-of-freedom adjustment. The residuals have only $n - 2$, not n, degrees of freedom. Therefore:

$$\widehat{\sigma}^2 = \frac{1}{n-2} SSR \qquad (2.71)$$

We now propose the following theorem. The unbiased estimator of σ^2 under the first five assumptions is:

$$E(\widehat{\sigma}^2) = \sigma^2 \qquad (2.72)$$

In most software packages, regression output will include:

$$\widehat{\sigma} = \sqrt{\widehat{\sigma}^2} \qquad (2.73)$$

$$= \sqrt{\frac{SSR}{(n-2)}} \qquad (2.74)$$

This is an estimator of $sd(u)$, the standard deviation of the population error. One small glitch is that $\widehat{\sigma}$ is not unbiased for σ.[26] This will not matter for our purposes: $\widehat{\sigma}$ is called the standard error of the regression, which means that it is an estimate of the standard deviation of the error in the regression. The software package Stata calls it the root mean squared error.

Given $\widehat{\sigma}$, we can now estimate $sd(\widehat{\beta}_1)$ and $sd(\widehat{\beta}_0)$. The estimates of these are called the standard errors of the $\widehat{\beta}_j$. We will use these *a lot*. Almost all regression packages report the standard errors in a column next to the coefficient estimates. We can just plug $\widehat{\sigma}$ in for σ:

---

26 There does exist an unbiased estimator of σ, but it's tedious and hardly anyone in economics seems to use it. See Holtzman [1950].

$$se(\widehat{\beta_1}) = \frac{\widehat{\sigma}}{\sqrt{SST_x}} \qquad (2.75)$$

where both the numerator and the denominator are computed from the data. For reasons we will see, it is useful to report the standard errors below the corresponding coefficient, usually in parentheses.

Robust standard errors. How realistic is it that the variance in the errors is the same for all slices of the explanatory variable, x? The short answer here is that it is probably unrealistic. Heterogeneity is just something I've come to accept as the rule, not the exception, so if anything, we should be opting in to believing in homoskedasticity, not opting out. You can just take it as a given that errors are never homoskedastic and move forward to the solution.

This isn't completely bad news, because the unbiasedness of our regressions based on repeated sampling never depended on assuming anything about the variance of the errors. Those four assumptions, and particularly the zero conditional mean assumption, guaranteed that the central tendency of the coefficients under repeated sampling would equal the true parameter, which for this book is a causal parameter. The problem is with the spread of the coefficients. Without homoskedasticity, OLS no longer has the minimum mean squared errors, which means that the estimated standard errors are biased. Using our sampling metaphor, then, the distribution of the coefficients is probably larger than we thought. Fortunately, there is a solution. Let's write down the variance equation under heterogeneous variance terms:

$$\text{Var}\,(\widehat{\beta_1}) = \frac{\sum_{i=1}^{n}(x_i - \overline{x})^2 \sigma_i^2}{SST_x^2} \qquad (2.76)$$

Notice the i subscript in our σ_i^2 term; that means variance is not a constant. When $\sigma_i^2 = \sigma^2$ for all i, this formula reduces to the usual form, $\frac{\sigma^2}{SST_x^2}$. But when that isn't true, then we have a problem called heteroskedastic errors. A valid estimator of $\text{Var}(\widehat{\beta_1})$ for heteroskedasticity of any form (including homoskedasticity) is

$$\text{Var}(\widehat{\beta}_1) = \frac{\sum_{i=1}^{n}(x_i - \overline{x})^2 \widehat{u}_i^2}{SST_x^2}$$

which is easily computed from the data after the OLS regression. We have Friedhelm Eicker, Peter J. Huber, and Halbert White to thank for this solution (White [1980]).[27] The solution for heteroskedasticity goes by several names, but the most common is "robust" standard error.

Cluster robust standard errors. People will try to scare you by challenging how you constructed your standard errors. Heteroskedastic errors, though, aren't the only thing you should be worried about when it comes to inference. Some phenomena do not affect observations individually, but they do affect groups of observations that involve individuals. And then they affect those individuals within the group in a common way. Say you want to estimate the effect of class size on student achievement, but you know that there exist unobservable things (like the teacher) that affect all the students equally. If we can commit to independence of these unobservables across classes, but individual student unobservables are correlated within a class, then we have a situation in which we need to cluster the standard errors. Before we dive into an example, I'd like to start with a simulation to illustrate the problem.

As a baseline for this simulation, let's begin by simulating nonclustered data and analyze least squares estimates of that nonclustered data. This will help firm up our understanding of the problems that occur with least squares when data is clustered.[28]

---

27 No one even bothers to cite White [1980] anymore, just like how no one cites Leibniz or Newton when using calculus. Eicker, Huber, and White created a solution so valuable that it got separated from the original papers when it was absorbed into the statistical toolkit.

28 Hat tip to Ben Chidmi, who helped create this simulation in Stata.

STATA

cluster1.do

```
1   clear all
2   set seed 20140
3   * Set the number of simulations
4   local n_sims  = 1000
5   set obs `n_sims'
6
7   * Create the variables that will contain the results of each simulation
8   generate beta_0 = .
9   generate beta_0_l = .
10  generate beta_0_u = .
11  generate beta_1 = .
12  generate beta_1_l = .
13  generate beta_1_u = .
14
15
16  * Provide the true population parameters
17  local beta_0_true = 0.4
18  local beta_1_true = 0
19  local rho = 0.5
20
21  * Run the linear regression 1000 times and save the parameters beta_0 and
    ↪  beta_1
22  quietly {
23      forvalues i = 1(1) `n_sims' {
24          preserve
25          clear
26          set obs 100
27          generate x = rnormal(0,1)
28          generate e = rnormal(0, sqrt(1 - `rho'))
29          generate y = `beta_0_true' + `beta_1_true'*x + e
30          regress y x
31          local b0 = _b[_cons]
32          local b1 = _b[x]
33          local df = e(df_r)
34          local critical_value = invt(`df', 0.975)
35          restore
```

(continued)

STATA *(continued)*

```
36         replace beta_0 = `b0' in `i'
37         replace beta_0_l = beta_0 - `critical_value'*_se[_cons]
38         replace beta_0_u = beta_0 + `critical_value'*_se[_cons]
39         replace beta_1 = `b1' in `i'
40         replace beta_1_l = beta_1 - `critical_value'*_se[x]
41         replace beta_1_u = beta_1 + `critical_value'*_se[x]
42
43      }
44  }
45  gen false = (beta_1_l > 0 )
46  replace false = 2 if beta_1_u < 0
47  replace false = 3 if false == 0
48  tab false
49
50  * Plot the parameter estimate
51  hist beta_1, frequency addplot(pci 0 0 100 0) title("Least squares estimates of
    ↪    non-clustered data") subtitle(" Monte Carlo simulation of the slope")
    ↪    legend(label(1 "Distribution of least squares estimates") label(2 "True
    ↪    population parameter")) xtitle("Parameter estimate")
52
53  sort beta_1
54  gen int sim_ID = _n
55  gen beta_1_True = 0
56  * Plot of the Confidence Interval
57  twoway rcap beta_1_l beta_1_u sim_ID if beta_1_l > 0 | beta_1_u < 0 , horizontal
    ↪    lcolor(pink) || || ///
58  rcap beta_1_l beta_1_u sim_ID if beta_1_l < 0 & beta_1_u > 0 , horizontal ysc(r(0))
    ↪    || || ///
59  connected sim_ID beta_1 || || ///
60  line sim_ID beta_1_True, lpattern(dash) lcolor(black) lwidth(1) ///
61  title("Least squares estimates of non-clustered data") subtitle(" 95% Confidence
    ↪    interval of the slope") ///
62  legend(label(1 "Missed") label(2 "Hit") label(3 "OLS estimates") label(4 "True
    ↪    population parameter")) xtitle("Parameter estimates") ///
63  ytitle("Simulation")
```

R

cluster1.R

```
1   #- Analysis of Clustered Data
2   #- Courtesy of Dr. Yuki Yanai,
3   #- http://yukiyanai.github.io/teaching/rm1/contents/R/clustered-data-
    ↪  analysis.html
4
5   library('arm')
6   library('mvtnorm')
7   library('lme4')
8   library('multiwayvcov')
9   library('clusterSEs')
10  library('ggplot2')
11  library('dplyr')
12  library('haven')
13
14  gen_cluster <- function(param = c(.1, .5), n = 1000, n_cluster = 50, rho = .5) {
15    # Function to generate clustered data
16    # Required package: mvtnorm
17
18    # individual level
19    Sigma_i <- matrix(c(1, 0, 0, 1 - rho), ncol = 2)
20    values_i <- rmvnorm(n = n, sigma = Sigma_i)
21
22    # cluster level
23    cluster_name <- rep(1:n_cluster, each = n / n_cluster)
24    Sigma_cl <- matrix(c(1, 0, 0, rho), ncol = 2)
25    values_cl <- rmvnorm(n = n_cluster, sigma = Sigma_cl)
26
27    # predictor var consists of individual- and cluster-level components
28    x <- values_i[ , 1] + rep(values_cl[ , 1], each = n / n_cluster)
29
30    # error consists of individual- and cluster-level components
31    error <- values_i[ , 2] + rep(values_cl[ , 2], each = n / n_cluster)
32
33    # data generating process
34    y <- param[1] + param[2]*x + error
35
36    df <- data.frame(x, y, cluster = cluster_name)
37    return(df)
38  }
```

(continued)

R *(continued)*

```
39
40   # Simulate a dataset with clusters and fit OLS
41   # Calculate cluster-robust SE when cluster_robust = TRUE
42   cluster_sim <- function(param = c(.1, .5), n = 1000, n_cluster = 50,
43                   rho = .5, cluster_robust = FALSE) {
44   # Required packages: mvtnorm, multiwayvcov
45   df <- gen_cluster(param = param, n = n , n_cluster = n_cluster, rho = rho)
46   fit <- lm(y ~ x, data = df)
47   b1 <- coef(fit)[2]
48   if (!cluster_robust) {
49     Sigma <- vcov(fit)
50     se <- sqrt(diag(Sigma)[2])
51     b1_ci95 <- confint(fit)[2, ]
52   } else { # cluster-robust SE
53     Sigma <- cluster.vcov(fit, ~ cluster)
54     se <- sqrt(diag(Sigma)[2])
55     t_critical <- qt(.025, df = n - 2, lower.tail = FALSE)
56     lower <- b1 - t_critical*se
57     upper <- b1 + t_critical*se
58     b1_ci95 <- c(lower, upper)
59   }
60   return(c(b1, se, b1_ci95))
61   }
62
63   # Function to iterate the simulation. A data frame is returned.
64   run_cluster_sim <- function(n_sims = 1000, param = c(.1, .5), n = 1000,
65                   n_cluster = 50, rho = .5, cluster_robust = FALSE) {
66     # Required packages: mvtnorm, multiwayvcov, dplyr
67     df <- replicate(n_sims, cluster_sim(param = param, n = n, rho = rho,
68                     n_cluster = n_cluster,
69                     cluster_robust = cluster_robust))
70     df <- as.data.frame(t(df))
71     names(df) <- c('b1', 'se_b1', 'ci95_lower', 'ci95_upper')
72     df <- df %>%
73       mutate(id = 1:n(),
74           param_caught = ci95_lower <= param[2] & ci95_upper >= param[2])
75     return(df)
76   }
```

(continued)

R (continued)

```
77
78   # Distribution of the estimator and confidence intervals
79   sim_params <- c(.4, 0)   # beta1 = 0: no effect of x on y
80   sim_nocluster <- run_cluster_sim(n_sims = 10000, param = sim_params, rho = 0)
81   hist_nocluster <- ggplot(sim_nocluster, aes(b1)) +
82     geom_histogram(color = 'black') +
83     geom_vline(xintercept = sim_params[2], color = 'red')
84   print(hist_nocluster)
85
86   ci95_nocluster <- ggplot(sample_n(sim_nocluster, 100),
87                     aes(x = reorder(id, b1), y = b1,
88                         ymin = ci95_lower, ymax = ci95_upper,
89                         color = param_caught)) +
90     geom_hline(yintercept = sim_params[2], linetype = 'dashed') +
91     geom_pointrange() +
92     labs(x = 'sim ID', y = 'b1', title = 'Randomly Chosen 100 95% CIs') +
93     scale_color_discrete(name = 'True param value', labels = c('missed', 'hit')) +
94     coord_flip()
95   print(ci95_nocluster)
96
97   sim_nocluster %>% summarize(type1_error = 1 - sum(param_caught)/n())
98
99
```

As we can see in Figure 7, the least squares estimate is centered on its true population parameter.

Setting the significance level at 5%, we should incorrectly reject the null that $\beta_1 = 0$ about 5% of the time in our simulations. But let's check the confidence intervals. As can be seen in Figure 8, about 95% of the 95% confidence intervals contain the true value of β_1, which is zero. In words, this means that we incorrectly reject the null about 5% of the time.

But what happens when we use least squares with *clustered* data? To see that, let's resimulate our data with observations that are no longer independent draws in a given cluster of observations.

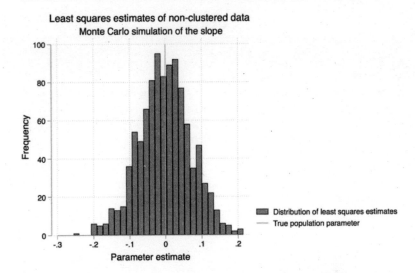

Figure 7. Distribution of the least squares estimator over 1,000 random draws.

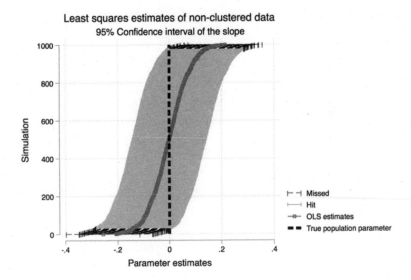

Figure 8. Distribution of the 95% confidence intervals with shading showing those that are incorrectly rejecting the null.

STATA

cluster2.do

```
1   clear all
2   set seed 20140
3   local n_sims = 1000
4   set obs `n_sims'
5
6   * Create the variables that will contain the results of each simulation
7   generate beta_0 = .
8   generate beta_0_l = .
9   generate beta_0_u = .
10  generate beta_1 = .
11  generate beta_1_l = .
12  generate beta_1_u = .
13
14
15  * Provide the true population parameters
16  local beta_0_true = 0.4
17  local beta_1_true = 0
18  local rho = 0.5
19
20  * Simulate a linear regression. Clustered data (x and e are clustered)
21
22
23  quietly {
24  forvalues i = 1(1) `n_sims' {
25      preserve
26      clear
27      set obs 50
28
29      * Generate cluster level data: clustered x and e
30      generate int cluster_ID = _n
31      generate x_cluster = rnormal(0,1)
32      generate e_cluster = rnormal(0, sqrt(`rho'))
33      expand 20
34      bysort cluster_ID : gen int ind_in_clusterID = _n
35
```

(continued)

STATA *(continued)*

```
36          * Generate individual level data
37          generate x_individual = rnormal(0,1)
38          generate e_individual = rnormal(0,sqrt(1 - `rho'))
39
40          * Generate x and e
41          generate x = x_individual + x_cluster
42          generate e = e_individual + e_cluster
43          generate y = `beta_0_true' + `beta_1_true'*x + e
44
45      * Least Squares Estimates
46          regress y x
47          local b0 = _b[_cons]
48          local b1 = _b[x]
49          local df = e(df_r)
50          local critical_value = invt(`df', 0.975)
51          * Save the results
52          restore
53          replace beta_0 = `b0' in `i'
54          replace beta_0_l = beta_0 - `critical_value'*_se[_cons]
55          replace beta_0_u = beta_0 + `critical_value'*_se[_cons]
56          replace beta_1 = `b1' in `i'
57          replace beta_1_l = beta_1 - `critical_value'*_se[x]
58          replace beta_1_u = beta_1 + `critical_value'*_se[x]
59      }
60  }
61
62  gen false = (beta_1_l > 0 )
63  replace false = 2 if beta_1_u < 0
64  replace false = 3 if false == 0
65  tab false
66
67  * Plot the parameter estimate
68  hist beta_1, frequency addplot(pci 0 0 100 0) title("Least squares estimates of
    ↪   clustered Data") subtitle(" Monte Carlo simulation of the slope")
    ↪   legend(label(1 "Distribution of least squares estimates") label(2 "True
    ↪   population parameter")) xtitle("Parameter estimate")
69
```

```
                                      R
                              cluster2.R
1   #- Analysis of Clustered Data - part 2
2   #- Courtesy of Dr. Yuki Yanai,
3   #- http://yukiyanai.github.io/teaching/rm1/contents/R/clustered-data-
    ↪  analysis.html
4
5   library('arm')
6   library('mvtnorm')
7   library('lme4')
8   library('multiwayvcov')
9   library('clusterSEs')
10  library('ggplot2')
11  library('dplyr')
12  library('haven')
13
14  #Data with clusters
15  sim_params <- c(.4, 0)   # beta1 = 0: no effect of x on y
16  sim_cluster_ols <- run_cluster_sim(n_sims = 10000, param = sim_params)
17  hist_cluster_ols <- hist_nocluster %+% sim_cluster_ols
18  print(hist_cluster_ols)
```

As can be seen in Figure 9, the least squares estimate has a narrower spread than that of the estimates when the data isn't clustered.
But to see this a bit more clearly, let's look at the confidence intervals
again.

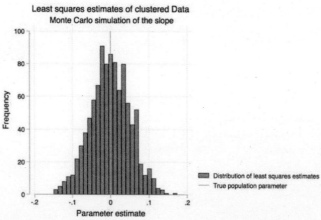

Figure 9. Distribution of the least squares estimator over 1,000 random draws.

STATA
cluster3.do

```stata
1   sort beta_1
2   gen int sim_ID = _n
3   gen beta_1_True = 0
4
5   * Plot of the Confidence Interval
6   twoway rcap beta_1_l beta_1_u sim_ID if beta_1_l > 0 | beta_1_u < 0 , horizontal
    ↪ lcolor(pink) || || ///
7   rcap beta_1_l beta_1_u sim_ID if beta_1_l < 0 & beta_1_u > 0 , horizontal ysc(r(0))
    ↪ || || ///
8   connected sim_ID beta_1 || || ///
9   line sim_ID beta_1_True, lpattern(dash) lcolor(black) lwidth(1) ///
10  title("Least squares estimates of clustered data") subtitle(" 95% Confidence
    ↪ interval of the slope") ///
11  legend(label(1 "Missed") label(2 "Hit") label(3 "OLS estimates") label(4 "True
    ↪ population parameter")) xtitle("Parameter estimates") ///
12  ytitle("Simulation")
13
```

R
cluster3.R

```r
1   #- Analysis of Clustered Data - part 3
2   #- Courtesy of Dr. Yuki Yanai,
3   #- http://yukiyanai.github.io/teaching/rm1/contents/R/clustered-data-
    ↪ analysis.html
4
5   library('arm')
6   library('mvtnorm')
7   library('lme4')
8   library('multiwayvcov')
9   library('clusterSEs')
10  library('ggplot2')
11  library('dplyr')
12  library('haven')
13
14  #Confidence interval
15  ci95_cluster_ols <- ci95_nocluster %+% sample_n(sim_cluster_ols, 100)
16  print(ci95_cluster_ols)
17
18  sim_cluster_ols %>% summarize(type1_error = 1 - sum(param_caught)/n())
```

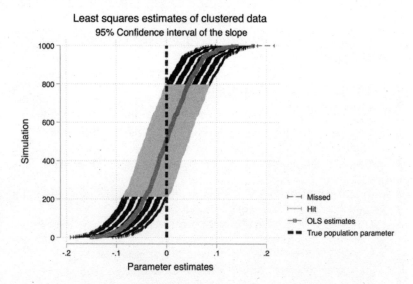

Figure 10. Distribution of 1,000 95% confidence intervals, with darker region representing those estimates that incorrectly reject the null.

Figure 10 shows the distribution of 95% confidence intervals from the least squares estimates. As can be seen, a much larger number of estimates incorrectly rejected the null hypothesis when the data was clustered. The standard deviation of the estimator shrinks under clustered data, causing us to reject the null incorrectly too often. So what can we do?

STATA
cluster4.do
```
1   * Robust Estimates
2   clear all
3   local n_sims = 1000
4   set obs `n_sims'
5
6   * Create the variables that will contain the results of each simulation
7   generate beta_0_robust = .
8   generate beta_0_l_robust = .
9   generate beta_0_u_robust = .
10  generate beta_1_robust = .
```

(continued)

STATA (continued)

```
11   generate beta_1_l_robust = .
12   generate beta_1_u_robust = .
13
14   * Provide the true population parameters
15   local beta_0_true = 0.4
16   local beta_1_true = 0
17   local rho = 0.5
18
19   quietly {
20   forvalues i = 1(1) `n_sims' {
21       preserve
22       clear
23       set obs 50
24
25       * Generate cluster level data: clustered x and e
26       generate int cluster_ID = _n
27       generate x_cluster = rnormal(0,1)
28       generate e_cluster = rnormal(0, sqrt(`rho'))
29       expand 20
30       bysort cluster_ID : gen int ind_in_clusterID = _n
31
32       * Generate individual level data
33       generate x_individual = rnormal(0,1)
34       generate e_individual = rnormal(0,sqrt(1 - `rho'))
35
36   * Robust Estimates
37   clear all
38   local n_sims = 1000
39   set obs `n_sims'
40
41   * Create the variables that will contain the results of each simulation
42   generate beta_0_robust = .
43   generate beta_0_l_robust = .
44   generate beta_0_u_robust = .
45   generate beta_1_robust = .
46   generate beta_1_l_robust = .
47   generate beta_1_u_robust = .
48
```

(continued)

STATA *(continued)*

```
49   * Provide the true population parameters
50   local beta_0_true = 0.4
51   local beta_1_true = 0
52   local rho = 0.5
53
54   quietly {
55   forvalues i = 1(1) `n_sims' {
56       preserve
57       clear
58       set obs 50
59
60       * Generate cluster level data: clustered x and e
61       generate int cluster_ID = _n
62       generate x_cluster = rnormal(0,1)
63       generate e_cluster = rnormal(0, sqrt(`rho'))
64       expand 20
65       bysort cluster_ID : gen int ind_in_clusterID = _n
66
67       * Generate individual level data
68       generate x_individual = rnormal(0,1)
69       generate e_individual = rnormal(0,sqrt(1 - `rho'))
70
71       * Generate x and e
72       generate x = x_individual + x_cluster
73       generate e = e_individual + e_cluster
74       generate y = `beta_0_true' + `beta_1_true'*x + e
75       regress y x, cl(cluster_ID)
76       local b0_robust = _b[_cons]
77       local b1_robust = _b[x]
78       local df = e(df_r)
79       local critical_value = invt(`df', 0.975)
80       * Save the results
81       restore
82       replace beta_0_robust = `b0_robust' in `i'
83       replace beta_0_l_robust = beta_0_robust - `critical_value'*_se[_cons]
84       replace beta_0_u_robust = beta_0_robust + `critical_value'*_se[_cons]
```

(continued)

STATA *(continued)*

```
85       replace beta_1_robust = `b1_robust' in `i'
86       replace beta_1_l_robust = beta_1_robust - `critical_value'*_se[x]
87       replace beta_1_u_robust = beta_1_robust + `critical_value'*_se[x]
88
89  }
90  }
91
92  * Plot the histogram of the parameters estimates of the robust least squares
93  gen false = (beta_1_l_robust > 0 )
94  replace false = 2 if beta_1_u_robust < 0
95  replace false = 3 if false == 0
96  tab false
97
98  * Plot the parameter estimate
99  hist beta_1_robust, frequency addplot(pci 0 0 110 0) title("Robust least
    ↪  squares estimates of clustered data") subtitle(" Monte Carlo simulation of
    ↪  the slope") legend(label(1 "Distribution of robust least squares
    ↪  estimates") label(2 "True population parameter")) xtitle("Parameter
    ↪  estimate")
100
101  sort beta_1_robust
102  gen int sim_ID = _n
103  gen beta_1_True = 0
104
105  * Plot of the Confidence Interval
106  twoway rcap beta_1_l_robust beta_1_u_robust sim_ID if beta_1_l_robust > 0 |
    ↪  beta_1_u_robust < 0, horizontal lcolor(pink) || || rcap beta_1_l_robust
    ↪  beta_1_u_robust sim_ID if beta_1_l_robust < 0 & beta_1_u_robust > 0 ,
    ↪  horizontal ysc(r(0)) || || connected sim_ID beta_1_robust || || line sim_ID
    ↪  beta_1_True, lpattern(dash) lcolor(black) lwidth(1) title("Robust least
    ↪  squares estimates of clustered data") subtitle(" 95% Confidence interval
    ↪  of the slope") legend(label(1 "Missed") label(2 "Hit") label(3 "Robust
    ↪  estimates") label(4 "True population parameter")) xtitle("Parameter
    ↪  estimates") ytitle("Simulation")
```

R
cluster4.R

```
1   #- Analysis of Clustered Data - part 4
2   #- Courtesy of Dr. Yuki Yanai,
3   #- http://yukiyanai.github.io/teaching/rm1/contents/R/clustered-data-
    ↳ analysis.html
4
5   library('arm')
6   library('mvtnorm')
7   library('lme4')
8   library('multiwayvcov')
9   library('clusterSEs')
10  library('ggplot2')
11  library('dplyr')
12  library('haven')
13
14  #clustered robust
15  sim_params <- c(.4, 0)   # beta1 = 0: no effect of x on y
16  sim_cluster_robust <- run_cluster_sim(n_sims = 10000, param = sim_params,
17                        cluster_robust = TRUE)
18
19  hist_cluster_robust <- hist_nocluster %+% sim_cluster_ols
20  print(hist_cluster_robust)
21
22  #Confidence Intervals
23  ci95_cluster_robust <- ci95_nocluster %+% sample_n(sim_cluster_robust, 100)
24  print(ci95_cluster_robust)
25
26  sim_cluster_robust %>% summarize(type1_error = 1 - sum(param_caught)/n())
```

Now in this case, notice that we included the ", cluster(cluster_ID)" syntax in our regression command. Before we dive in to what this syntax did, let's look at how the confidence intervals changed. Figure 11 shows the distribution of the 95% confidence intervals where, again, the darkest region represents those estimates that incorrectly rejected the null. Now, when there are observations whose errors are correlated within a cluster, we find that estimating the model using least squares leads us back to a situation in which the type I error has decreased considerably.

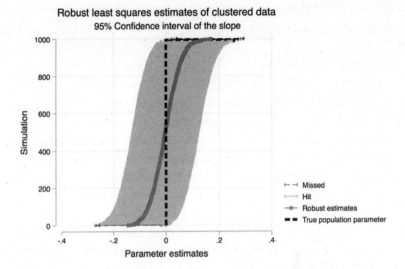

Figure 11. Distribution of 1,000 95% confidence intervals from a clustered robust least squares regression, with dashed region representing those estimates that incorrectly reject the null.

This leads us to a natural question: what did the adjustment of the estimator's variance do that caused the type I error to decrease by so much? Whatever it's doing, it sure seems to be working! Let's dive in to this adjustment with an example. Consider the following model:

$$y_{ig} = x'_{ig}\beta + u_{ig} \quad \text{where } 1,\ldots,G$$

and

$$E[u_{ig}u'_{jg}]$$

which equals zero if $g = g'$ and equals $\sigma_{(ij)g}$ if $g \neq g'$.

Let's stack the data by cluster first.

$$y_g = x'_g\beta + u_g$$

The OLS estimator is still $\widehat{\beta} = E[X'X]^{-1}X'Y$. We just stacked the data, which doesn't affect the estimator itself. But it does change the variance.

$$V(\beta) = E\Big[[X'X]^{-1}X'\Omega X[X'X]^{-1}\Big]$$

With this in mind, we can now write the variance-covariance matrix for clustered data as

$$\widehat{V}(\widehat{\beta}) = [X'X]^{-1} \left[\sum_{i=1}^{G} x_g' \widehat{u}_g \widehat{u}_g' \right] [X'X]^{-1}$$

Adjusting for clustered data will be quite common in your applied work given the ubiquity of clustered data in the first place. It's absolutely essential for working in the panel contexts, or in repeated cross-sections like the difference-in-differences design. But it also turns out to be important for experimental design, because often, the treatment will be at a higher level of aggregation than the microdata itself. In the real world, though, you can never assume that errors are independent draws from the same distribution. You need to know how your variables were constructed in the first place in order to choose the correct error structure for calculating your standard errors. If you have aggregate variables, like class size, then you'll need to cluster at that level. If some treatment occurred at the state level, then you'll need to cluster at that level. There's a large literature available that looks at even more complex error structures, such as multi-way clustering [Cameron et al., 2011].

But even the concept of the sample as the basis of standard errors may be shifting. It's becoming increasingly less the case that researchers work with random samples; they are more likely working with administrative data containing the population itself, and thus the concept of sampling uncertainty becomes strained.[29] For instance, Manski and Pepper [2018] wrote that "random sampling assumptions ... are not natural when considering states or counties as units of observation." So although a metaphor of a superpopulation may be useful for extending these classical uncertainty concepts, the ubiquity of digitized administrative data sets has led econometricians and statisticians to think about uncertainty in other ways.

New work by Abadie et al. [2020] explores how sampling-based concepts of the standard error may not be the right way to think about

---

[29] Usually we appeal to superpopulations in such situations where the observed population is simply itself a draw from some "super" population.

uncertainty in the context of causal inference, or what they call *design-based uncertainty*. This work in many ways anticipates the next two chapters because of its direct reference to the concept of the counterfactual. Design-based uncertainty is a reflection of not knowing which values would have occurred had some intervention been different in counterfactual. And Abadie et al. [2020] derive standard errors for design-based uncertainty, as opposed to sampling-based uncertainty. As luck would have it, those standard errors are usually *smaller*.

Let's now move into these fundamental concepts of causality used in applied work and try to develop the tools to understand how counterfactuals and causality work together.

Directed Acyclic Graphs

*Everyday it rains, so
everyday the pain
Went ignored and I'm sure
ignorance was to blame
But life is a chain, cause and
effected.*

Jay-Z

The history of graphical causal modeling goes back to the early twentieth century and Sewall Wright, one of the fathers of modern genetics and son of the economist Philip Wright. Sewall developed path diagrams for genetics, and Philip, it is believed, adapted them for econometric identification [Matsueda, 2012].[1]

But despite that promising start, the use of graphical modeling for causal inference has been largely ignored by the economics profession, with a few exceptions [Heckman and Pinto, 2015; Imbens, 2019]. It was revitalized for the purpose of causal inference when computer scientist and Turing Award winner Judea Pearl adapted them for his work on artificial intelligence. He explained this in his mangum opus, which is a general theory of causal inference that expounds on the usefulness of his directed graph notation [Pearl, 2009]. Since graphical models are immensely helpful for designing a credible identification strategy, I have chosen to include them for your consideration. Let's

---

1 I will discuss the Wrights again in the chapter on instrumental variables. They were an interesting pair.

review graphical models, one of Pearl's contributions to the theory of causal inference.[2]

Introduction to DAG Notation

Using directed acyclic graphical (DAG) notation requires some up-front statements. The first thing to notice is that in DAG notation, causality runs in one direction. Specifically, it runs forward in time. There are no cycles in a DAG. To show reverse causality, one would need to create multiple nodes, most likely with two versions of the same node separated by a time index. Similarly, simultaneity, such as in supply and demand models, is not straightforward with DAGs [Heckman and Pinto, 2015]. To handle either simultaneity or reverse causality, it is recommended that you take a completely different approach to the problem than the one presented in this chapter. Third, DAGs explain causality in terms of counterfactuals. That is, a causal effect is defined as a comparison between two states of the world—one state that actually happened when some intervention took on some value and another state that didn't happen (the "counterfactual") under some other intervention.

Think of a DAG as like a graphical representation of a chain of causal effects. The causal effects are themselves based on some underlying, unobserved structured process, one an economist might call the equilibrium values of a system of behavioral equations, which are themselves nothing more than a *model* of the world. All of this is captured efficiently using graph notation, such as nodes and arrows. Nodes represent random variables, and those random variables are assumed to be created by some data-generating process.[3] Arrows represent a causal effect between two random variables moving in the intuitive direction of the arrow. The direction of the arrow captures the direction of causality.

---

2 If you find this material interesting, I highly recommend Morgan and Winship [2014], an all-around excellent book on causal inference, and especially on graphical models.

3 I leave out some of those details, though, because their presence (usually just error terms pointing to the variables) clutters the graph unnecessarily.

Causal effects can happen in two ways. They can either be direct (e.g., $D \to Y$), or they can be mediated by a third variable (e.g., $D \to X \to Y$). When they are mediated by a third variable, we are capturing a sequence of events originating with D, which may or may not be important to you depending on the question you're asking.

A DAG is meant to describe all causal relationships relevant to the effect of D on Y. What makes the DAG distinctive is both the explicit commitment to a causal effect pathway and the complete commitment to the *lack of* a causal pathway represented by missing arrows. In other words, a DAG will contain both arrows connecting variables and choices to exclude arrows. And the lack of an arrow necessarily means that you think there is no such relationship in the data—this is one of the strongest beliefs you can hold. A complete DAG will have all direct causal effects among the variables in the graph as well as all common causes of any pair of variables in the graph.

At this point, you may be wondering where the DAG comes from. It's an excellent question. It may be *the* question. A DAG is supposed to be a theoretical representation of the state-of-the-art knowledge about the phenomena you're studying. It's what an expert would say is the thing itself, and that expertise comes from a variety of sources. Examples include economic theory, other scientific models, conversations with experts, your own observations and experiences, literature reviews, as well as your own intuition and hypotheses.

I have included this material in the book because I have found DAGs to be useful for understanding the critical role that prior knowledge plays in identifying causal effects. But there are other reasons too. One, I have found that DAGs are very helpful for communicating research designs and estimators if for no other reason than pictures speak a thousand words. This is, in my experience, especially true for instrumental variables, which have a very intuitive DAG representation. Two, through concepts such as the backdoor criterion and collider bias, a well-designed DAG can help you develop a credible research design for identifying the causal effects of some intervention. As a bonus, I also think a DAG provides a bridge between various empirical schools, such as the structural and reduced form groups. And finally, DAGs drive home the point that assumptions are necessary for any and all identification of causal effects, which economists have been hammering at for years [Wolpin, 2013].

A simple DAG. Let's begin with a simple DAG to illustrate a few basic ideas. I will expand on it to build slightly more complex ones later.

In this DAG, we have three random variables: X, D, and Y. There is a direct *path* from D to Y, which represents a causal effect. That path is represented by $D \to Y$. But there is also a second path from D to Y called the *backdoor path*. The backdoor path is $D \leftarrow X \to Y$. While the direct path is a causal effect, the backdoor path is not causal. Rather, it is a process that creates spurious correlations between D and Y that are driven solely by fluctuations in the X random variable.

The idea of the backdoor path is one of the most important things we can learn from the DAG. It is similar to the notion of omitted variable bias in that it represents a variable that determines the outcome and the treatment variable. Just as not controlling for a variable like that in a regression creates omitted variable bias, leaving a backdoor open creates bias. The backdoor path is $D \leftarrow X \to Y$. We therefore call X a *confounder* because it jointly determines D and Y, and so confounds our ability to discern the effect of D on Y in naïve comparisons.

Think of the backdoor path like this: Sometimes when D takes on different values, Y takes on different values because D causes Y. But sometimes D and Y take on different values because X takes on different values, and that bit of the correlation between D and Y is purely spurious. The existence of two causal pathways is contained within the correlation between D and Y.

Let's look at a second DAG, which is subtly different from the first. In the previous example, X was observed. We know it was observed because the direct edges from X to D and Y were solid lines. But sometimes there exists a confounder that is unobserved, and when there is, we represent its direct edges with dashed lines. Consider the following DAG:

Same as before, U is a noncollider along the backdoor path from D to Y, but unlike before, U is unobserved to the researcher. It exists, but it may simply be missing from the data set. In this situation, there are two pathways from D to Y. There's the direct pathway, $D \to Y$, which is the causal effect, and there's the backdoor pathway, $D \gets U \to Y$. And since U is unobserved, that backdoor pathway is *open*.

Let's now move to another example, one that is slightly more realistic. A classical question in labor economics is whether college education increases earnings. According to the Becker human capital model [Becker, 1994], education increases one's marginal product, and since workers are paid their marginal product in competitive markets, education also increases their earnings. But college education is not random; it is optimally chosen given an individual's subjective preferences and resource constraints. We represent that with the following DAG. As always, let D be the treatment (e.g., college education) and Y be the outcome of interest (e.g., earnings). Furthermore, let PE be parental education, I be family income, and B be unobserved background factors, such as genetics, family environment, and mental ability.

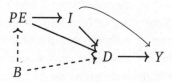

This DAG is telling a story. And one of the things I like about DAGs is that they invite everyone to listen to the story together. Here is my interpretation of the story being told. Each person has some background. It's not contained in most data sets, as it measures things like intelligence, contentiousness, mood stability, motivation, family dynamics, and other environmental factors—hence, it is unobserved in the picture. Those environmental factors are likely correlated between parent and child and therefore subsumed in the variable B.

Background causes a child's parent to choose her own optimal level of education, and that choice also causes the child to choose their level of education through a variety of channels. First, there is the shared background factors, B. Those background factors cause the child to choose a level of education, just as her parent had. Second, there's a direct effect, perhaps through simple modeling of achievement or setting expectations, a kind of peer effect. And third, there's the effect that parental education has on family earnings, I, which in turn affects how much schooling the child receives. Family earnings may itself affect the child's future earnings through bequests and other transfers, as well as external investments in the child's productivity.

This is a simple story to tell, and the DAG tells it well, but I want to alert your attention to some subtle points contained in this DAG. The DAG is actually telling two stories. It is telling what is happening, and it is telling what is *not* happening. For instance, notice that B has no direct effect on the child's earnings except through its effect on schooling. Is this realistic, though? Economists have long maintained that unobserved ability both determines how much schooling a child gets and directly affects the child's future earnings, insofar as intelligence and motivation can influence careers. But in this DAG, there is no relationship between background and earnings, which is itself an *assumption*. And you are free to call foul on this assumption if you think that background factors affect both schooling and the child's own productivity, which itself should affect wages. So what if you think that there should be an arrow from B to Y? Then you would draw one and rewrite all the backdoor paths between D and Y.

Now that we have a DAG, what do we do? I like to list out all direct and indirect paths (i.e., backdoor paths) between D and Y. Once I have all those, I have a better sense of where my problems are. So:

1. $D \rightarrow Y$ (the causal effect of education on earnings)
2. $D \leftarrow I \rightarrow Y$ (backdoor path 1)
3. $D \leftarrow PE \rightarrow I \rightarrow Y$ (backdoor path 2)
4. $D \leftarrow B \rightarrow PE \rightarrow I \rightarrow Y$ (backdoor path 3)

So there are four paths between D and Y: one direct causal effect (which arguably is the important one if we want to know the return on schooling) and three backdoor paths. And since none of the variables along the backdoor paths is a collider, each of the backdoors paths is *open*. The problem, though, with open backdoor paths is that they create systematic and independent correlations between D and Y. Put a different way, the presence of open backdoor paths introduces bias when comparing educated and less-educated workers.

Colliding. But what is this collider? It's an unusual term, one you may have never seen before, so let's introduce it with another example. I'm going to show you what a collider is graphically using a simple DAG, because it's an easy thing to see and a slightly more complicated phenomenon to explain. So let's work with a new DAG. Pay careful attention to the directions of the arrows, which have changed.

As before, let's list all paths from D to Y:

1. $D \rightarrow Y$ (causal effect of D on Y)
2. $D \rightarrow X \leftarrow Y$ (backdoor path 1)

Just like last time, there are two ways to get from D to Y. You can get from D to Y using the direct (causal) path, $D \rightarrow Y$. Or you can use the backdoor path, $D \rightarrow X \leftarrow Y$. But something is different about this backdoor path; do you see it? This time the X has two arrows pointing to it, not away from it. When two variables cause a third variable along some path, we call that third variable a "collider." Put differently, X is a collider along this backdoor path because D and the causal effects of Y collide at X. But so what? What makes a collider so special? Colliders are special in part because when they appear along a backdoor path, that backdoor path is *closed* simply because of their presence. Colliders, when they are left alone, always close a specific backdoor path.

Backdoor criterion. We care about open backdoor paths because they they create systematic, noncausal correlations between the causal variable of interest and the outcome you are trying to study. In regression terms, open backdoor paths introduce omitted variable bias, and for all you know, the bias is so bad that it flips the sign entirely. Our goal, then, is to close these backdoor paths. And if we can close all of the otherwise open backdoor paths, then we can isolate the causal effect of D on Y using one of the research designs and identification strategies discussed in this book. So how do we close a backdoor path?

There are two ways to close a backdoor path. First, if you have a confounder that has created an open backdoor path, then you can close that path by *conditioning* on the confounder. Conditioning requires holding the variable fixed using something like subclassification, matching, regression, or another method. It is equivalent to "controlling for" the variable in a regression. The second way to close a backdoor path is the appearance of a collider along that backdoor path. Since colliders always close backdoor paths, and conditioning on a collider always opens a backdoor path, choosing to ignore the colliders is part of your overall strategy to estimate the causal effect itself. By not conditioning on a collider, you will have closed that backdoor path and that takes you closer to your larger ambition to isolate some causal effect.

When all backdoor paths have been closed, we say that you have come up with a research design that satisfies the *backdoor criterion*. And if you have satisfied the backdoor criterion, then you have in effect isolated some causal effect. But let's formalize this: a set of variables X satisfies the backdoor criterion in a DAG if and only if X blocks every path between confounders that contain an arrow from D to Y. Let's review our original DAG involving parental education, background and earnings.

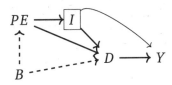

The minimally sufficient conditioning strategy necessary to achieve the backdoor criterion is the control for I, because I appeared as a noncollider along every backdoor path (see earlier). It might literally be no simpler than to run the following regression:

$$Y_i = \alpha + \delta D_i + \beta I_i + \varepsilon_i$$

By simply conditioning on *I*, your estimated $\widehat{\delta}$ takes on a causal interpretation.[4]

But maybe in hearing this story, and studying it for yourself by reviewing the literature and the economic theory surrounding it, you are skeptical of this DAG. Maybe this DAG has really bothered you from the moment you saw me produce it because you are skeptical that *B* has no relationship to *Y* except through *D* or *PE*. That skepticism leads you to believe that there should be a *direct* connection from *B* to *Y*, not merely one mediated through own education.

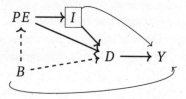

Note that including this new backdoor path has created a problem because our conditioning strategy no longer satisfies the backdoor criterion. Even controlling for *I*, there still exist spurious correlations between *D* and *Y* due to the $D \leftarrow B \rightarrow Y$ backdoor path. Without more information about the nature of $B \rightarrow Y$ and $B \rightarrow D$, we cannot say much more about the partial correlation between *D* and *Y*. We just are not legally allowed to interpret $\widehat{\delta}$ from our regression as the causal effect of *D* on *Y*.

More examples of collider bias. The issue of conditioning on a collider is important, so how do we know if we have that problem or not? No data set comes with a flag saying "collider" and "confounder." Rather, the only way to know whether you have satisfied the backdoor criterion is with a DAG, and a DAG requires a model. It requires in-depth knowledge of the data-generating process for the variables in your DAG,

---

4 Subsequent chapters discuss other estimators, such as matching.

but it also requires ruling out pathways. And the only way to rule out pathways is through logic and models. There is no way to avoid it—all empirical work requires theory to guide it. Otherwise, how do you know if you've conditioned on a collider or a noncollider? Put differently, you cannot identify treatment effects without making assumptions.

In our earlier DAG with collider bias, we conditioned on some variable X that was a collider—specifically, it was a descendent of D and Y. But that is just one example of a collider. Oftentimes, colliders enter into the system in very subtle ways. Let's consider the following scenario: Again, let D and Y be child schooling and child future earnings. But this time we introduce three new variables—$U1$, which is father's unobserved genetic ability; $U2$, which is mother's unobserved genetic ability; and I, which is joint family income. Assume that I is observed but that U_i is unobserved for both parents.

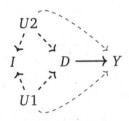

Notice in this DAG that there are several backdoor paths from D to Y. They are as follows:

1. $D \leftarrow U2 \rightarrow Y$
2. $D \leftarrow U1 \rightarrow Y$
3. $D \leftarrow U1 \rightarrow I \leftarrow U2 \rightarrow Y$
4. $D \leftarrow U2 \rightarrow I \leftarrow U1 \rightarrow Y$

Notice, the first two are open-backdoor paths, and as such, they cannot be closed, because $U1$ and $U2$ are not observed. But what if we controlled for I anyway? Controlling for I only makes matters worse, because it opens the third and fourth backdoor paths, as I was a collider along both of them. It does not appear that *any* conditioning strategy could meet the backdoor criterion in this DAG. And any strategy controlling for I would actually make matters worse. Collider bias is a difficult concept to understand at first, so I've included a couple of examples to help you sort through it.

Discrimination and collider bias. Let's examine a real-world example around the problem of gender discrimination in labor-markets. It is common to hear that once occupation or other characteristics of a job are conditioned on, the wage disparity between genders disappears or gets smaller. For instance, critics once claimed that Google systematically underpaid its female employees. But Google responded that its data showed that when you take "location, tenure, job role, level and performance" into consideration, women's pay is basically identical to that of men. In other words, controlling for characteristics of the job, women received the same pay.

But what if one of the ways gender discrimination creates gender disparities in earnings is through occupational sorting? If discrimination happens via the occupational match, then naïve contrasts of wages by gender controlling for occupation characteristics will likely understate the presence of discrimination in the marketplace. Let me illustrate this with a DAG based on a simple occupational sorting model with unobserved heterogeneity.

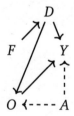

Notice that there is in fact no effect of female gender on earnings; women are assumed to have productivity identical to that of men. Thus, if we could control for discrimination, we'd get a coefficient of zero as in this example because women are, initially, just as productive as men.[5]

---

5 Productivity could diverge, though, if women systematically sort into lower-quality occupations in which human capital accumulates over time at a lower rate.

But in this example, we aren't interested in estimating the effect of being female on earnings; we are interested in estimating the effect of discrimination itself. Now you can see several noticeable paths between discrimination and earnings. They are as follows:

1. $D \to O \to Y$
2. $D \to O \leftarrow A \to Y$

The first path is not a backdoor path; rather, it is a path whereby discrimination is mediated by occupation before discrimination has an effect on earnings. This would imply that women are discriminated against, which in turn affects which jobs they hold, and as a result of holding marginally worse jobs, women are paid less. The second path relates to that channel but is slightly more complicated. In this path, unobserved ability affects both which jobs people get and their earnings.

So let's say we regress Y onto D, our discrimination variable. This yields the total effect of discrimination as the weighted sum of both the direct effect of discrimination on earnings and the mediated effect of discrimination on earnings through occupational sorting. But say that we want to control for occupation because we want to compare men and women in similar jobs. Well, controlling for occupation in the regression closes down the mediation channel, but it then opens up the second channel. Why? Because $D \to O \leftarrow A \to Y$ has a collider O. So when we control for occupation, we open up this second path. It had been closed because colliders close backdoor paths, but since we conditioned on it, we actually opened it instead. This is the reason we cannot merely control for occupation. Such a control ironically introduces new patterns of bias.[6]

---

6 Angrist and Pischke [2009] talk about this problem in a different way using language called "bad controls." Bad controls are not merely conditioning on outcomes. Rather, they are any situation in which the outcome had been a collider linking the treatment to the outcome of interest, like $D \to O \leftarrow A \to Y$.

What is needed is to control for occupation and ability, but since ability is unobserved, we cannot do that, and therefore we do not possess an identification strategy that satisfies the backdoor criterion. Let's now look at code to illustrate this DAG.[7]

```
                            STATA
                collider_discrimination.do
1    clear all
2    set obs 10000
3
4    * Half of the population is female.
5    generate female = runiform()>=0.5
6
7    * Innate ability is independent of gender.
8    generate ability = rnormal()
9
10   * All women experience discrimination.
11   generate discrimination = female
12
13   * Data generating processes
14   generate occupation = (1) + (2)*ability + (0)*female + (-2)*discrimination +
     ↪  rnormal()
15   generate wage = (1) + (-1)*discrimination + (1)*occupation + 2*ability + rnormal()
16
17   * Regressions
18   regress wage female
19   regress wage female occupation
20   regress wage female occupation ability
21
22
```

7 Erin Hengel is a professor of economics at the University of Liverpool. She and I were talking about this on Twitter one day, and she and I wrote down the code describing this problem. Her code was better, so I asked if I could reproduce it here, and she said yes. Erin's work partly focuses on gender discrimination. You can see some of that work on her website at http://www.erinhengel.com.

```R
                              R
              collider_discrimination.R
1   library(tidyverse)
2   library(stargazer)
3
4   tb <- tibble(
5     female = ifelse(runif(10000)>=0.5,1,0),
6     ability = rnorm(10000),
7     discrimination = female,
8     occupation = 1 + 2*ability + 0*female - 2*discrimination + rnorm(10000),
9     wage = 1 - 1*discrimination + 1*occupation + 2*ability + rnorm(10000)
10  )
11
12  lm_1 <- lm(wage ~ female, tb)
13  lm_2 <- lm(wage ~ female + occupation, tb)
14  lm_3 <- lm(wage ~ female + occupation + ability, tb)
15
16  stargazer(lm_1,lm_2,lm_3, type = "text",
17          column.labels = c("Biased Unconditional",
18                  "Biased",
19                  "Unbiased Conditional"))
```

This simulation hard-codes the data-generating process represented by the previous DAG. Notice that ability is a random draw from the standard normal distribution. Therefore it is independent of female preferences. And then we have our last two generated variables: the heterogeneous occupations and their corresponding wages. Occupations are increasing in unobserved ability but decreasing in discrimination. Wages are decreasing in discrimination but increasing in higher-quality jobs and higher ability. Thus, we know that discrimination exists in this simulation because we are hard-coding it that way with the negative coefficients both the occupation and wage processes.

The regression coefficients from the three regressions at the end of the code are presented in Table 9. First note that when we simply regress wages onto gender, we get a large negative effect, which is the combination of the direct effect of discrimination on earnings and the

Table 9. Regressions illustrating confounding bias with simulated gender disparity.

Covariates:	Biased unconditional	Biased biased	Unbiased conditional
Female	−3.074***	0.601***	−0.994***
	(0.000)	(0.000)	(0.000)
Occupation		1.793***	0.991***
		(0.000)	(0.000)
Ability			2.017***
			(0.000)
N	10,000	10,000	10,000
Mean of dependent variable	0.45	0.45	0.45

indirect effect via occupation. But if we run the regression that Google and others recommend wherein we control for occupation, the sign on gender changes. It becomes positive! We know this is wrong because we hard-coded the effect of gender to be −1! The problem is that occupation is a collider. It is caused by ability and discrimination. If we control for occupation, we open up a backdoor path between discrimination and earnings that is spurious and so strong that it perverts the entire relationship. So only when we control for occupation and ability can we isolate the direct causal effect of gender on wages.

Sample selection and collider bias. Bad controls are not the only kind of collider bias to be afraid of, though. Collider bias can also be baked directly into the sample if the sample itself was a collider. That's no doubt a strange concept to imagine, so I have a funny illustration to clarify what I mean.

A 2009 CNN blog post reported that Megan Fox, who starred in the movie *Transformers*, was voted the worst and most attractive actress of 2009 in some survey about movie stars [Piazza, 2009]. The implication could be taken to be that talent and beauty are negatively correlated. But are they? And why might they be? What if they are independent of each other in reality but negatively correlated in a sample of movie stars because of collider bias? Is that even possible?[8]

To illustrate, we will generate some data based on the following DAG:

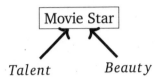

Let's illustrate this with a simple program.

STATA
moviestar.do

```
1   clear all
2   set seed 3444
3
4   * 2500 independent draws from standard normal distribution
5   set obs 2500
6   generate beauty=rnormal()
7   generate talent=rnormal()
8
9   * Creating the collider variable (star)
10  gen score=(beauty+talent)
11  egen c85=pctile(score), p(85)
12  gen star=(score>=c85)
13  label variable star "Movie star"
14
15  * Conditioning on the top 15\%
16  twoway (scatter beauty talent, mcolor(black) msize(small) msymbol(smx)),
    ↪  ytitle(Beauty) xtitle(Talent) subtitle(Aspiring actors and actresses) by(star,
    ↪  total)
```

---

8 I *wish* I had thought of this example, but alas the sociologist Gabriel Rossman gets full credit.

R

moviestar.R

```
1    library(tidyverse)
2
3    set.seed(3444)
4
5    star_is_born <- tibble(
6      beauty = rnorm(2500),
7      talent = rnorm(2500),
8      score = beauty + talent,
9      c85 = quantile(score, .85),
10     star = ifelse(score>=c85,1,0)
11   )
12
13   star_is_born %>%
14     lm(beauty ~ talent, .) %>%
15     ggplot(aes(x = talent, y = beauty)) +
16     geom_point(size = 0.5, shape=23) + xlim(-4, 4) + ylim(-4, 4)
17
18   star_is_born %>%
19     filter(star == 1) %>%
20     lm(beauty ~ talent, .) %>%
21     ggplot(aes(x = talent, y = beauty)) +
22     geom_point(size = 0.5, shape=23) + xlim(-4, 4) + ylim(-4, 4)
23
24   star_is_born %>%
25     filter(star == 0) %>%
26     lm(beauty ~ talent, .) %>%
27     ggplot(aes(x = talent, y = beauty)) +
28     geom_point(size = 0.5, shape=23) + xlim(-4, 4) + ylim(-4, 4)
29
```

Figure 12 shows the output from this simulation. The bottom left panel shows the scatter plot between talent and beauty. Notice that the two variables are independent, random draws from the standard normal distribution, creating an oblong data cloud. But because "movie star" is in the top 85th percentile of the distribution of a linear combination of talent and beauty, the sample consists of people whose combined score is in the top right portion of the joint distribution. This frontier has a negative slope and is in the upper right portion of the data

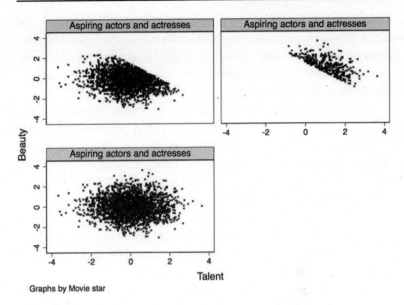

Figure 12. Aspiring actors and actresses.
   Note: Top left: Non-star sample scatter plot of beauty (vertical axis) and talent (horizontal axis). Top right: Star sample scatter plot of beauty and talent. Bottom left: Entire (stars and non-stars combined) sample scatter plot of beauty and talent.

cloud, creating a negative correlation between the observations in the movie-star sample. Likewise, the collider bias has created a negative correlation between talent and beauty in the non-movie-star sample as well. Yet we know that there is in fact *no* relationship between the two variables. This kind of sample selection creates spurious correlations. A random sample of the full population would be sufficient to show that there is no relationship between the two variables, but splitting the sample into movie stars only, we introduce spurious correlations between the two variables of interest.

Collider bias and police use of force. We've known about the problems of nonrandom sample selection for decades [Heckman, 1979]. But DAGs may still be useful for helping spot what might be otherwise subtle cases of conditioning on colliders [Elwert and Winship, 2014].

And given the ubiquitous rise in researcher access to large administrative databases, it's also likely that some sort of theoretically guided reasoning will be needed to help us determine whether the databases we have are themselves rife with collider bias. A contemporary debate could help illustrate what I mean.

Public concern about police officers systematically discriminating against minorities has reached a breaking point and led to the emergence of the Black Lives Matter movement. "Vigilante justice" episodes such as George Zimmerman's killing of teenage Trayvon Martin, as well as police killings of Michael Brown, Eric Garner, and countless others, served as catalysts to bring awareness to the perception that African Americans face enhanced risks for shootings. Fryer [2019] attempted to ascertain the degree to which there was racial bias in the use of force by police. This is perhaps one of the most important questions in policing as of this book's publication.

There are several critical empirical challenges in studying racial biases in police use of force, though. The main problem is that all data on police-citizen interactions are conditional on an interaction having already occurred. The data themselves were generated as a function of earlier police-citizen interactions. In this sense, we can say that the data itself are endogenous. Fryer [2019] collected several databases that he hoped would help us better understand these patterns. Two were public-use data sets—the New York City Stop and Frisk database and the Police-Public Contact Survey. The former was from the New York Police Department and contained data on police stops and questioning of pedestrians; if the police wanted to, they could frisk them for weapons or contraband. The latter was a survey of civilians describing interactions with the police, including the use of force.

But two of the data sets were administrative. The first was a compilation of event summaries from more than a dozen large cities and large counties across the United States from all incidents in which an officer discharged a weapon at a civilian. The second was a random sample of police-civilian interactions from the Houston Police Department. The accumulation of these databases was by all evidence a gigantic empirical task. For instance, Fryer [2019] notes that the Houston data was based on arrest narratives that ranged from two to one hundred pages in length. From these arrest narratives, a team of researchers collected almost three hundred variables relevant to the police use of force on the incident. This is the world in which we now live, though. Administrative databases can be accessed more easily than ever, and they are helping break open the black box of many opaque social processes.

A few facts are important to note. First, using the stop-and-frisk data, Fryer finds that blacks and Hispanics were more than 50 percent more likely to have an interaction with the police in the raw data. The racial difference survives conditioning on 125 baseline characteristics, encounter characteristics, civilian behavior, precinct, and year fixed effects. In his full model, blacks are 21 percent more likely than whites to be involved in an interaction with police in which a weapon is drawn (which is statistically significant). These racial differences show up in the Police-Public Contact Survey as well, only here the racial differences are considerably larger. So the first thing to note is that the actual stop itself appears to be larger for minorities, which I will come back to momentarily.

Things become surprising when Fryer moves to his rich administrative data sources. He finds that conditional on a police interaction, there are no racial differences in officer-involved shootings. In fact, controlling for suspect demographics, officer demographics, encounter characteristics, suspect weapon, and year fixed effects, blacks are 27 percent less likely to be shot at by police than are non-black non-Hispanics. The coefficient is not significant, and it shows up across alternative specifications and cuts of the data. Fryer is simply unable with these data to find evidence for racial discrimination in officer-involved shootings.

One of the main strengths of Fryer's study are the shoe leather he used to accumulate the needed data sources. Without data, one cannot study the question of whether police shoot minorities more than they shoot whites. And the extensive coding of information from the narratives is also a strength, for it afforded Fryer the ability to control for observable confounders. But the study is not without issues that could cause a skeptic to take issue. Perhaps the police departments most willing to cooperate with a study of this kind are the ones with the least racial bias, for instance. In other words, maybe these are not the departments with the racial bias to begin with.[9] Or perhaps a more sinister explanation exists, such as records being unreliable because administrators scrub out the data on racially motivated shootings before handing them over to Fryer altogether.

But I would like to discuss a more innocent possibility, one that requires no conspiracy theories and yet is so basic a problem that it is in fact more worrisome. Perhaps the administrative datasource is endogenous because of conditioning on a collider. If so, then the administrative data itself may have the racial bias baked into it from the start. Let me explain with a DAG.

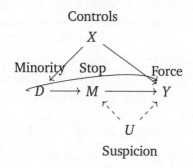

Fryer showed that minorities were more likely to be stopped using both the stop-and-frisk data and the Police-Public Contact Survey. So we know already that the $D \rightarrow M$ pathway exists. In fact, it was a very robust correlation across multiple studies. Minorities are more likely to have an encounter with the police. Fryer's study introduces extensive controls about the nature of the interaction, time of day, and hundreds of factors that I've captured with X. Controlling for X allows Fryer to shut this backdoor path.

But notice M—the stop itself. All the administrative data is conditional on a stop. Fryer [2019] acknowledges this from the outset: "Unless otherwise noted, all results are conditional on an interaction. Understanding potential selection into police data sets due to bias in who police interacts with is a difficult endeavor" (3). Yet what this DAG

9 I am not sympathetic to this claim. The administrative data comes from large Texas cities, a large county in California, the state of Florida, and several other cities and counties racial bias has been reported.

shows is that if police stop people who they believe are suspicious *and* use force against people they find suspicious, then conditioning on the stop is *equivalent* to conditioning on a collider. It opens up the $D \rightarrow M \leftarrow U \rightarrow Y$ mediated path, which introduces spurious patterns into the data that, depending on the signs of these causal associations, may distort any true relationship between police and racial differences in shootings.

Dean Knox, Will Lowe, and Jonathan Mummolo are a talented team of political scientists who study policing, among other things. They produced a study that revisited Fryer's question and in my opinion both yielded new clues as to the role of racial bias in police use of force and the challenges of using administrative data sources to do so. I consider Knox et al. [2020] one of the more methodologically helpful studies for understanding this problem and attempting to solve it. The study should be widely read by *every* applied researcher whose day job involves working with proprietary administrative data sets, because this DAG may in fact be a more general problem. After all, administrative data sources are already select samples, and depending on the study question, they may constitute a collider problem of the sort described in this DAG. The authors develop a bias correction procedure that places bounds on the severity of the selection problems. When using this bounding approach, they find that even lower-bound estimates of the incidence of police violence against civilians is as much as five times higher than a traditional approach that ignores the sample selection problem altogether.

It is incorrect to say that sample selection problems were unknown without DAGs. We've known about them and have had some limited solutions to them since at least Heckman [1979]. What I have tried to show here is more general. An atheoretical approach to empiricism will simply fail. Not even "big data" will solve it. Causal inference is not solved with more data, as I argue in the next chapter. Causal inference requires knowledge about the behavioral processes that structure equilibria in the world. Without them, one cannot hope to devise a credible identification strategy. Not even data is a substitute for deep institutional knowledge about the phenomenon you're studying. That, strangely enough, even includes the behavioral processes that generated the *samples* you're using in the first place. You simply must take seriously the behavioral theory that is behind the phenomenon you're studying if you hope to obtain believable estimates of causal effects. And DAGs are a helpful tool for wrapping your head around and expressing those problems.

Conclusion. In conclusion, DAGs are powerful tools.[10] They are helpful at both clarifying the relationships between variables and guiding you in a research design that has a shot at identifying a causal effect. The two concepts we discussed in this chapter—the backdoor criterion and collider bias—are but two things I wanted to bring to your attention. And since DAGs are themselves based on counterfactual forms of reasoning, they fit well with the potential outcomes model that I discuss in the next chapter.

---

10 There is far more to DAGs than I have covered here. If you are interested in learning more about them, then I encourage you to carefully read Pearl [2009], which is his magnum opus and a major contribution to the theory of causation.

Potential Outcomes Causal Model

It's like the more money we come across, the more problems we see.

Notorious B.I.G.

Practical questions about causation have been a preoccupation of economists for several centuries. Adam Smith wrote about the causes of the wealth of nations [Smith, 2003]. Karl Marx was interested in the transition of society from capitalism to socialism [Needleman and Needleman, 1969]. In the twentieth century the Cowles Commission sought to better understand identifying causal parameters [Heckman and Vytlacil, 2007].[1] Economists have been wrestling with both the big ideas around causality and the development of useful empirical tools from day one.

We can see the development of the modern concepts of causality in the writings of several philosophers. Hume [1993] described causation as a sequence of temporal events in which, had the first event not occurred, subsequent ones would not either. For example, he said:

---

1 This brief history will focus on the development of the potential outcomes model. See Morgan [1991] for a more comprehensive history of econometric ideas.

We may define a cause to be an object, followed by another, and where all the objects similar to the first are followed by objects similar to the second. Or in other words where, if the first object had not been, the second never had existed.

Mill [2010] devised five methods for inferring causation. Those methods were (1) the method of agreement, (2) the method of difference, (3) the joint method, (4) the method of concomitant variation, and (5) the method of residues. The second method, the method of difference, is most similar to the idea of causation as a comparison among counterfactuals. For instance, he wrote:

If a person eats of a particular dish, and dies in consequence, that is, would not have died if he had not eaten it, people would be apt to say that eating of that dish was the source of his death. [399]

Statistical inference. A major jump in our understanding of causation occurs coincident with the development of modern statistics. Probability theory and statistics revolutionized science in the nineteenth century, beginning with the field of astronomy. Giuseppe Piazzi, an early nineteenth-century astronomer, discovered the dwarf planet Ceres, located between Jupiter and Mars, in 1801. Piazzi observed it 24 times before it was lost again. Carl Friedrich Gauss proposed a method that could successfully predict Ceres's next location using data on its prior location. His method minimized the sum of the squared errors; in other words, the ordinary least squares method we discussed earlier. He discovered OLS at age 18 and published his derivation of OLS in 1809 at age 24 [Gauss, 1809].[2] Other scientists who contributed to our understanding of OLS include Pierre-Simon LaPlace and Adrien-Marie Legendre.

The statistician G. Udny Yule made early use of regression analysis in the social sciences. Yule [1899] was interested in the causes of poverty in England. Poor people depended on either poorhouses or the local authorities for financial support, and Yule wanted to know if

---

2 Around age 20, I finally beat *Tomb Raider 2* on the Sony PlayStation. So yeah, I can totally relate to Gauss's accomplishments at such a young age.

public assistance increased the number of paupers, which is a causal question. Yule used least squares regression to estimate the partial correlation between public assistance and poverty. His data was drawn from the English censuses of 1871 and 1881, and I have made his data available at my website for Stata or the Mixtape library for R users. Here's an example of the regression one might run using these data:

$$\text{Pauper} = \alpha + \delta\text{Outrelief} + \beta_1\text{Old} + \beta_2\text{Pop} + u$$

Let's run this regression using the data.

STATA
yule.do

```
1   use https://github.com/scunning1975/mixtape/raw/master/yule.dta, clear
2   regress paup outrelief old pop
```

R
yule.R

```
1   library(tidyverse)
2   library(haven)
3
4   read_data <- function(df)
5   {
6     full_path <- paste("https://raw.github.com/scunning1975/mixtape/master/",
7                 df, sep = "")
8     df <- read_dta(full_path)
9     return(df)
10  }
11
12  yule <- read_data("yule.dta") %>%
13    lm(paup ~ outrelief + old + pop, .)
14  summary(yule)
```

Each row in this data set is a particular location in England (e.g., Chelsea, Strand). So, since there are 32 rows, that means the data set contains 32 English locations. Each of the variables is expressed as

Table 10. Estimated association between pauperism growth rates and public assistance.

Covariates	Dependent variable Pauperism growth
Out-relief	0.752
	(0.135)
Old	0.056
	(0.223)
Pop	−0.311
	(0.067)

an annual growth rate. As a result, each regression coefficient has elasticity interpretations, with one caveat—technically, as I explained at the beginning of the book, elasticities are actually *causal* objects, not simply correlations between two variables. And it's unlikely that the conditions needed to interpret these as causal relationships are met in Yule's data. Nevertheless, let's run the regression and look at the results, which I report in Table 10.

In words, a 10-percentage-point change in the out-relief growth rate is associated with a 7.5-percentage-point increase in the pauperism growth rate, or an elasticity of 0.75. Yule used his regression to crank out the correlation between out-relief and pauperism, from which he concluded that public assistance increased pauper growth rates.

But what might be wrong with this reasoning? How convinced are you that all backdoor paths between pauperism and out-relief are blocked once you control for two covariates in a cross-sectional database for all of England? Could there be unobserved determinants of both poverty and public assistance? After all, he does not control for any economic factors, which surely affect both poverty and the amount of resources allocated to out-relief. Likewise, he may have the causality backwards—perhaps increased poverty causes communities to increase relief, and not merely the other way around. The earliest adopters of some new methodology or technique are often the ones who get the most criticism, despite being pioneers of the methods themselves. It's trivially easy to beat up on a researcher from one hundred years ago, working at a time when the alternative to regression was ideological make-believe. Plus he isn't here to reply. I merely want to note that the naïve use of regression to estimate correlations as a way of making causal claims that inform important policy questions has been the norm for a very long time, and it likely isn't going away any time soon.

Physical Randomization

The notion of physical randomization as the foundation of causal inference was in the air in the nineteenth and twentieth centuries, but it was not until Fisher [1935] that it crystallized. The first historically recognized randomized experiment had occurred fifty years earlier in psychology [Peirce and Jastrow, 1885]. But interestingly, in that experiment, the reason for randomization was *not* as the basis for causal inference. Rather, the researchers proposed randomization as a way of fooling subjects in their experiments. Peirce and Jastrow [1885] used several treatments, and they used physical randomization so that participants couldn't guess what would happen next. Unless I'm mistaken, recommending physical randomization of treatments to units as a basis for causal inference is based on Splawa-Neyman [1923] and Fisher [1925]. More specifically, Splawa-Neyman [1923] developed the powerful potential outcomes notation (which we will discuss soon), and while he proposed randomization, it was not taken to be literally necessary until Fisher [1925]. Fisher [1925] proposed the explicit use of randomization in experimental design for causal inference.[3]

Physical randomization was largely the domain of agricultural experiments until the mid-1950s, when it began to be used in medical trials. Among the first major randomized experiments in medicine—in fact, ever attempted—were the Salk polio vaccine field trials. In 1954, the Public Health Service set out to determine whether the Salk vaccine prevented polio. Children in the study were assigned *at random* to receive the vaccine or a placebo.[4] Also, the doctors making the

---

3 For more on the transition from Splawa-Neyman [1923] to Fisher [1925], see Rubin [2005].

4 In the placebo, children were injected with a saline solution.

diagnoses of polio did not know whether the child had received the vaccine or the placebo. The polio vaccine trial was called a *double-blind, randomized controlled trial* because neither the patient nor the administrator of the vaccine knew whether the treatment was a placebo or a vaccine. It was necessary for the field trial to be very large because the rate at which polio occurred in the population was 50 per 100,000. The treatment group, which contained 200,745 individuals, saw 33 polio cases. The control group had 201,229 individuals and saw 115 cases. The probability of seeing such a big difference in rates of polio because of chance alone is about one in a billion. The only plausible explanation, it was argued, was that the polio vaccine caused a reduction in the risk of polio.

A similar large-scale randomized experiment occurred in economics in the 1970s. Between 1971 and 1982, the RAND Corporation conducted a large-scale randomized experiment studying the causal effect of health-care insurance on health-care utilization. For the study, Rand recruited 7,700 individuals younger than age 65. The experiment was somewhat complicated, with multiple treatment arms. Participants were randomly assigned to one of five health insurance plans: free care, three plans with varying levels of cost sharing, and an HMO plan. Participants with cost sharing made fewer physician visits and had fewer hospitalizations than those with free care. Other declines in health-care utilization, such as fewer dental visits, were also found among the cost-sharing treatment groups. Overall, participants in the cost-sharing plans tended to spend less on health because they used fewer services. The reduced use of services occurred mainly because participants in the cost-sharing treatment groups were opting not to initiate care.[5]

But the use of randomized experiments has exploded since that health-care experiment. There have been multiple Nobel Prizes given to those who use them: Vernon Smith for his pioneering of the laboratory experiments in 2002, and more recently, Abhijit Bannerjee, Esther Duflo, and Michael Kremer in 2019 for their leveraging of field

---

[5] More information about this fascinating experiment can be found in Newhouse [1993].

experiments at the service of alleviating global poverty.[6] The experimental design has become a hallmark in applied microeconomics, political science, sociology, psychology, and more. But why is it viewed as important? Why is randomization such a key element of this design for isolating causal effects? To understand this, we need to learn more about the powerful notation that Splawa-Neyman [1923] developed, called "potential outcomes."

Potential outcomes. While the potential outcomes notation goes back to Splawa-Neyman [1923], it got a big lift in the broader social sciences with Rubin [1974].[7] As of this book's writing, potential outcomes is more or less the lingua franca for thinking about and expressing causal statements, and we probably owe Rubin [1974] for that as much as anyone.

In the potential outcomes tradition [Rubin, 1974; Splawa-Neyman, 1923], a causal effect is defined as a comparison between two states of the world. Let me illustrate with a simple example. In the first state of the world (sometimes called the "actual" state of the world), a man takes aspirin for his headache and one hour later reports the severity of his headache. In the second state of the world (sometimes called the "counterfactual" state of the world), that same man takes nothing for his headache and one hour later reports the severity of his headache. What was the causal effect of the aspirin? According to the potential outcomes tradition, the causal effect of the aspirin is the difference in the severity of his headache between two states of the world: one where he took the aspirin (the actual state of the world) and one where he never took the aspirin (the counterfactual state of the world). The difference in headache severity between these two states of the world, measured at what is otherwise the same point in time, is the causal effect of aspirin on his headache. Sounds easy!

---

6 If I were a betting man—and I am—then I would bet we see at least one more experimental prize given out. The most likely candidate being John List, for his work using field experiments.

7 Interestingly, philosophy as a field undertakes careful consideration of counterfactuals at the same time as Rubin's early work with the great metaphysical philosopher David Lewis [Lewis, 1973]. This stuff was apparently in the air, which makes tracing the causal effect of scientific ideas tough.

To even ask questions like this (let alone attempt to answer them) is to engage in storytelling. Humans have always been interested in stories exploring counterfactuals. What if Bruce Wayne's parents had never been murdered? What if that waitress had won the lottery? What if your friend from high school had never taken that first drink? What if in *The Matrix* Neo had taken the blue pill? These are fun hypotheticals to entertain, but they are still ultimately storytelling. We need Doctor Strange to give us the Time Stone to answer questions like these.

You can probably see where this is going. The potential outcomes notation expresses causality in terms of counterfactuals, and since counterfactuals do not exist, confidence about causal effects must to some degree be unanswerable. To wonder how life would be different had one single event been different is to indulge in counterfactual reasoning, and counterfactuals are not realized in history because they are hypothetical states of the world. Therefore, if the answer requires data on those counterfactuals, then the question cannot be answered. History is a sequence of observable, *factual* events, one after another. We don't know what would have happened had one event changed because we are missing data on the *counterfactual outcome*.[8] Potential outcomes exist ex ante as a set of possibilities, but once a decision is made, all but one outcome disappears.[9]

To make this concrete, let's introduce some notation and more specific concepts. For simplicity, we will assume a *binary* variable that takes on a value of 1 if a particular unit *i* receives the *treatment* and a 0 if it does not.[10] Each unit will have two *potential outcomes*, but only

---

[8] Counterfactual reasoning can be helpful, but it can also be harmful, particularly when it is the source of regret. There is likely a counterfactual version of the sunk-cost fallacy wherein, since we cannot know with certainty what would've happened had we made a different decision, we must accept a certain amount of basic uncertainty just to move on and get over it. Ultimately, no one can say that an alternative decision would've had a better outcome. You cannot know, and that can be difficult sometimes. It has been and will continue to be, for me at least.

[9] As best I can tell, the philosopher I mentioned earlier, David Lewis, believed that potential outcomes were actually separate worlds—just as real as our world. That means that, according to Lewis, there is a very real, yet inaccessible, world in which Kanye released *Yandhi* instead of *Jesus Is King*, I find extremely frustrating.

[10] A couple of things. First, this analysis can be extended to more than two potential outcomes, but as a lot of this book focuses on program evaluation, I am sticking

one observed outcome. Potential outcomes are defined as Y_i^1 if unit i received the treatment and as Y_i^0 if the unit did not. Notice that both potential outcomes have the same i subscript—this indicates two separate states of the world for the exact same person in our example at the exact same moment in time. We'll call the state of the world where no treatment occurred the *control* state. Each unit i has exactly two potential outcomes: a potential outcome under a state of the world where the treatment occurred (Y^1) and a potential outcome where the treatment did not occur (Y^0).

Observable or "actual" outcomes, Y_i, are distinct from potential outcomes. First, notice that actual outcomes do not have a superscript. That is because they are not potential outcomes—they are the realized, actual, historical, empirical—however you want to say it—outcomes that unit i experienced. Whereas potential outcomes are hypothetical random variables that differ across the population, observable outcomes are factual random variables. How we get from potential outcomes to actual outcomes is a major philosophical move, but like any good economist, I'm going to make it seem simpler than it is with an equation. A unit's observable outcome is a function of its potential outcomes determined according to the *switching equation*:

$$Y_i = D_i Y_i^1 + (1 - D_i) Y_i^0 \tag{4.1}$$

where D_i equals 1 if the unit received the treatment and 0 if it did not. Notice the logic of the equation. When $D_i = 1$, then $Y_i = Y_i^1$ because the second term zeroes out. And when $D_i = 0$, the first term zeroes out and therefore $Y_i = Y_i^0$. Using this notation, we define the unit-specific treatment effect, or causal effect, as the difference between the two states of the world:

$$\delta_i = Y_i^1 - Y_i^0$$

Immediately we are confronted with a problem. If a treatment effect requires knowing two states of the world, Y_i^1 and Y_i^0, but by the switching equation we observe only one, then we cannot calculate the treatment effect. Herein lies the fundamental problem of causal

with just two. Second, the treatment here is any particular intervention that can be manipulated, such as the taking of aspirin or not. In the potential outcomes tradition, manipulation is central to the concept of causality.

inference—*certainty* around causal effects requires access to data that is and always will be missing.

Average treatment effects. From this simple definition of a treatment effect come three different parameters that are often of interest to researchers. They are all population means. The first is called the *average treatment effect*:

$$ATE = E[\delta_i]$$
$$= E[Y_i^1 - Y_i^0]$$
$$= E[Y_i^1] - E[Y_i^0] \qquad (4.2)$$

Notice, as with our definition of individual-level treatment effects, that the average treatment effect requires both potential outcomes for each *i* unit. Since we only know one of these by the switching equation, the average treatment effect, or the *ATE*, is inherently unknowable. Thus, the ATE, like the individual treatment effect, is not a quantity that can be calculated. But it can be *estimated*.

The second parameter of interest is the *average treatment effect for the treatment group*. That's a mouthful, but let me explain. There exist two groups of people in this discussion we've been having: a treatment group and a control group. The average treatment effect for the treatment group, or *ATT* for short, is simply that population mean treatment effect for the group of units that had been assigned the treatment in the first place according to the switching equation. Insofar as δ_i differs across the population, the ATT will likely differ from the ATE. In observational data involving human beings, it almost always will be different from the ATE, and that's because individuals will be endogenously sorting into some treatment based on the gains they expect from it. Like the ATE, the ATT is unknowable, because like the ATE, it also requires two observations per treatment unit *i*. Formally we write the ATT as:

$$ATT = E[\delta_i \mid D_i = 1]$$
$$= E[Y_i^1 - E_i^0 \mid D_i = 1]$$
$$= E[Y_i^1 \mid D_i = 1] - E[Y_i^0 \mid D_i = 1] \qquad (4.3)$$

The final parameter of interest is called the average treatment effect for the control group, or *untreated* group. It's shorthand is *ATU*, which stands for average treatment effect for the untreated. And like ATT, the ATU is simply the population mean treatment effect for those units who sorted into the control group.[11] Given heterogeneous treatment effects, it's probably the case that the $ATT \neq ATU$, especially in an observational setting. The formula for the ATU is as follows:

$$ATU = E[\delta_i \mid D_i = 0]$$
$$= E[Y_i^1 - Y_i^0 \mid D_i = 0]$$
$$= E[Y_i^1 \mid D_i = 0] - E[Y_i^0 \mid D_i = 0] \qquad (4.4)$$

Depending on the research question, one, or all three, of these parameters is interesting. But the two most common ones of interest are the ATE and the ATT.

Simple difference in means decomposition. This discussion has been somewhat abstract, so let's be concrete. Let's assume there are ten patients *i* who have cancer, and two medical procedures or treatments. There is a surgery intervention, $D_i = 1$, and there is a chemotherapy intervention, $D_i = 0$. Each patient has the following two potential outcomes where a potential outcome is defined as post-treatment life span in years: a potential outcome in a world where they received surgery and a potential outcome where they had instead received chemo. We use the notation Y^1 and Y^0, respectively, for these two states of the world.

We can calculate the average treatment effect if we have this matrix of data, because the average treatment effect is simply the mean difference between columns 2 and 3. That is, $E[Y^1] = 5.6$, and $E[Y^0] = 5$, which means that $ATE = 0.6$. In words, the average treatment effect of surgery across these specific patients is 0.6 additional years (compared to chemo).

But that is just the average. Notice, though: not everyone benefits from surgery. Patient 7, for instance, lives only one additional year post-surgery versus ten additional years post-chemo. But the ATE is simply the average over these heterogeneous treatment effects.

---

11 This can happen because of preferences, but it also can happen because of constraints. Utility maximization, remember, is a constrained optimization process, and therefore value and obstacles both play a role in sorting.

Table 11. Potential outcomes for ten patients receiving surgery Y^1 or chemo Y^0.

Patient	Y^1	Y^0	δ
1	7	1	6
2	5	6	−1
3	5	1	4
4	7	8	−1
5	4	2	2
6	10	1	9
7	1	10	−9
8	5	6	−1
9	3	7	−4
10	9	8	1

To maintain this fiction, let's assume that there exists the perfect doctor who knows each person's potential outcomes and chooses whichever treatment that maximizes a person's post-treatment life span.[12] In other words, the doctor chooses to put a patient in surgery or chemotherapy depending on whichever treatment has the longer post-treatment life span. Once he makes that treatment assignment, the doctor observes their post-treatment actual outcome according to the switching equation mentioned earlier.

Table 12 shows only the observed outcome for treatment and control group. Table 12 differs from Table 11, which shows each unit's potential outcomes. Once treatment has been assigned, we can calculate the average treatment effect for the surgery group (ATT) versus the chemo group (ATU). The ATT equals 4.4, and the ATU equals −3.2. That means that the average post-surgery life span for the surgery group is 4.4 additional years, whereas the average post-surgery life span for the chemotherapy group is 3.2 fewer years.[13]

---

12 Think of the "perfect doctor" as like a Harry Potter–style Sorting Hat. I first learned of this "perfect doctor" illustration from Rubin himself.

13 The reason that the ATU is negative is because the treatment here is the surgery, which did not perform as well as chemotherapy-untreated units. But you could just as easily interpret this as 3.2 *additional* years of life if they had received chemo instead of surgery.

Table 12. Post-treatment observed life spans in years for surgery $D = 1$ versus chemotherapy $D = 0$.

Patients	Y	D
1	7	1
2	6	0
3	5	1
4	8	0
5	4	1
6	10	1
7	10	0
8	6	0
9	7	0
10	9	1

Now the ATE is 0.6, which is just a weighted average between the ATT and the ATU.[14] So we know that the overall effect of surgery is positive, although the effect for some is negative. There exist heterogeneous treatment effects, in other words, but the net effect is positive. What if we were to simply compare the average post-surgery life span for the two groups? This simplistic estimator is called the simple difference in means, and it is an *estimate* of the ATE equal to

$$E[Y^1 \mid D = 1] - E[Y^0 \mid D = 0]$$

which can be estimated using samples of data:

$$SDO = E[Y^1 \mid D = 1] - E[Y^0 \mid D = 0]$$

$$= \frac{1}{N_T} \sum_{i=1}^{n} (y_i \mid d_i = 1) - \frac{1}{N_C} \sum_{i=1}^{n} (y_i \mid d_i = 0) \qquad (4.5)$$

which in this situation is equal to $7 - 7.4 = -0.4$. That means that the treatment group lives 0.4 fewer years post-surgery than the chemo group when the perfect doctor assigned each unit to its best treatment. While the statistic is true, notice how misleading it is. This statistic without proper qualification could easily be used to claim that, on average, surgery is harmful, when we know that's not true.

---

14 $ATE = p \times ATT + (1 - p) \times ATU = 0.5 \times 4.4 + 0.5 \times -3.2 = 0.6$.

It's biased because the individuals units were optimally sorting into their best treatment option, creating fundamental differences between treatment and control group that are a direct function of the potential outcomes themselves. To make this as clear as I can make it, we will decompose the simple difference in means into three parts. Those three parts are listed below:

$$E[Y^1 \mid D = 1] - E[Y^0 \mid D = 0] = ATE$$
$$+ E[Y^0 \mid D = 1] - E[Y^0 \mid D = 0]$$
$$+ (1 - \pi)(ATT - ATU) \tag{4.6}$$

To understand where these parts on the right-hand side originate, we need to start over and decompose the parameter of interest, *ATE*, into its basic building blocks. ATE is equal to the weighted sum of conditional average expectations, *ATT* and *ATU*.

$$ATE = \pi ATT + (1 - \pi)ATU$$
$$= \pi E[Y^1 \mid D = 1] - \pi E[Y^0 \mid D = 1]$$
$$+ (1 - \pi)E[Y^1 \mid D = 0] - (1 - \pi)E[Y^0 \mid D = 0]$$
$$= \left\{ \pi E[Y^1 \mid D = 1] + (1 - \pi)E[Y^1 \mid D = 0] \right\}$$
$$- \left\{ \pi E[Y^0 \mid D = 1] + (1 - \pi)E[Y^0 \mid D = 0] \right\}$$

where π is the share of patients who received surgery and $1 - \pi$ is the share of patients who received chemotherapy. Because the conditional expectation notation is a little cumbersome, let's exchange each term on the left side, *ATE*, and right side with some letters. This will make the proof a little less cumbersome:

$$E[Y^1 \mid D = 1] = a$$
$$E[Y^1 \mid D = 0] = b$$
$$E[Y^0 \mid D = 1] = c$$
$$E[Y^0 \mid D = 0] = d$$
$$ATE = e$$

Now that we have made these substitutions, let's rearrange the letters by redefining ATE as a weighted average of all conditional expectations

$$e = \{\pi a + (1-\pi)b\} - \{\pi c + (1-\pi)d\}$$

$$e = \pi a + b - \pi b - \pi c - d + \pi d$$

$$e = \pi a + b - \pi b - \pi c - d + \pi d + (\mathbf{a} - \mathbf{a}) + (\mathbf{c} - \mathbf{c}) + (\mathbf{d} - \mathbf{d})$$

$$0 = e - \pi a - b + \pi b + \pi c + d - \pi d - \mathbf{a} + \mathbf{a} - \mathbf{c} + \mathbf{c} - \mathbf{d} + \mathbf{d}$$

$$\mathbf{a} - \mathbf{d} = e - \pi a - b + \pi b + \pi c + d - \pi d + \mathbf{a} - \mathbf{c} + \mathbf{c} - \mathbf{d}$$

$$\mathbf{a} - \mathbf{d} = e + (\mathbf{c} - \mathbf{d}) + \mathbf{a} - \pi a - b + \pi b - \mathbf{c} + \pi c + d - \pi d$$

$$\mathbf{a} - \mathbf{d} = e + (\mathbf{c} - \mathbf{d}) + (1-\pi)a - (1-\pi)b + (1-\pi)d - (1-\pi)c$$

$$\mathbf{a} - \mathbf{d} = e + (\mathbf{c} - \mathbf{d}) + (1-\pi)(a-c) - (1-\pi)(b-d)$$

Now, substituting our definitions, we get the following:

$$E[Y^1 \mid D=1] - E[Y^0 \mid D=0] = ATE$$
$$+ \left(E[Y^0 \mid D=1] - E[Y^0 \mid D=0] \right)$$
$$+ (1-\pi)(ATT - ATU) \qquad (4.7)$$

And the decomposition ends. Now the fun part—let's think about what we just made! The left side can be estimated with a sample of data, as both of those potential outcomes become actual outcomes under the switching equation. That's just the simple difference in mean outcomes. It's the right side that is more interesting because it tells us what the simple difference in mean outcomes is by definition. Let's put some labels to it.

$$\underbrace{\frac{1}{N_T}\sum_{i=1}^{n}(y_i \mid d_i = 1) - \frac{1}{N_C}\sum_{i=1}^{n}(y_i \mid d_i = 0)}_{\text{Simple Difference in Outcomes}} = \underbrace{E[Y^1] - E[Y^0]}_{\text{Average Treatment Effect}}$$

$$+ \underbrace{E[Y^0 \mid D=1] - E[Y^0 \mid D=0]}_{\text{Selection bias}}$$

$$+ \underbrace{(1-\pi)(ATT - ATU)}_{\text{Heterogeneous treatment effect bias}}$$

Let's discuss each of these in turn. The left side is the simple difference in mean outcomes, and we already know it is equal to -0.4. Since this is a decomposition, it must be the case that the right side also equals -0.4.

The first term is the average treatment effect, which is the parameter of interest, and we know that it is equal to 0.6. Thus, the remaining two terms must be the source of the bias that is causing the simple difference in means to be negative.

The second term is called the *selection bias*, which merits some unpacking. In this case, the selection bias is the inherent difference between the two groups if both received chemo. Usually, though, it's just a description of the differences between the two groups if there had never been a treatment in the first place. There are in other words two groups: a surgery group and a chemo group. How do their potential outcomes under control differ? Notice that the first is a counterfactual, whereas the second is an observed outcome according to the switching equation. We can calculate this difference here because we have the complete potential outcomes in Table 11. That difference is equal to -4.8.

The third term is a lesser-known form of bias, but it's interesting. Plus, if the focus is the ATE, then it is always present.[15] The *heterogeneous treatment effect bias* is simply the different returns to surgery for the two groups multiplied by the share of the population that is in the chemotherapy group at all. This final term is $0.5 \times (4.4 - (-3.2))$ or 3.8. Note in case it's not obvious that the reason $\pi = 0.5$ is because 5 out of 10 units are in the chemotherapy group.

Now that we have all three parameters on the right side, we can see why the simple difference in mean outcomes is equal to -0.4.

$$-0.4 = 0.6 - 4.8 + 3.8$$

What I find interesting—hopeful even—in this decomposition is that it shows that a contrast between treatment and control group technically "contains" the parameter of interest. I placed "contains" in quotes

---

15 Note that Angrist and Pischke [2009] have a slightly different decomposition where the $SDO = ATT +$ selection bias, but that is because their parameter of interest is the ATT, and therefore the third term doesn't appear.

because while it is clearly visible in the decomposition, the simple difference in outcomes is ultimately not laid out as the sum of three parts. Rather, the simple difference in outcomes is nothing more than a number. The number is the sum of the three parts, but we cannot calculate each individual part because we do not have data on the underlying counterfactual outcomes needed to make the calculations. The problem is that that parameter of interest has been masked by two forms of bias, the selection bias and the heterogeneous treatment effect bias. If we knew those, we could just subtract them out, but ordinarily we don't know them. We develop strategies to negate these biases, but we cannot directly calculate them any more than we can directly calculate the ATE, as these biases depend on unobservable counterfactuals.

The problem isn't caused by assuming heterogeneity either. We can make the strong assumption that treatment effects are constant, $\delta_i = \delta \ \forall i$, which will cause $ATU = ATT$ and make $SDO = ATE$ + selection bias. But we'd still have that nasty selection bias screwing things up. One could argue that the entire enterprise of causal inference is about developing a reasonable strategy for negating the role that selection bias is playing in estimated causal effects.

Independence assumption. Let's start with the most credible situation for using SDO to estimate ATE: when the treatment itself (e.g., surgery) has been assigned to patients *independent* of their potential outcomes. But what does this word "independence" mean anyway? Well, notationally, it means:

$$(Y^1, Y^0) \perp\!\!\!\perp D \tag{4.8}$$

What this means is that surgery was assigned to an individual for reasons that had *nothing* to do with the gains to surgery.[16] Now in our example, we already know that this is violated because the perfect doctor specifically chose surgery or chemo based on potential outcomes. Specifically, a patient received surgery if $Y^1 > Y^0$ and chemo if $Y^1 < Y^0$. Thus, in our case, the perfect doctor ensured that D *depended* on Y^1 *and* Y^0. All forms of human-based sorting—probably as a rule

---

16 Why do I say "gains"? Because the gain to surgery is $Y_i^1 - Y_i^0$. Thus, if we say it's independent of gains, we are saying it's independent of Y^1 and Y^0.

to be honest—violate independence, which is the main reason naïve observational comparisons are almost always incapable of recovering causal effects.[17]

But what if he hadn't done that? What if he had chosen surgery in such a way that did not depend on Y^1 or Y^0? How does one choose surgery independent of the expected gains of the surgery? For instance, maybe he alphabetized them by last name, and the first five received surgery and the last five received chemotherapy. Or maybe he used the second hand on his watch to assign surgery to them: if it was between 1 and 30 seconds, he gave them surgery, and if it was between 31 and 60 seconds, he gave them chemotherapy.[18] In other words, let's say that he chose some method for assigning treatment that did not depend on the values of potential outcomes under either state of the world. What would that mean in this context? Well, it would mean:

$$E[Y^1 \mid D = 1] - E[Y^1 \mid D = 0] = 0 \qquad (4.9)$$

$$E[Y^0 \mid D = 1] - E[Y^0 \mid D = 0] = 0 \qquad (4.10)$$

In other words, it would mean that the mean potential outcome for Y^1 or Y^0 is the same (in the population) for either the surgery group or the chemotherapy group. This kind of *randomization* of the treatment assignment would eliminate both the selection bias and the heterogeneous treatment effect bias. Let's take it in order. The selection bias zeroes out as follows:

$$E[Y^0 \mid D = 1] - E[Y^0 \mid D = 0] = 0$$

And thus the *SDO* no longer suffers from selection bias. How does randomization affect heterogeneity treatment bias from the third line?

---

17 This is actually where economics is helpful in my opinion. Economics emphasizes that observed values are equilibria based on agents engaging in constrained optimization and that all but guarantees that independence is violated in observational data. Rarely are human beings making important life choices by flipping coins.

18 In Craig [2006], a poker-playing banker used the second hand on his watch as a random number generator to randomly bluff when he had a weak hand.

Rewrite definitions for ATT and ATU:

$$ATT = E[Y^1 \mid D = 1] - E[Y^0 \mid D = 1]$$
$$ATU = E[Y^1 \mid D = 0] - E[Y^0 \mid D = 0]$$

Rewrite the third row bias after $1 - \pi$:

$$ATT - ATU = \mathbf{E[Y^1 \mid D = 1]} - E[Y^0 \mid D = 1]$$
$$- \mathbf{E[Y^1 \mid D = 0]} + E[Y^0 \mid D = 0]$$
$$= 0$$

If treatment is independent of potential outcomes, then:

$$\frac{1}{N_T} \sum_{i=1}^{n} (y_i \mid d_i = 1) - \frac{1}{N_C} \sum_{i=1}^{n} (y_i \mid d_i = 0) = E[Y^1] - E[Y^0]$$

$$SDO = ATE$$

What's necessary in this situation is simply (a) data on observable outcomes, (b) data on treatment assignment, and (c) $(Y^1, Y^0) \perp\!\!\!\perp D$. We call (c) the independence assumption. To illustrate that this would lead to the SDO, we use the following Monte Carlo simulation. Note that *ATE* in this example is equal to 0.6.

STATA		
	independence.do	

```
1   clear all
2   program define gap, rclass
3
4       version 14.2
5       syntax [, obs(integer 1) mu(real 0) sigma(real 1) ]
6       clear
7       drop _all
8       set obs 10
9       gen     y1 = 7 in 1
10      replace y1 = 5 in 2
```

(continued)

STATA (*continued*)

```
11      replace y1 = 5 in 3
12      replace y1 = 7 in 4
13      replace y1 = 4 in 5
14      replace y1 = 10 in 6
15      replace y1 = 1 in 7
16      replace y1 = 5 in 8
17      replace y1 = 3 in 9
18      replace y1 = 9 in 10
19
20      gen      y0 = 1 in 1
21      replace y0 = 6 in 2
22      replace y0 = 1 in 3
23      replace y0 = 8 in 4
24      replace y0 = 2 in 5
25      replace y0 = 1 in 6
26      replace y0 = 10 in 7
27      replace y0 = 6 in 8
28      replace y0 = 7 in 9
29      replace y0 = 8 in 10
30      drawnorm random
31      sort random
32
33      gen      d=1 in 1/5
34      replace d=0 in 6/10
35      gen      y=d*y1 + (1-d)*y0
36      egen sy1 = mean(y) if d==1
37      egen sy0 = mean(y) if d==0
38      collapse (mean) sy1 sy0
39      gen sdo = sy1 - sy0
40      keep sdo
41      summarize sdo
42      gen mean = r(mean)
43      end
44
45   simulate mean, reps(10000): gap
46   su _sim_1
47
```

R

independence.R

```r
1    library(tidyverse)
2
3    gap <- function()
4    {
5      sdo <- tibble(
6        y1 = c(7,5,5,7,4,10,1,5,3,9),
7        y0 = c(1,6,1,8,2,1,10,6,7,8),
8        random = rnorm(10)
9      ) %>%
10      arrange(random) %>%
11      mutate(
12        d = c(rep(1,5), rep(0,5)),
13        y = d * y1 + (1 - d) * y0
14      ) %>%
15      pull(y)
16
17      sdo <- mean(sdo[1:5]-sdo[6:10])
18
19      return(sdo)
20    }
21
22    sim <- replicate(10000, gap())
23    mean(sim)
```

This Monte Carlo runs 10,000 times, each time calculating the average SDO under independence—which is ensured by the random number sorting that occurs. In my running of this program, the ATE is 0.6, and the SDO is on average equal to 0.59088.[19]

Before we move on from the SDO, let's just emphasize something that is often lost on students first learning the independence concept and notation. Independence does not imply that $E[Y^1 \mid D = 1] - E[Y^0 \mid D = 0] = 0$. Nor does it imply that $E[Y^1 \mid D = 1] - E[Y^0 \mid D = 1] = 0$. Rather, it implies

$$E[Y^1 \mid D = 1] - E[Y^1 \mid D = 0] = 0$$

19 Because it's not seeded, when you run it, your answer will be close but slightly different because of the randomness of the sample drawn.

in a large population.[20] That is, independence implies that the two groups of units, surgery and chemo, have the same potential outcome on average in the population.

How realistic is independence in observational data? Economics—maybe more than any other science—tells us that independence is unlikely to hold observationally. Economic actors are always attempting to achieve some optima. For instance, parents are putting kids in what they perceive to be the best school for them, and that is based on potential outcomes. In other words, people are *choosing* their interventions, and most likely their decisions are related to the potential outcomes, which makes simple comparisons improper. Rational choice is always pushing against the independence assumption, and therefore simple comparison in means will not approximate the true causal effect. We need unit randomization for simple comparisons to help us understand the causal effects at play.

SUTVA. Rubin argues that there are a bundle of assumptions behind this kind of calculation, and he calls these assumptions the *stable unit treatment value assumption*, or SUTVA for short. That's a mouthful, but here's what it means: our potential outcomes framework places limits on us for calculating treatment effects. When those limits do not credibly hold in the data, we have to come up with a new solution. And those limitations are that each unit receives the same sized dose, no spillovers ("externalities") to other units' potential outcomes when a unit is exposed to some treatment, and no general equilibrium effects.

First, this implies that the treatment is received in homogeneous doses to all units. It's easy to imagine violations of this, though—for instance, if some doctors are better surgeons than others. In which case, we just need to be careful what we are and are not defining as the treatment.

Second, this implies that there are no externalities, because by definition, an externality spills over to other untreated units. In other words, if unit 1 receives the treatment, and there is some externality,

---

20 Here's a simple way to remember what equality we get with independence. The term before the vertical bar is the same, but the term after the vertical bar is different. So independence guarantees that in the population Y^1 is the same on average, for each group.

then unit 2 will have a different Y^0 value than if unit 1 had not received the treatment. We are assuming away this kind of spillover. When there are such spillovers, though, such as when we are working with social network data, we will need to use models that can explicitly account for such SUTVA violations, such as that of Goldsmith-Pinkham and Imbens [2013].

Related to this problem of spillovers is the issue of general equilibrium. Let's say we are estimating the causal effect of returns to schooling. The increase in college education would in general equilibrium cause a change in relative wages that is different from what happens under partial equilibrium. This kind of scaling-up issue is of common concern when one considers extrapolating from the experimental design to the large-scale implementation of an intervention in some population.

Replicating "demand for learning HIV status." Rebecca Thornton is a prolific, creative development economist. Her research has spanned a number of topics in development and has evaluated critically important questions regarding optimal HIV policy, demand for learning, circumcision, education, and more. Some of these papers have become major accomplishments. Meticulous and careful, she has become a leading expert on HIV in sub-Saharan Africa. I'd like to discuss an ambitious project she undertook as a grad student in rural Malawi concerning whether cash incentives caused people to learn their HIV status and the cascading effect of that learning on subsequent risky sexual behavior [Thornton, 2008].

Thornton's study emerges in a policy context where people believed that HIV testing could be used to fight the epidemic. The idea was simple: if people learned their HIV status, then maybe learning they were infected would cause them to take precautions, thus slowing the rate of infection. For instance, they might seek medical treatment, thus prolonging their life and the quality of life they enjoyed. But upon learning their HIV status, maybe finding out they were HIV-positive would cause them to decrease high-risk behavior. If so, then increased testing could create frictions throughout the sexual network itself that would slow an epidemic. So commonsense was this policy that the assumptions on which it rested were not challenged until Thornton [2008] did an ingenious field experiment in rural

Malawi. Her results were, like many studies, a mixture of good news and bad.

Attempting to understand the demand for HIV status, or the effect of HIV status on health behaviors, is generally impossible without an experiment. Insofar as individuals are optimally choosing to learn about their type or engaging in health behaviors, then it is unlikely that knowledge about HIV status is independent of potential outcomes. Almost certainly, it is those very potential outcomes that shape the decisions both to acquire that information and to engage in risky behaviors of any sort. Thus, a field experiment would be needed if we were to test the underlying assumptions behind this commonsense policy to use testing to fight the epidemic.

How did she do this, though? Respondents in rural Malawi were offered a free door-to-door HIV test and randomly assigned no voucher or vouchers ranging from $1–$3. These vouchers were redeemable once they visited a nearby voluntary counseling and testing center (VCT). The most encouraging news was that monetary incentives were highly effective in causing people to seek the results of tests. On average, respondents who received any cash-value voucher were two times as likely to go to the VCT center to get their test results compared to those individuals who received no compensation. How big was this incentive? Well, the average incentive in her experiment was worth about a day's wage. But she found positive status-seeking behavior even for the smallest incentive, which was worth only one-tenth a day's wage. Thornton showed that even small monetary nudges could be used to encourage people to learn their HIV type, which has obvious policy implications.

The second part of the experiment threw cold water on any optimism from her first results. Several months after the cash incentives were given to respondents, Thornton followed up and interviewed them about their subsequent health behaviors. Respondents were also given the opportunity to purchase condoms. Using her randomized assignment of incentives for learning HIV status, she was able to isolate the causal effect of learning itself on condom purchase her proxy for engaging in risky sex. She finds that conditional on learning one's HIV status from the randomized incentives, HIV-positive individuals did increase their condom usage over those HIV-positive individuals who had not learned their results *but only in the form of buying two additional condoms.* This study suggested that some kinds of outreach, such as door-to-door testing, may cause people to learn their type—particularly when bundled with incentives—but simply having been incentivized to learn one's HIV status may not itself lead HIV-positive individuals to reduce any engagement in high-risk sexual behaviors, such as having sex without a condom.

Thorton's experiment was more complex than I am able to represent here, and also, I focus now on only the cash-transfer aspect of the experiment, in the form of vouchers. but I am going to focus purely on her incentive results. But before I do so, let's take a look at what she found. Table 13 shows her findings.

Since her project uses randomized assignment of cash transfers for identifying causal effect on learning, she mechanically creates a treatment assignment that is independent of the potential outcomes under consideration. We know this even though we cannot directly test it (i.e., potential outcomes are unseen) because we know how the science works. Randomization, in other words, by design assigns treatment independent of potential outcomes. And as a result, simple differences in means are sufficient for getting basic estimates of causal effects.

But Thornton is going to estimate a linear regression model with controls instead of using a simple difference in means for a few reasons. One, doing so allows her to include a variety of controls that can reduce the residual variance and thus improve the precision of her estimates. This has value because in improving precision, she is able to rule out a broader range of treatment effects that are technically contained by her confidence intervals. Although probably in this case, that's not terribly important given, as we will see, that her standard errors are miniscule.

But the inclusion of controls has other value. For instance, if assignment was conditional on observables, or if the assignment was done at different times, then including these controls (such as district fixed effects) is technically needed to isolate the causal effects

Table 13. Impact of monetary incentives and distance on learning HIV results [Thornton, 2008].

	1	2	3	4	5
Any incentive	0.431***	0.309***	0.219***	0.220***	0.219***
	(0.023)	(0.026)	(0.029)	(0.029)	(0.029)
Amount of incentive		0.091***	0.274***	0.274***	0.273***
		(0.012)	(0.036)	(0.035)	(0.036)
Amount of ^{2} incentive			−0.063***	−0.063***	−0.063***
			(0.011)	(0.011)	(0.011)
HIV	−0.055*	−0.052	−0.05	−0.058*	−0.055*
	(0.031)	(0.032)	(0.032)	(0.031)	(0.031)
Distance (km)				−0.076***	
				(0.027)	
Distance^2				0.010**	
				(0.005)	
Controls	Yes	Yes	Yes	Yes	Yes
Sample size	2,812	2,812	2,812	2,812	2,812
Average attendance	0.69	0.69	0.69	0.69	0.69

Note: Columns 1–5 represent OLS coefficients; robust standard errors clustered by village (for 119 villages) with district fixed effects in parentheses. All specifications also include a term for age-squared. "Any incentive" is an indicator if the respondent received any nonzero monetary incentive. "HIV" is an indicator of being HIV-positive. "Simulated average distance" is an average distance of respondents' households to simulated randomized locations of HIV results centers. Distance is measured as a straight-line spherical distance from a respondent's home to a randomly assigned VCT center from geospatial coordinates and is measured in kilometers. *** Significantly different from zero at 99 percent confidence level. ** Significantly different from zero at 95 percent confidence level. * Significantly different from zero at 90 percent confidence level.

themselves. And finally, regression generates nice standard errors, and maybe for that alone, we should give it a chance.[21]

So what did Thornton find? She uses least squares as her primary model, represented in columns 1–5. The effect sizes that she finds

21 She also chose to cluster those standard errors by village for 119 villages. In doing so, she addresses the over-rejection problem that we saw earlier when discussing clustering in the probability and regression chapter.

could be described as gigantic. Because only 34 percent of the control group participants went to a center to learn their HIV status, it is impressive that receiving any money caused a 43-percentage-point increase in learning one's HIV status. Monetary incentives—even very small ones—are enough to push many people over the hump to go collect health data.

Columns 2–5 are also interesting, but I won't belabor them here. In short, column 2 includes a control for the amount of the incentive, which ranged from US$0 to US$3. This allows us to estimate the linear impact of each additional dollar on learning, which is relatively steep. Columns 3–5 include a quadratic and as a result we see that while each additional dollar increases learning, it does so only at a decreasing rate. Columns 4 and 5 include controls for distance to the VCT center, and as with other studies, distance itself is a barrier to some types of health care [Lindo et al., 2019].

Thornton also produces a simple graphic of her results, showing box plots with mean and confidence intervals for the treatment and control group. As we will continually see throughout the book, the best papers estimating causal effects will always summarize their main results in smart and effective pictures, and this study is no exception. As this figure shows, the effects were huge.

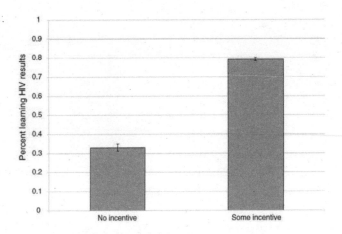

Figure 13. Visual representation of cash transfers on learning HIV test results [Thornton, 2008].

While learning one's own HIV status is important, particularly if it leads to medical care, the gains to policies that nudge learning are particularly higher if they lead to changes in high-risk sexual behavior among HIV-positive individuals. In fact, given the multiplier effects associated with introducing frictions into the sexual network via risk-mitigating behavior (particularly if it disrupts concurrent partnerships), such efforts may be so beneficial that they justify many types of programs that otherwise may not be cost-effective.

Thornton examines in her follow-up survey where she asked all individuals, regardless of whether they learned their HIV status, the effect of a cash transfer on condom purchases. Let's first see her main results in Figure 14.

It is initially encouraging to see that the effects on condom purchases are large for the HIV-positive individuals who, as a result of the incentive, got their test results. Those who bought *any* condoms increases from a baseline that's a little over 30 percent to a whopping 80 percent with any incentive. But where things get discouraging is when we examine *how many* additional condoms this actually entailed. In columns 3 and 4 of Table 14, we see the problem.

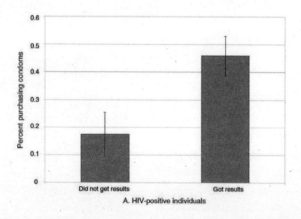

Figure 14. Visual representation of cash transfers on condom purchases for HIV-positive individuals [Thornton, 2008].

Table 14. Reactions to learning HIV results among sexually active at baseline [Thornton, 2008].

Dependent variables	Bought condoms		Number of condoms bought	
	OLS	IV	OLS	IV
Got results	−0.022	−0.069	−0.193	−0.303
	(0.025)	(0.062)	(0.148)	(0.285)
Got results × HIV	0.418***	0.248	1.778***	1.689**
	(0.143)	(0.169)	(0.564)	(0.784)
HIV	−0.175**	−0.073	−0.873	−0.831
	(0.085)	(0.123)	(0.275)	(0.375)
Controls	Yes	Yes	Yes	Yes
Sample size	1,008	1,008	1,008	1,008
Mean	0.26	0.26	0.95	0.95

Note: Sample includes individuals who tested for HIV and have demographic data.

Now Thornton wisely approaches the question in two ways for the sake of the reader and for the sake of accuracy. She wants to know the effect of getting results, but the results only matter (1) for those who got their status and (2) for those who were HIV-positive. The effects shouldn't matter if they were HIV-negative. And ultimately that is what she finds, but how is she going to answer the first? Here she examines the effect for those who got their results and who were HIV-positive using an interaction. And that's column 1: individuals who got their HIV status and who learned they were HIV positive were 41% more likely to buy condoms several months later. This result shrinks, though, once she utilizes the *randomization* of the incentives in an instrumental variables framework, which we will discuss later in the book. The coefficient is almost cut in half and her confidence intervals are so large that we can't be sure the effects are nonexistent.

But let's say that the reason she failed to find an effect on any purchasing behavior is because the sample size is just small enough that to pick up the effect with IV is just asking too much of the data. What if we used something that had a little more information, like number of condoms bought? And that's where things get pessimistic. Yes, Thornton does find evidence that the HIV-positive individuals were buying more condoms, but when see how many, we learn that it is only around 2 more condoms at the follow-up visit (columns 3–4). And the effect on sex itself (not shown) was negative, small (4% reduction), and not precise enough to say either way anyway.

In conclusion, Thornton's study is one of those studies we regularly come across in causal inference, a mixture of positive and negative. It's positive in that nudging people with small incentives leads them to collecting information about their own HIV status. But our enthusiasm is muted when we learn the effect on actual risk behaviors is not very large—a mere two additional condoms bought several months later for the HIV-positive individuals is likely not going to generate large positive externalities unless it falls on the highest-risk HIV-positive individuals.

Randomization Inference

Athey and Imbens [2017a], in their chapter on randomized experiments, note that "in randomization-based inference, uncertainty in estimates arises naturally from the random assignment of the treatments, rather than from hypothesized sampling from a large population" (73). Athey and Imbens are part of a growing trend of economists using randomization-based methods for inferring the probability that an estimated coefficient is not simply a result of change. This growing trend uses randomization-based methods to construct exact p-values that reflect the likelihood that chance could've produced the estimate.

Why has randomization inference become so population now? Why not twenty years ago or more? It's not clear why randomization-based inference has become so popular in recent years, but a few possibilities could explain the trend. It may be the rise in the randomized controlled trials within economics, the availability of large-scale administrative databases that are not samples of some larger population but rather represent "all the data," or it may be that computational power has improved so much that randomization inference has become trivially simple to implement when working with thousands of observations. But whatever the reason, randomization inference has become a very common way to talk about the uncertainty around one's estimates.

There are at least three reasons we might conduct randomization inference. First, it may be because we aren't working with samples, and since standard errors are often justified on the grounds that they reflect sampling uncertainty, traditional methods may not be as meaningful. The core uncertainty within a causal study is not based on sampling uncertainty, but rather on the fact that we do not know the counterfactual [Abadie et al., 2020, 2010]. Second, it may be that we are uncomfortable appealing to the large sample properties of an estimator in a particular setting, such as when we are working with a small number of treatment units. In such situations, maybe assuming the number of units increases to infinity stretches credibility [Buchmueller et al., 2011]. This can be particularly problematic in practice. Young [2019] shows that in finite samples, it is common for some observations to experience concentrated leverage. Leverage causes standard errors and estimates to become volatile and can lead to overrejection. Randomization inference can be more robust to such outliers. Finally, there seems to be some aesthetic preference for these types of placebo-based inference, as many people find them intuitive. While this is not a sufficient reason to adopt a methodological procedure, it is nonetheless very common to hear someone say that they used randomization inference because it makes sense. I figured it was worth mentioning since you'll likely run into comments like that as well. But before we dig into it, let's discuss its history, which dates back to Ronald Fisher in the early twentieth century.

Lady tasting tea. Fisher [1935] described a thought experiment in which a woman claims she can discern whether milk or tea was poured first into a cup of tea. While he does not give her name, we now know that the woman in the thought experiment was Muriel Bristol and that the thought experiment in fact did happen.[22] Muriel Bristol was a PhD scientist back in the days when women rarely were able to become PhD scientists. One day during afternoon tea, Muriel claimed that

---

22 Apparently, Bristol correctly guessed all four cups of tea.

she could tell whether the milk was added to the cup before or after the tea. Incredulous, Fisher hastily devised an experiment to test her self-proclaimed talent.

The hypothesis, properly stated, is that, given a cup of tea with milk, a woman can discern whether milk or tea was first added to the cup. To test her claim, eight cups of tea were prepared; in four the milk was added first, and in four the tea was added first. How many cups does she have to correctly identify to convince us of her uncanny ability?

Fisher [1935] proposed a kind of permutation-based inference—a method we now call the Fisher's exact test. The woman possesses the ability probabilistically, not with certainty, if the likelihood of her guessing all four correctly was sufficiently low. There are $8 \times 7 \times 6 \times 5 = 1,680$ ways to choose a first cup, a second cup, a third cup, and a fourth cup, in order. There are $4 \times 3 \times 2 \times 1 = 24$ ways to order four cups. So the number of ways to choose four cups out of eight is $\frac{1680}{24} = 70$. Note, the woman performs the experiment by selecting four cups. The probability that she would correctly identify all four cups is $\frac{1}{70}$, which is $p = 0.014$.

Maybe you would be more convinced of this method if you could see a simulation, though. So let's conduct a simple combination exercise. You can with the following code.

STATA

tea.do

```
1   clear
2   capture log close
3
4   * Create the data. 4 cups with tea, 4 cups with milk.
5
6   set obs 8
7   gen cup = _n
8
```

(continued)

STATA *(continued)*

```
9    * Assume she guesses the first cup (1), then the second cup (2), and so forth
10   gen      guess = 1 in 1
11   replace guess = 2 in 2
12   replace guess = 3 in 3
13   replace guess = 4 in 4
14   replace guess = 0 in 5
15   replace guess = 0 in 6
16   replace guess = 0 in 7
17   replace guess = 0 in 8
18   label variable guess "1: she guesses tea before milk then stops"
19
20   tempfile correct
21   save "`correct'", replace
22
23   * ssc install percom
24   combin cup, k(4)
25   gen permutation = _n
26   tempfile combo
27   save "'combo'", replace
28
29   destring cup*, replace
30   cross using 'correct'
31   sort permutation cup
32
33   gen      correct = 0
34   replace correct = 1 if cup_1 == 1 & cup_2 == 2 & cup_3 == 3 & cup_4 == 4
35
36   * Calculation p-value
37   count if correct==1
38   local correct `r(N)'
39   count
40   local total `r(N)'
41   di 'correct'/`total'
42   gen pvalue = (`correct')/(`total')
43   su pvalue
44
45   * pvalue equals 0.014
46
47   capture log close
48   exit
```

R
tea.R

```
1   library(tidyverse)
2   library(utils)
3
4   correct <- tibble(
5     cup   = c(1:8),
6     guess = c(1:4,rep(0,4))
7   )
8
9   combo <- correct %$% as_tibble(t(combn(cup, 4))) %>%
10    transmute(
11      cup_1 = V1, cup_2 = V2,
12      cup_3 = V3, cup_4 = V4) %>%
13    mutate(permutation = 1:70) %>%
14    crossing(., correct) %>%
15    arrange(permutation, cup) %>%
16    mutate(correct = case_when(cup_1 == 1 & cup_2 == 2 &
17                        cup_3 == 3 & cup_4 == 4 ~ 1,
18                        TRUE ~ 0))
19  sum(combo$correct == 1)
20  p_value <- sum(combo$correct == 1)/nrow(combo)
```

Notice, we get the same answer either way—0.014. So, let's return to Dr. Bristol. Either she has no ability to discriminate the order in which the tea and milk were poured, and therefore chose the correct four cups by random chance, or she (like she said) has the ability to discriminate the order in which ingredients were poured into a drink. Since choosing correctly is highly unlikely (1 chance in 70), it is reasonable to believe she has the talent that she claimed all along that she had.

So what exactly have we done? Well, what we have done is provide an exact probability value that the observed phenomenon was merely the product of chance. You can never let the fundamental problem of causal inference get away from you: we never *know* a causal effect. We only estimate it. And then we rely on other procedures to give us reasons to believe the number we calculated is probably a causal effect. Randomization inference, like all inference, is epistemological scaffolding for a particular kind of belief—specifically, the likelihood that chance created this observed value through a particular kind of procedure.

But this example, while it motivated Fisher to develop this method, is not an experimental design wherein causal effects are estimated. So now I'd like to move beyond it. Here, I hope, the randomization inference procedure will become a more interesting and powerful tool for making credible causal statements.

Methodology of Fisher's sharp null. Let's discuss more of what we mean by randomization inference in a context that is easier to understand—a literal experiment or quasi-experiment. We will conclude with code that illustrates how we might implement it. The main advantage of randomization inference is that it allows us to make probability calculations revealing whether the data are likely a draw from a truly random distribution or not.

The methodology can't be understood without first understanding the concept of *Fisher's sharp null*. Fisher's sharp null is a claim we make wherein no unit in our data, when treated, had a causal effect. While that is a subtle concept and maybe not readily clear, it will be much clearer once we work through some examples. The value of Fisher's sharp null is that it allows us to make an "exact" inference that does not depend on hypothesized distributions (e.g., Gaussian) or large sample approximations. In this sense, it is *nonparametric*.[23]

Some, when first confronted with the concept of randomization inference, think, "Oh, this sounds like bootstrapping," but the two are in fact completely different. Bootstrapped *p*-values are random draws from the sample that are then used to conduct inference. This means that bootstrapping is primarily about uncertainty in the observations used in the sample itself. But randomization inference *p*-values are not about uncertainty in the sample; rather, they are based on uncertainty over *which units* within a sample are assigned to the treatment itself.

To help you understand randomization inference, let's break it down into a few methodological steps. You could say that there are six steps to randomization inference: (1) the choice of the sharp null,

---

23 I simply mean that the inference does not depend on asymptotics or a type of distribution in the data-generating process.

(2) the construction of the null, (3) the picking of a different treatment vector, (4) the calculation of the corresponding test statistic for that new treatment vector, (5) the randomization over step 3 as you cycle through a number of new treatment vectors (ideally all possible combinations), and (6) the calculation the exact p-value.

Steps to a p value. Fisher and Neyman debated about this first step. Fisher's "sharp" null was the assertion that *every single unit* had a treatment effect of zero, which leads to an easy statement that the ATE is also zero. Neyman, on the other hand, started at the other direction and asserted that there was no *average* treatment effect, not that each unit had a zero treatment effect. This is an important distinction. To see this, assume that your treatment effect is a 5, but my treatment effect is -5. Then the $ATE = 0$ which was Neyman's idea. But Fisher's idea was to say that my treatment effect was zero, and your treatment effect was zero. This is what "sharp" means—it means literally that no single unit has a treatment effect. Let's express this using potential outcomes notation, which can help clarify what I mean.

$$H_0 : \delta_i = Y_i^1 - Y_i^0 = 0 \forall i$$

Now, it may not be obvious how this is going to help us, but consider this—since we know all observed values, if there is no treatment effect, *then* we also know each unit's counterfactual. Let me illustrate my point using the example in Table 15.

Table 15. Example of made-up data for eight people with missing counterfactuals.

Name	D	Y	Y^0	Y^1
Andy	1	10	.	10
Ben	1	5	.	5
Chad	1	16	.	16
Daniel	1	3	.	3
Edith	0	5	5	.
Frank	0	7	7	.
George	0	8	8	.
Hank	0	10	10	.

If you look closely at Table 15, you will see that for each unit, we only observe one potential outcome. But under the sharp null, we can infer the other missing counterfactual. We only have information on observed outcomes based on the switching equation. So if a unit is treated, we know its Y^1 but not its Y^0.

The second step is the construction of what is called a "test statistic." What is this? A test statistic $t(D,Y)$ is simply a known, *scalar* quantity calculated from the treatment assignments and the observed outcomes. It is often simply nothing more than a measurement of the relationship between the Y values by D. In the rest of this section, we will build out a variety of ways that people construct test statistics, but we will start with a fairly straightforward measurement—the simple difference in mean outcome.

Test statistics ultimately help us distinguish between the sharp null itself and some other hypothesis. And if you want a test statistic with high statistical power, then you need the test statistic to take on "extreme" values (i.e., large in absolute values) when the null is false, and you need for these large values to be unlikely when the null is true.[24]

As we said, there are a number of ways to estimate a test statistic, and we will be discussing several of them, but let's start with the simple difference in mean outcomes. The average values for the treatment group are 34/4, the average values for the control group are 30/4, and the difference between these two averages is 1. So given this *particular* treatment assignment in our sample—the true assignment, mind you—there is a corresponding test statistic (the simple difference in mean outcomes) that is equal to 1.

Now, what is implied by Fisher's sharp null is one of the more interesting parts of this method. While historically we do not know each unit's counterfactual, under the sharp null we *do* know each unit's counterfactual. How is that possible? Because if none of the units has nonzero treatment effects, then it must be that each counterfactual

---

24 It's kind of interesting what precisely the engine of this method is—it's actually not designed to pick up small treatment effects because often those small values will be swamped by the randomization process. There's no philosophical reason to believe, though, that average treatment effects have to be relatively "large." It's just that randomization inference *does* require that so as to distinguish the true effect from that of the sharp null.

Table 16. Example of made-up data for eight people with filled-in counterfactuals according to Fisher's sharp null hypothesis.

Name	D	Y	Y^0	Y^1
Andy	1	10	**10**	10
Ben	1	5	**5**	5
Chad	1	16	**16**	16
Daniel	1	3	**3**	3
Edith	0	5	5	**5**
Frank	0	7	7	**7**
George	0	8	8	**8**
Hank	0	10	10	**10**

Note: Under the sharp null, we can infer the missing counterfactual, which I have represented with bold face.

is equal to its observed outcome. This means that we can fill in those missing counterfactuals *with the observed values* (Table 16).

With these missing counterfactuals replaced by the corresponding observed outcome, there's no treatment effect at the unit level and therefore a zero ATE. So why did we find earlier a simple difference in mean outcomes of 1 if in fact there was no average treatment effect? Simple—it was just noise, pure and simple. It was simply a reflection of some arbitrary treatment assignment under Fisher's sharp null, and through random chance it just so happens that this assignment generated a test statistic of 1.

So, let's summarize. We have a particular treatment assignment and a corresponding test statistic. If we assume Fisher's sharp null, that test statistic is simply a draw from some random process. And if that's true, then we can shuffle the treatment assignment, calculate a new test statistic and ultimately compare this "fake" test statistic with the real one.

The key insight of randomization inference is that under the sharp null, the treatment assignment ultimately does not matter. It explicitly assumes as we go from one assignment to another that the counterfactuals aren't changing—they are always just equal to the observed outcomes. So the randomization distribution is simply a set of all possible test statistics for each possible treatment assignment vector.

The third and fourth steps extend this idea by *literally* shuffling the treatment assignment and calculating the unique test statistic for each assignment. And as you do this repeatedly (step 5), in the limit you will eventually cycle through all possible combinations that will yield a distribution of test statistics under the sharp null.

Once you have the entire distribution of test statistics, you can calculate the exact *p*-value. How? Simple—you rank these test statistics, fit the true effect into that ranking, count the number of fake test statistics that dominate the real one, and divide that number by all possible combinations. Formally, that would be this:

$$\Pr\left(t(D, Y) \geq t(D, Y \mid \delta = 0)\right) = \frac{\sum_{D \in \Omega} I(t(D, Y) \geq t(D, Y)}{K}$$

Again, we see what is meant by "exact." These *p*-values are exact, not approximations. And with a rejection threshold of α—for instance, 0.05—then a randomization inference test will falsely reject the sharp null less than $100 \times \alpha$ percent of the time.

Example. I think this has been kind of abstract, and when things are abstract, it's easy to be confused, so let's work through an example with some new data. Imagine that you work for a homeless shelter with a cognitive behavioral therapy (CBT) program for treating mental illness and substance abuse. You have enough funding to enlist four people into the study, but you have eight residents. Therefore, there are four in treatment and four in control. After concluding the CBT, residents are interviewed to determine the severity of their mental illness symptoms. The therapist records their mental health on a scale of 0 to 20. With the following information, we can both fill in missing counterfactuals so as to satisfy Fisher's sharp null and calculate a corresponding test statistic based on this treatment assignment. Our test statistic will be the absolute value of the simple difference in mean outcomes for simplicity. The test statistic for this particular treatment assignment is simply $|34/4 - 30/4| = 8.5 - 7.5 = 1$, using the data in Table 17.

Now we move to the randomization stage. Let's shuffle the treatment assignment and calculate the new test statistic for that new treatment vector. Table 18 shows this permutation. But first, one thing. We are going to keep the number of treatment units fixed throughout this example. But if treatment assignment had followed some random

Table 17. Self-reported mental health for eight residents in a homeless shelter (treatment and control).

Name	D_1 ($15)	Y (health)	Y^0	Y^1
Andy	1	10	**10**	10
Ben	1	5	**5**	5
Chad	1	16	**16**	16
Daniel	1	3	**3**	3
Edith	0	5	5	**5**
Frank	0	7	7	**7**
George	0	8	8	**8**
Hank	0	10	10	**10**

Table 18. First permutation holding the number of treatment units fixed.

Name	$\tilde{D}_2$	Y	Y^0	Y^1
Andy	1	10	**10**	10
Ben	0	5	**5**	5
Chad	1	16	**16**	16
Daniel	1	3	**3**	3
Edith	0	5	5	**5**
Frank	1	7	7	**7**
George	0	8	8	**8**
Hank	0	10	10	**10**

process, like the Bernoulli, then the number of treatment units would be random and the randomized treatment assignment would be larger than what we are doing here. Which is right? Neither is right in itself. Holding treatment units fixed is ultimately a reflection of whether it had been fixed in the original treatment assignment. That means that you need to know your data and the process by which units were assigned to treatment to know how to conduct the randomization inference.

With this shuffling of the treatment assignment, we can calculate a new test statistic, which is $|36/4 - 28/4| = 9 - 7 = 2$. Now before we move on, look at this test statistic: that test statistic of 2 is "fake" because it is not the true treatment assignment. But under the null, the treatment assignment, was already meaningless, since there were no nonzero treatment effects anyway. The point is that even when null

Table 19. Second permutation holding the number of treatment units fixed.

Name	$\tilde{D}_3$	Y	Y^0	Y^1
Andy	1	10	**10**	10
Ben	0	5	**5**	5
Chad	1	16	**16**	16
Daniel	1	3	**3**	3
Edith	0	5	5	**5**
Frank	0	7	7	**7**
George	1	8	8	**8**
Hank	0	10	10	**10**

Table 20. The first few permutations for a randomization of treatment assignments.

Assignment	D_1	D_2	D_3	D_4	D_5	D_6	D_7	D_8	$|T_i|$
True D	1	1	1	1	0	0	0	0	1
$\tilde{D}_2$	1	0	1	1	0	1	0	0	2
$\tilde{D}_3$	1	0	1	1	0	0	1	0	2.25
...

of no effect holds, it can and usually will yield a nonzero effect for no other reason than finite sample properties.

Let's write that number 2 down and do another permutation, by which I mean, let's shuffle the treatment assignment again. Table 19 shows this second permutation, again holding the number of treatment units fixed at four in treatment and four in control.

The test statistic associated with this treatment assignment is $|36/4 - 27/4| = 9 - 6.75 = 2.25$. Again, 2.25 is a draw from a random treatment assignment where each unit has no treatment effect.

Each time you randomize the treatment assignment, you calculate a test statistic, store that test statistic somewhere, and then go onto the next combination. You repeat this over and over until you have exhausted all possible treatment assignments. Let's look at the first iterations of this in Table 20.

The final step is the calculation of the exact p-value. To do this, we have a couple of options. We can either use software to do it, which is a fine way to do it, or we can manually do it ourselves. And for pedagogical reasons, I am partial to doing this manually. So let's go.

STATA
ri.do

```
1   use https://github.com/scunning1975/mixtape/raw/master/ri.dta, clear
2
3   tempfile ri
4   gen id = _n
5   save "`ri'", replace
6
7   * Create combinations
8   * ssc install percom
9   combin id, k(4)
10  gen permutation = _n
11  tempfile combo
12  save "`combo'", replace
13
14  forvalue i =1/4 {
15      ren id_`i' treated`i'
16  }
17
18
19  destring treated*, replace
20  cross using `ri'
21  sort permutation name
22  replace d = 1 if id == treated1 | id == treated2 | id == treated3 | id == treated4
23  replace d = 0 if ~(id == treated1 | id == treated2 | id == treated3 | id == treated4)
24
25  * Calculate true effect using absolute value of SDO
26  egen    te1 = mean(y) if d==1, by(permutation)
27  egen    te0 = mean(y) if d==0, by(permutation)
28
29  collapse (mean) te1 te0, by(permutation)
30  gen     ate = te1 - te0
31  keep    ate permutation
32
33  sort ate
34  gen rank = _n
35  su rank if permutation==1
36  gen pvalue = (`r(mean)'/70)
37  list pvalue if permutation==1
38  * pvalue equals 0.6
```

R
ri.R

```
1   library(tidyverse)
2   library(magrittr)
3   library(haven)
4
5   read_data <- function(df)
6   {
7    full_path <- paste("https://raw.github.com/scunning1975/mixtape/master/",
8            df, sep = "")
9    df <- read_dta(full_path)
10   return(df)
11  }
12
13  ri <- read_data("ri.dta") %>%
14    mutate(id = c(1:8))
15
16  treated <- c(1:4)
17
18  combo <- ri %$% as_tibble(t(combn(id, 4))) %>%
19    transmute(
20     treated1 = V1, treated2 = V2,
21     treated3 = V3, treated4 = V4) %>%
22    mutate(permutation = 1:70) %>%
23    crossing(., ri) %>%
24    arrange(permutation, name) %>%
25    mutate(d = case_when(id == treated1 | id == treated2 |
26             id == treated3 | id == treated4 ~ 1,
27             TRUE ~ 0))
28
29  te1 <- combo %>%
30    group_by(permutation) %>%
31    filter(d == 1) %>%
32    summarize(te1 = mean(y, na.rm = TRUE))
33
34  te0 <- combo %>%
35    group_by(permutation) %>%
36    filter(d == 0) %>%
37    summarize(te0 = mean(y, na.rm = TRUE))
38
```

(continued)

R *(continued)*
39   n <- nrow(inner_join(te1, te0, by = "permutation"))
40
41   p_value <- inner_join(te1, te0, by = "permutation") %>%
42   mutate(ate = te1 - te0) %>%
43   select(permutation, ate) %>%
44   arrange(ate) %>%
45   mutate(rank = 1:nrow(.)) %>%
46   filter(permutation == 1) %>%
47   pull(rank)/n

This program was fairly straightforward because the number of possible combinations was so small. Out of eight observations, then four choose eight equals 70. We just had to manipulate the data to get to that point, but once we did, the actual calculation was straighforward. So we can see that the estimated ATE cannot reject the null in the placebo distribution.

But often the data sets we work with will be much larger than eight observations. In those situations, we cannot use this method, as the sheer volume of combination grows very fast as *n* increases. We will hold off for now reviewing this inference method when *n* is too large until we've had a chance to cover more ground.

Other test statistics. Recall that the second step in this methodology was selection of the test statistic.[25] We chose the simple difference in mean outcomes (or the absolute value of such), which is fine when effects are additive and there are few outliers in the data. But outliers create problems for that test statistic because of the variation that gets introduced in the randomization distribution. So other alternative test statistics become more attractive.

One transformation that handles outliers and skewness more generally is the log transformation. Imbens and Rubin [2015] define this as the average difference on a log scale by treatment status, or

---

25 For more in-depth discussion of the following issues, I highly recommend the excellent Imbens and Rubin [2015], chapter 5 in particular.

Table 21. Illustrating ranks using the example data.

Name	D	Y	Y^0	Y^1	Rank	R_i
Andy	1	10	**10**	10	6.5	2
Ben	1	5	**5**	5	2.5	−2
Chad	1	16	**16**	16	8	3.5
Daniel	1	3	**3**	3	1	−3.5
Edith	0	5	5	**5**	2.5	−1
Frank	0	7	7	**7**	4	−0.5
George	0	8	8	**8**	5	0.5
Hank	0	10	10	**10**	6.5	2

$$T_{\log} = \left| \frac{1}{N_T} \sum_{i=1}^{N} D_i \ln(Y_i) - \frac{1}{N_C} \sum_{i=1}^{N} (1 - D_i) \ln(Y_i) \right|$$

This makes sense when the raw data is skewed, which happens for positive values like earnings and in instances when treatment effects are multiplicative rather than additive.

Another test statistic seen is the absolute value in the difference in quantiles. This also protects against outliers and is represented as

$$T_{\text{median}} = \left| \text{median}(Y_T) - \text{median}(Y_C) \right|$$

We could look at the median, the 25th quantile, the 75th quantile, or anything along the unit interval.

The issue of outliers also leads us to consider a test statistic that uses ranks rather than differences. This again is useful when there are large numbers of outliers, when outcomes are continuous or data sets are small. Rank statistics transform outcomes to ranks and then conduct analysis on the ranks themselves. The basic idea is to rank the outcomes and then compare the average rank of the treated and control groups. Let's illustrate this with an example first (Table 21).

As before, we only observe one half of the potential outcomes given the switching equation which assigns potential outcomes to actual outcomes. But under Fisher's sharp null, we can impute the missing counterfactual so as to ensure no treatment effect. To calculate ranks, we simply count the number of units with higher values of Y, including the unit in question. And in instances of ties, we simply take the average over all tied units.

For instance, consider Andy. Andy has a value of 10. Andy is as large as himself (1); larger than Ben (2), Daniel (3), Edith (4), Frank (5), and George (6); and tied with Hank (7). Since he is tied with Hank, we average the two, which brings his rank to 6.5. Now consider Ben. Ben has a value of 5. He is as large as himself (1), larger than Daniel (2), and tied with Edith (3). Therefore, we average Edith and himself to get 0.5, bringing us to a rank of 2.

It is common, though, to normalize the ranks to have mean 0, which is done according to the following formula:

$$\widetilde{R}_i = \widetilde{R}_i(Y_1, \ldots, Y_N) = \sum_{j=1}^{N} I(Y_j \leq Y_i) - \frac{N+1}{2}$$

This gives us the final column, which we will now use to calculate the test statistic. Let's use the absolute value of the simple difference in mean outcomes on the normalized rank, which here is

$$T_{\text{rank}} = |0 - 1/4| = 1/4$$

To calculate the exact p-value, we would simply conduct the same randomization process as earlier, only instead of calculating the simple difference in mean outcomes, we would calculate the absolute value of the simpler difference in mean *rank*.

But all of these test statistics we've been discussing have been *differences* in the outcomes by treatment status. We considered simple differences in averages, simple differences in log averages, differences in quantiles, and differences in ranks. Imbens and Rubin [2015] note that there are shortcomings that come from focusing solely on a few features of the data (e.g., skewness), as it can cause us to miss differences in other aspects. This specifically can be problematic if the variance in potential outcomes for the treatment group differs from that of the control group. Focusing only on the simple average differences we discussed may not generate p-values that are "extreme" enough to reject the null even when the null in fact does not hold. So we may be interested in a test statistic that can detect differences in *distributions* between the treatment and control units. One such test statistic is the Kolmogorov-Smirnov test statistic (see figure 15).

Let's first define the empirical cumulative distribution function (CDF) as:

$$\widehat{F}_C(Y) = \frac{1}{N_C} \sum_{i:D_i=0} 1(Y_i \leq Y)$$

$$\widehat{F}_T(Y) = \frac{1}{N_T} \sum_{i:D_i=1} 1(Y_i \leq Y)$$

If two distributions are the same, then their empirical CDF is the same. But note, empirical CDFs are functions, and test statistics are *scalars*. So how will we take differences between two functions and turn that into a single scalar quantity? Easy—we will use the *maximum* difference between the two empirical CDFs. Visually, it will literally be the greatest vertical distance between the two empirical CDFs. That vertical distance will be our test statistic. Formally it is:

$$T_{KS} = \max \left| \widehat{F}_T(Y_i) - \widehat{F}_C(Y_i) \right|$$

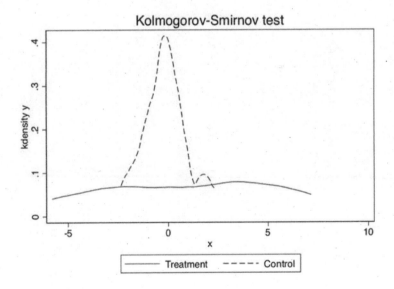

Figure 15. Visualization of distributions by treatment status using a kernel density.

STATA
ks.do

```
1   clear
2   input d y
3   0    0.22
4   0   -0.87
5   0   -2.39
6   0   -1.79
7   0    0.37
8   0   -1.54
9   0    1.28
10  0   -0.31
11  0   -0.74
12  0    1.72
13  0    0.38
14  0   -0.17
15  0   -0.62
16  0   -1.10
17  0    0.30
18  0    0.15
19  0    2.30
20  0    0.19
21  0   -0.50
22  0   -0.09
23  1   -5.13
24  1   -2.19
25  1   -2.43
26  1   -3.83
27  1    0.50
28  1   -3.25
29  1    4.32
30  1    1.63
31  1    5.18
32  1   -0.43
33  1    7.11
34  1    4.87
35  1   -3.10
36  1   -5.81
```

(continued)

STATA *(continued)*

```
37  1     3.76
38  1     6.31
39  1     2.58
40  1     0.07
41  1     5.76
42  1     3.50
43  end
44
45  twoway (kdensity y if d==1) (kdensity y if d==0, lcolor(blue) lwidth(medium)
    ↪  lpattern(dash)), \\\
46  title(Kolmogorov-Smirnov test) legend(order(1 ``Treatment'' 2 ``Control''))
```

R
ks.R

```
1   library(tidyverse)
2   library(stats)
3
4   tb <- tibble(
5    d = c(rep(0, 20), rep(1, 20)),
6    y = c(0.22, -0.87, -2.39, -1.79, 0.37, -1.54,
7        1.28, -0.31, -0.74, 1.72,
8        0.38, -0.17, -0.62, -1.10, 0.30,
9        0.15, 2.30, 0.19, -0.50, -0.9,
10       -5.13, -2.19, 2.43, -3.83, 0.5,
11       -3.25, 4.32, 1.63, 5.18, -0.43,
12       7.11, 4.87, -3.10, -5.81, 3.76,
13       6.31, 2.58, 0.07, 5.76, 3.50)
14  )
15
16  kdensity_d1 <- tb %>%
17    filter(d == 1) %>%
18    pull(y)
19  kdensity_d1 <- density(kdensity_d1)
20
21  kdensity_d0 <- tb %>%
22    filter(d == 0) %>%
23    pull(y)
24  kdensity_d0 <- density(kdensity_d0)
25
```

(continued)

R *(continued)*

```
26   kdensity_d0 <- tibble(x = kdensity_d0$x, y = kdensity_d0$y, d = 0)
27   kdensity_d1 <- tibble(x = kdensity_d1$x, y = kdensity_d1$y, d = 1)
28
29   kdensity <- full_join(kdensity_d1, kdensity_d0)
30   kdensity$d <- as_factor(kdensity$d)
31
32   ggplot(kdensity)+
33     geom_point(size = 0.3, aes(x,y, color = d))+
34     xlim(-7, 8)+
35     labs(title = "Kolmogorov-Smirnov Test")+
36     scale_color_discrete(labels = c("Control", "Treatment"))
```

And to calculate the *p*-value, you repeat what we did in earlier examples. Specifically, drop the treatment variable, re-sort the data, reassign new (fixed) treatment values, calculate T_{KS}, save the coefficient, and repeat a thousand or more times until you have a distribution that you can use to calculate an empirical *p*-value.

Randomization inference with large n. What did we do when the number of observations is very large? For instance, Thornton's total sample was 2,901 participants. Of those, 2,222 received any incentive at all. Wolfram Alpha is an easy to use online calculator for more complicated calculations and easy to use interface. If you go to the website and type "2901 choose 2222" you get the following truncated number of combinations:

6150566109498251513699280333307718471623795043419269261826403 1826638575892109580799569314255435267978378517415493374384524 5116605236515180505177864028242897940877670928487172011882232 1888594251573599135614428312093501743827746469215584985879012 3688111563011540267646207996405072248645607065160780040934113 0655444540016312151177000750339179099962167196885539725968603 1228687680364730936480933074665307...

Good luck calculating those combinations. So clearly, exact *p*-values using all of the combinations won't work. So instead, we are going estimate approximate *p*-values. To do that, we will need to randomly assign the treatment, estimate a test statistic satisfying the sharp null for that sample, repeating that thousands of times, and then calculate the *p*-value associated with this treatment assignment based on its ranked position in the distribution.

STATA

thornton_ri.do

```
1   use https://github.com/scunning1975/mixtape/raw/master/thornton_hiv.dta,
    ↪  clear
2
3   tempfile hiv
4   save "`hiv'", replace
5
6   * Calculate true effect using absolute value of SDO
7   egen      te1 = mean(got) if any==1
8   egen      te0 = mean(got) if any==0
9
10  collapse (mean) te1 te0
11  gen       ate = te1 - te0
12  keep      ate
13  gen iteration = 1
14
15  tempfile permute1
16  save "`permute1'", replace
17
18  * Create a hundred datasets
19
20  forvalues i = 2/1000 {
21
22  use "`hiv'", replace
23
24  drop any
25  set seed `i'
26  gen random_`i' = runiform()
27  sort random_`i'
28  gen one=_n
29  drop random*
30  sort one
31
32  gen       any = 0
33  replace any = 1 in 1/2222
34
```

(continued)

STATA (continued)

```
35   * Calculate test statistic using absolute value of SDO
36   egen     te1 = mean(got) if any==1
37   egen     te0 = mean(got) if any==0
38
39   collapse (mean) te1 te0
40   gen      ate = te1 - te0
41   keep     ate
42
43   gen      iteration = `i'
44   tempfile permute`i'
45   save "`permute`i''", replace
46
47   }
48
49   use "`permute1'", replace
50   forvalues i = 2/1000 {
51      append using "`permute`i''"
52   }
53
54   tempfile final
55   save "`final'", replace
56
57   * Calculate exact p-value
58   gsort -ate
59   gen rank = _n
60   su rank if iteration==1
61   gen pvalue = (`r(mean)'/1000)
62   list if iteration==1
63
```

R

thornton_ri.R

```
1   library(tidyverse)
2   library(haven)
3
4   read_data <- function(df)
5   {
6    full_path <- paste("https://raw.github.com/scunning1975/mixtape/master/",
7              df, sep = "")
```

(continued)

	R *(continued)*

```
 8    df <- read_dta(full_path)
 9    return(df)
10   }
11
12   hiv <- read_data("thornton_hiv.dta")
13
14
15   # creating the permutations
16
17   tb <- NULL
18
19   permuteHIV <- function(df, random = TRUE){
20     tb <- df
21     first_half <- ceiling(nrow(tb)/2)
22     second_half <- nrow(tb) - first_half
23
24     if(random == TRUE){
25       tb <- tb %>%
26         sample_frac(1) %>%
27         mutate(any = c(rep(1, first_half), rep(0, second_half)))
28     }
29
30     te1 <- tb %>%
31       filter(any == 1) %>%
32       pull(got) %>%
33       mean(na.rm = TRUE)
34
35     te0 <- tb %>%
36       filter(any == 0) %>%
37       pull(got) %>%
38       mean(na.rm = TRUE)
39
40     ate <- te1 - te0
41     return(ate)
42   }
43
44   permuteHIV(hiv, random = FALSE)
45
```

(continued)

R *(continued)*

```
46   iterations <- 1000
47
48   permutation <- tibble(
49     iteration = c(seq(iterations)),
50     ate = as.numeric(
51       c(permuteHIV(hiv, random = FALSE), map(seq(iterations-1), ~permuteHIV(hiv,
         ↪ random = TRUE)))
52     )
53   )
54
55   #calculating the p-value
56
57   permutation <- permutation %>%
58     arrange(-ate) %>%
59     mutate(rank = seq(iterations))
60
61   p_value <- permutation %>%
62     filter(iteration == 1) %>%
63     pull(rank)/iterations
```

Quite impressive. Table 22 shows Thornton's experiment under Fisher's sharp null with between 100 and 1,000 repeated draws yields highly significant *p*-values. In fact, it is always the highest-ranked ATE in a one-tailed test.

So what I have done here is obtain an approximation of the *p*-value associated with our test statistic and the sharp null hypothesis. In practice, if the number of draws is large, the *p*-value based on this random sample will be fairly accurate [Imbens and Rubin, 2015]. I wanted to illustrate this randomization method because in reality this is exactly

Table 22. Estimated *p*-value using different number of trials.

ATE	Iteration	Rank	*p*	No. trials
0.45	1	1	0.01	100
0.45	1	1	0.002	500
0.45	1	1	0.001	1000

what you will be doing most of the time since the number of combinations with any reasonably sized data set will be computationally prohibitive.

Now, in some ways, this randomization exercise didn't reveal a whole lot, and that's probably because Thornton's original findings were just so precise to begin with (0.4 with a standard error of 0.02). We could throw atom bombs at this result and it won't go anywhere. But the purpose here is primarily to show its robustness under different ways of generating those precious *p*-values, as well as provide you with a map for programming this yourself and for having an arguably separate intuitive way of thinking about significance itself.

Leverage. Before we conclude, I'd like to go back to something I said earlier regarding *leverage*. A recent provocative study by Young [2019] has woken us up to challenges we may face when using traditional inference for estimating the uncertainty of some point estimate, such as robust standard errors. He finds practical problems with our traditional forms of inference, which while previously known, had not been made as salient as they were made by his study. The problem that he highlights is one of concentrated leverage. Leverage is a measure of the degree to which a single observation on the right-hand-side variable takes on extreme values and is influential in estimating the slope of the regression line. A concentration of leverage in even a few observations can make coefficients and standard errors extremely volatile and even bias robust standard errors towards zero, leading to higher rejection rates.

To illustrate this problem, Young [2019] went through a simple exercise. He collected over fifty experimental (lab and field) articles from the American Economic Association's flagship journals: *American Economic Review, American Economic Journal: Applied*, and *American Economic Journal: Economic Policy*. He then reanalyzed these papers, using the authors' models, by dropping one observation or cluster and reestimating the entire model, repeatedly. What he found was shocking:

With the removal of just one observation, 35% of 0.01-significant reported results in the average paper can be rendered insignificant at that level. Conversely, 16% of 0.01-insignificant reported results can be found to be significant at that level. (567)

For evidence to be so dependent on just a few observations creates some doubt about the clarity of our work, so what are our alternatives? The randomization inference method based on Fisher's sharp null, which will be discussed in this section, can improve upon these problems of leverage, in addition to the aforementioned reasons to consider it. In the typical paper, randomization inference found individual treatment effects that were 13 to 22 percent fewer significant results than what the authors' own analysis had discovered. Randomization inference, it appears, is somewhat more robust to the presence of leverage in a few observations.

Conclusion

In conclusion, we have done a few things in this chapter. We've introduced the potential outcomes notation and used it to define various types of causal effects. We showed that the simple difference in mean outcomes was equal to the sum of the average treatment effect, or the selection bias, and the weighted heterogeneous treatment effect bias. Thus the simple difference-in-mean outcomes estimator is biased unless those second and third terms zero out. One situation in which they zero out is under *independence* of the treatment, which is when the treatment has been assigned independent of the potential outcomes. When does independence occur? The most commonly confronted situation is under physical randomization of the treatment to the units. Because physical randomization assigns the treatment for reasons that are *independent* of the potential outcomes, the selection bias zeroes out, as does the heterogeneous treatment effect bias. We now move to discuss a second situation where the two terms zero out: *conditional* independence.

Matching and Subclassification

Subclassification

One of the main things I wanted to cover in the chapter on directed acylical graphical models was the idea of the backdoor criterion. Specifically, insofar as there exists a conditioning strategy that will satisfy the backdoor criterion, then you can use that strategy to identify some causal effect. We now discuss three different kinds of conditioning strategies. They are subclassification, exact matching, and approximate matching.[1]

Subclassification is a method of satisfying the backdoor criterion by weighting differences in means by strata-specific weights. These strata-specific weights will, in turn, adjust the differences in means so that their distribution by strata is the same as that of the counterfactual's strata. This method implicitly achieves distributional *balance* between the treatment and control in terms of that known, observable confounder. This method was created by statisticians like Cochran [1968], who tried to analyze the causal effect of smoking on lung cancer, and while the methods today have moved beyond it, we include

---

1 Everything I know about matching I learned from the Northwestern causal inference workshops in lectures taught by the econometrician Alberto Abadie. I would like to acknowledge him as this chapter is heavily indebted to him and those lectures.

it because some of the techniques implicit in subclassification are present throughout the rest of the book.

One of the concepts threaded through this chapter is the conditional independence assumption, or *CIA*. Sometimes we know that randomization occurred only conditional on some observable characteristics. For instance, in Krueger [1999], Tennessee randomly assigned kindergarten students and their teachers to small classrooms, large classrooms, and large classrooms with an aide. But the state did this conditionally—specifically, schools were chosen, and then students were randomized. Krueger therefore estimated regression models that included a school fixed effect because he knew that the treatment assignment was only *conditionally* random.

This assumption is written as

$$(Y^1, Y^0) \perp\!\!\!\perp D \mid X \tag{5.1}$$

where again $\perp\!\!\!\perp$ is the notation for statistical independence and X is the variable we are conditioning on. What this means is that the expected values of Y^1 and Y^0 are equal for treatment and control group *for each value of X*. Written out, this means:

$$E[Y^1 \mid D = 1, X] = E[Y^1 \mid D = 0, X] \tag{5.2}$$

$$E[Y^0 \mid D = 1, X] = E[Y^0 \mid D = 0, X] \tag{5.3}$$

Let me link together some concepts. First, insofar as CIA is credible, then CIA means you have found a conditioning strategy that satisfies the backdoor criterion. Second, when treatment assignment had been conditional on observable variables, it is a situation of *selection on observables*. The variable X can be thought of as an $n \times k$ matrix of covariates that satisfy the CIA as a whole.

Some background. A major public health problem of the mid- to late twentieth century was the problem of rising lung cancer. For instance, the mortality rate per 100,000 from cancer of the lungs in males reached 80–100 per 100,000 by 1980 in Canada, England, and Wales. From 1860 to 1950, the incidence of lung cancer found in cadavers during autopsy grew from 0% to as high as 7%. The rate of lung cancer incidence appeared to be increasing.

Studies began emerging that suggested smoking was the cause since it was so highly correlated with incidence of lung cancer. For instance, studies found that the relationship between daily smoking and lung cancer in males was monotonically increasing in the number of cigarettes a male smoked per day. But some statisticians believed that scientists couldn't draw a causal conclusion because it was possible that smoking was not independent of potential health outcomes. Specifically, perhaps the people who smoked cigarettes differed from non-smokers in ways that were directly related to the incidence of lung cancer. After all, no one is flipping coins when deciding to smoke.

Thinking about the simple difference in means decomposition from earlier, we know that contrasting the incidence of lung cancer between smokers and non-smokers will be biased in observational data if the independence assumption does not hold. And because smoking is endogenous—that is, people choose to smoke—it's entirely possible that smokers differed from the non-smokers in ways that were directly related to the incidence of lung cancer.

Criticisms at the time came from such prominent statisticians as Joseph Berkson, Jerzy Neyman, and Ronald Fisher. They made several compelling arguments. First, they suggested that the correlation was spurious due to a non-random selection of subjects. Functional form complaints were also common. This had to do with people's use of risk ratios and odds ratios. The association, they argued, was sensitive to those kinds of functional form choices, which is a fair criticism. The arguments were really not so different from the kinds of arguments you might see today when people are skeptical of a statistical association found in some observational data set.

Probably most damning, though, was the hypothesis that there existed an unobservable genetic element that both caused people to smoke and independently caused people to develop lung cancer. This confounder meant that smokers and non-smokers differed from one another in ways that were directly related to their potential outcomes, and thus independence did not hold. And there was plenty of evidence that the two groups were different. For instance, smokers were more extroverted than non-smokers, and they also differed in age, income, education, and so on.

The arguments against the smoking cause mounted. Other criticisms included that the magnitudes relating smoking and lung cancer were implausibly large. And again, the ever-present criticism of observational studies: there did not exist any experimental evidence that could incriminate smoking as a cause of lung cancer.[2]

The theory that smoking causes lung cancer is now accepted science. I wouldn't be surprised if more people believe in a flat Earth than that smoking causes lung cancer. I can't think of a more well-known and widely accepted causal theory, in fact. So how did Fisher and others fail to see it? Well, in Fisher's defense, his arguments were based on sound causal logic. Smoking *was* endogenous. There *was* no experimental evidence. The two groups differed considerably on observables. And the decomposition of the simple difference in means shows that contrasts will be biased if there is selection bias. Nonetheless, Fisher was wrong, and his opponents were right. They just were right for the wrong reasons.

To motivate what we're doing in subclassification, let's work with Cochran [1968], which was a study trying to address strange patterns in smoking data by adjusting for a confounder. Cochran lays out mortality rates by country and smoking type (Table 23).

As you can see, the highest death rate for Canadians is among the cigar and pipe smokers, which is considerably higher than for non-smokers or for those who smoke cigarettes. Similar patterns show up

---

2 But think about the hurdle that the last criticism actually creates. Just imagine the hypothetical experiment: a large sample of people, with diverse potential outcomes, are assigned to a treatment group (smoker) and control (non-smoker). These people must be dosed with their corresponding treatments long enough for us to observe lung cancer develop—so presumably years of heavy smoking. How could anyone ever run an experiment like that? Who in their right mind would participate!? Just to describe the idealized experiment is to admit it's impossible. But how do we answer the causal question without independence (i.e., randomization)?

Table 23. Death rates per 1,000 person-years [Cochran, 1968].

Smoking group	Canada	UK	US
Non-smokers	20.2	11.3	13.5
Cigarettes	20.5	14.1	13.5
Cigars/pipes	35.5	20.7	17.4

in both countries, though smaller in magnitude than what we see in Canada.

This table suggests that pipes and cigars are more dangerous than cigarette smoking, which, to a modern reader, sounds ridiculous. The reason it sounds ridiculous is because cigar and pipe smokers often do not inhale, and therefore there is less tar that accumulates in the lungs than with cigarettes. And insofar as it's the tar that causes lung cancer, it stands to reason that we should see higher mortality rates among cigarette smokers.

But, recall the independence assumption. Do we really believe that:

$$E[Y^1 \mid \text{Cigarette}] = E[Y^1 \mid \text{Pipe}] = E[Y^1 \mid \text{Cigar}]$$
$$E[Y^0 \mid \text{Cigarette}] = E[Y^0 \mid \text{Pipe}] = E[Y^0 \mid \text{Cigar}]$$

Is it the case that factors related to these three states of the world are truly independent to the factors that determine death rates? Well, let's assume for the sake of argument that these independence assumptions held. What else would be true across these three groups? Well, if the mean potential outcomes are the same for each type of smoking category, then wouldn't we expect the observable characteristics of the smokers themselves to be as well? This connection between the independence assumption and the characteristics of the groups is called *balance*. If the means of the covariates are the same for each group, then we say those covariates are balanced and the two groups are exchangeable with respect to those covariates.

One variable that appears to matter is the age of the person. Older people were more likely at this time to smoke cigars and pipes, and

Table 24. Mean ages, years [Cochran, 1968].

Smoking group	Canada	UK	US
Non-smokers	54.9	49.1	57.0
Cigarettes	50.5	49.8	53.2
Cigars/pipes	65.9	55.7	59.7

without stating the obvious, older people were more likely to die. In Table 24 we can see the mean ages of the different groups.

The high means for cigar and pipe smokers are probably not terribly surprising. Cigar and pipe smokers are typically older than cigarette smokers, or at least they were in 1968 when Cochran was writing. And since older people die at a higher rate (for reasons other than just smoking cigars), maybe the higher death rate for cigar smokers is because they're older on average. Furthermore, maybe by the same logic, cigarette smoking has such a low mortality rate because cigarette smokers are younger on average. Note, using DAG notation, this simply means that we have the following DAG:

where D is smoking, Y is mortality, and A is age of the smoker. Insofar as CIA is violated, then we have a backdoor path that is open, which also means that we have omitted variable bias. But however we want to describe it, the common thing is that the distribution of age for each group will be different—which is what I mean by *covariate imbalance*. My first strategy for addressing this problem of covariate imbalance is to *condition* on age in such a way that the distribution of age is comparable for the treatment and control groups.[3]

So how does subclassification achieve covariate balance? Our first step is to divide age into strata: say, 20–40, 41–70, and 71

---

3 Interestingly, this issue of covariate balance weaves throughout nearly every identification strategy that we will discuss.

Table 25. Subclassification example.

| | Death rates | Number of | |
	Cigarette smokers	Cigarette smokers	Pipe or cigar smokers
Age 20–40	20	65	10
Age 41–70	40	25	25
Age ≥ 71	60	10	65
Total		100	100

and older. Then we can calculate the mortality rate for some treatment group (cigarette smokers) by strata (here, that is age). Next, weight the mortality rate for the treatment group by a strata-specific (or age-specific) weight that corresponds to the control group. This gives us the age-adjusted mortality rate for the treatment group. Let's explain with an example by looking at Table 25. Assume that age is the only relevant confounder between cigarette smoking and mortality.[4]

What is the average death rate for pipe smokers without subclassification? It is the weighted average of the mortality rate column where each weight is equal to $\frac{N_t}{N}$, and N_t and N are the number of people in each group and the total number of people, respectively. Here that would be

$$20 \times \frac{65}{100} + 40 \times \frac{25}{100} + 60 \times \frac{10}{100} = 29.$$

That is, the mortality rate of smokers in the population is 29 per 100,000.

But notice that the age distribution of cigarette smokers is the exact opposite (by construction) of pipe and cigar smokers. Thus the age distribution is imbalanced. Subclassification simply adjusts the mortality rate for cigarette smokers so that it has the same age distribution as the comparison group. In other words, we would multiply each age-specific mortality rate by the proportion of individuals in that age strata for the comparison group. That would be

$$20 \times \frac{10}{100} + 40 \times \frac{25}{100} + 60 \times \frac{65}{100} = 51$$

---

4 A truly hilarious assumption, but this is just illustrative.

Table 26. Adjusted mortality rates using three age groups [Cochran, 1968].

Smoking group	Canada	UK	US
Non-smokers	20.2	11.3	13.5
Cigarettes	29.5	14.8	21.2
Cigars/pipes	19.8	11.0	13.7

That is, when we adjust for the age distribution, the age-adjusted mortality rate for cigarette smokers (were they to have the same age distribution as pipe and cigar smokers) would be 51 per 100,000—almost twice as large as we got taking a simple naïve calculation unadjusted for the age confounder.

Cochran uses a version of this subclassification method in his paper and recalculates the mortality rates for the three countries and the three smoking groups (see Table 26). As can be seen, once we adjust for the age distribution, cigarette smokers have the highest death rates among any group.

This kind of adjustment raises a question—which variable(s) should we use for adjustment? First, recall what we've emphasized repeatedly. Both the backdoor criterion and CIA tell us precisely what we need to do. We need to choose a set of variables that satisfy the backdoor criterion. If the backdoor criterion is met, then all backdoor paths are closed, and if all backdoor paths are closed, then CIA is achieved. We call such a variable the *covariate*. A covariate is usually a random variable assigned to the individual units prior to treatment. This is sometimes also called exogenous. Harkening back to our DAG chapter, this variable must not be a collider as well. A variable is exogenous with respect to D if the value of X does not depend on the value of D. Oftentimes, though not always and not necessarily, this variable will be time-invariant, such as race. Thus, when trying to adjust for a confounder using subclassification, rely on a credible DAG to help guide the selection of variables. Remember—your goal is to meet the backdoor criterion.

Identifying assumptions. Let me now formalize what we've learned. In order to estimate a causal effect when there is a confounder, we need

(1) CIA and (2) the probability of treatment to be between 0 and 1 for each strata. More formally,

1. $(Y^1, Y^0) \perp\!\!\!\perp D \mid X$ (conditional independence)
2. $0 < Pr(D = 1 \mid X) < 1$ with probability one (common support)

These two assumptions yield the following identity

$$E[Y^1 - Y^0 \mid X] = E[Y^1 - Y^0 \mid X, D = 1]$$
$$= E[Y^1 \mid X, D = 1] - E[Y^0 \mid X, D = 0]$$
$$= E[Y \mid X, D = 1] - E[Y \mid X, D = 0]$$

where each value of Y is determined by the switching equation. Given common support, we get the following estimator:

$$\widehat{\delta_{ATE}} = \int \Big(E[Y \mid X, D = 1] - E[Y \mid X, D = 0] \Big) d\,Pr(X)$$

Whereas we need treatment to be conditionally independent of both potential outcomes to identify the ATE, we need only treatment to be conditionally independent of Y^0 to identify the ATT and the fact that there exist some units in the control group for each treatment strata. Note, the reason for the common support assumption is because we are weighting the data; without common support, we cannot calculate the relevant weights.

Subclassification exercise: Titanic data set. For what we are going to do next, I find it useful to move into actual data. We will use an interesting data set to help us better understand subclassification. As everyone knows, the *Titanic* ocean cruiser hit an iceberg and sank on its maiden voyage. Slightly more than 700 passengers and crew survived out of the 2,200 people on board. It was a horrible disaster. One of the things about it that was notable, though, was the role that wealth and norms played in passengers' survival.

Imagine that we wanted to know whether or not being seated in first class made someone more likely to survive. Given that the cruiser contained a variety of levels for seating and that wealth was highly concentrated in the upper decks, it's easy to see why wealth might have a leg up for survival. But the problem was that women and children were explicitly given priority for boarding the scarce lifeboats. If women and children were more likely to be seated in first class, then maybe differences in survival by first class is simply picking up the effect of that social norm. Perhaps a DAG might help us here, as a DAG can help us outline the sufficient conditions for identifying the causal effect of first class on survival.

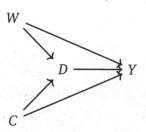

Now before we commence, let's review what this DAG is telling us. This says that being a female made you more likely to be in first class but also made you more likely to survive because lifeboats were more likely to be allocated to women. Furthermore, being a child made you more likely to be in first class and made you more likely to survive. Finally, there are no other confounders, observed or unobserved.[5]

Here we have one direct path (the causal effect) between first class (D) and survival (Y) and that's $D \rightarrow Y$. But, we have two backdoor paths. One travels through the variable Child (C): $D \leftarrow C \rightarrow Y$; the other travels through the variable Woman (W): $D \leftarrow W \rightarrow Y$. Fortunately for us, our data includes both age and gender, so it is possible to close each backdoor path and therefore satisfy the backdoor criterion. We will use subclassification to do that, but before we do, let's calculate a naïve simple difference in outcomes (SDO), which is just

$$E[Y \mid D = 1] - E[Y \mid D = 0]$$

for the sample.

---

5 I'm sure you can think of others, though, in which case this DAG is misleading.

STATA

titanic.do

```
1   use https://github.com/scunning1975/mixtape/raw/master/titanic.dta, clear
2   gen female=(sex==0)
3   label variable female "Female"
4   gen male=(sex==1)
5   label variable male "Male"
6   gen     s=1 if (female==1 & age==1)
7   replace s=2 if (female==1 & age==0)
8   replace s=3 if (female==0 & age==1)
9   replace s=4 if (female==0 & age==0)
10  gen     d=1 if class==1
11  replace d=0 if class!=1
12  summarize survived if d==1
13  gen ey1=r(mean)
14  summarize survived if d==0
15  gen ey0=r(mean)
16  gen sdo=ey1-ey0
17  su sdo
```

R

titanic.R

```
1   library(tidyverse)
2   library(haven)
3
4   read_data <- function(df)
5   {
6    full_path <- paste("https://raw.github.com/scunning1975/mixtape/master/",
7                df, sep = "")
8    df <- read_dta(full_path)
9    return(df)
10  }
11
12  titanic <- read_data("titanic.dta") %>%
13   mutate(d = case_when(class == 1 ~ 1, TRUE ~ 0))
14
```

(continued)

R *(continued)*

```
15   ey1 <- titanic %>%
16     filter(d == 1) %>%
17     pull(survived) %>%
18     mean()
19
20   ey0 <- titanic %>%
21     filter(d == 0) %>%
22     pull(survived) %>%
23     mean()
24
25   sdo <- ey1 - ey0
```

Using the data set on the *Titanic*, we calculate a simple difference in mean outcomes (SDO), which finds that being seated in first
class raised the probability of survival by 35.4%. But note, since this
does not adjust for observable confounders age and gender, it is a
biased estimate of the ATE. So next we use subclassification weighting to control for these confounders. Here are the steps that will
entail:

1. Stratify the data into four groups: young males, young females,
   old males, old females.
2. Calculate the difference in survival probabilities for each group.
3. Calculate the number of people in the non-first-class groups and
   divide by the total number of non-first-class population. These
   are our strata-specific weights.
4. Calculate the weighted average survival rate using the strata
   weights.

Let's review this with some code so that you can better understand
what these four steps actually entail.

STATA

titanic_subclassification.do

```
1   * Subclassification
2   cap n drop ey1 ey0
3   su survived if s==1 & d==1
4   gen ey11=r(mean)
5   label variable ey11 "Average survival for male child in treatment"
6   su survived if s==1 & d==0
7   gen ey10=r(mean)
8   label variable ey10 "Average survival for male child in control"
9   gen diff1=ey11-ey10
10  label variable diff1 "Difference in survival for male children"
11  su survived if s==2 & d==1
12  gen ey21=r(mean)
13  su survived if s==2 & d==0
14  gen ey20=r(mean)
15  gen diff2=ey21-ey20
16  su survived if s==3 & d==1
17  gen ey31=r(mean)
18  su survived if s==3 & d==0
19  gen ey30=r(mean)
20  gen diff3=ey31-ey30
21  su survived if s==4 & d==1
22  gen ey41=r(mean)
23  su survived if s==4 & d==0
24  gen ey40=r(mean)
25  gen diff4=ey41-ey40
26  count if s==1 & d==0
27  count if s==2 & d==0
28  count if s==3 & d==0
29  count if s==4 & d==0
30  count
31  gen wt1=425/2201
32  gen wt2=45/2201
33  gen wt3=1667/2201
34  gen wt4=64/2201
35  gen wate=diff1*wt1 + diff2*wt2 + diff3*wt3 + diff4*wt4
36  sum wate sdo
```

R

titanic_subclassification.R

```
1   library(stargazer)
2   library(magrittr) # for %$% pipes
3   library(tidyverse)
4   library(haven)
5
6   titanic <- read_data("titanic.dta") %>%
7     mutate(d = case_when(class == 1 ~ 1, TRUE ~ 0))
8
9
10  titanic %<>%
11    mutate(s = case_when(sex == 0 & age == 1 ~ 1,
12                 sex == 0 & age == 0 ~ 2,
13                 sex == 1 & age == 1 ~ 3,
14                 sex == 1 & age == 0 ~ 4,
15                 TRUE ~ 0))
16
17  ey11 <- titanic %>%
18    filter(s == 1 & d == 1) %$%
19    mean(survived)
20
21  ey10 <- titanic %>%
22    filter(s == 1 & d == 0) %$%
23    mean(survived)
24
25  ey21 <- titanic %>%
26    filter(s == 2 & d == 1) %$%
27    mean(survived)
28
29  ey20 <- titanic %>%
30    filter(s == 2 & d == 0) %$%
31    mean(survived)
32
33  ey31 <- titanic %>%
34    filter(s == 3 & d == 1) %$%
35    mean(survived)
36
```

(continued)

	R *(continued)*

```
37   ey30 <- titanic %>%
38     filter(s == 3 & d == 0) %$%
39     mean(survived)
40
41   ey41 <- titanic %>%
42     filter(s == 4 & d == 1) %$%
43     mean(survived)
44
45   ey40 <- titanic %>%
46     filter(s == 4 & d == 0) %$%
47     mean(survived)
48
49   diff1 = ey11 - ey10
50   diff2 = ey21 - ey20
51   diff3 = ey31 - ey30
52   diff4 = ey41 - ey40
53
54   obs = nrow(titanic)
55
56   wt1 <- titanic %>%
57     filter(s == 1 & d == 0) %$%
58     nrow(.)/obs
59
60   wt2 <- titanic %>%
61     filter(s == 2 & d == 0) %$%
62     nrow(.)/obs
63
64   wt3 <- titanic %>%
65     filter(s == 3 & d == 0) %$%
66     nrow(.)/obs
67
68   wt4 <- titanic %>%
69     filter(s == 4 & d == 0) %$%
70     nrow(.)/obs
71
72   wate = diff1*wt1 + diff2*wt2 + diff3*wt3 + diff4*wt4
73
74   stargazer(wate, sdo, type = "text")
75
```

Table 27. Subclassification example of *Titanic* survival for large *K*.

Age and Gender	Survival Prob. 1st Class	Controls	Diff.	Number of 1st Class	Controls
Male 11-yo	1.0	0	1	1	2
Male 12-yo	–	1	–	0	1
Male 13-yo	1.0	0	1	1	2
Male 14-yo	–	0.25	–	0	4
...					

Here we find that once we condition on the confounders gender and age, first-class seating has a much lower probability of survival associated with it (though frankly, still large). The weighted ATE is 16.1%, versus the SDO, which is 35.4%.

Curse of dimensionality. Here we've been assuming two covariates, each of which has two possible set of values. But this was for convenience. Our data set, for instance, only came to us with two possible values for age—child and adult. But what if it had come to us with multiple values for age, like specific age? Then once we condition on individual age and gender, it's entirely likely that we will not have the information necessary to calculate differences within strata, and therefore be unable to calculate the strata-specific weights that we need for subclassification.

For this next part, let's assume that we have precise data on *Titanic* survivor ages. But because this will get incredibly laborious, let's just focus on a few of them.

Here we see an example of the common support assumption being violated. The common support assumption requires that for each strata, there exist observations in both the treatment and control group, but as you can see, there are not any 12-year-old male passengers in first class. Nor are there any 14-year-old male passengers in first class. And if we were to do this for every combination of age and gender, we would find that this problem was quite common. Thus, we cannot estimate the ATE using subclassification. The problem is that our stratifying variable has too many dimensions, and as a result, we have sparseness in some cells because the sample is too small.

But let's say that the problem was always on the treatment group, not the control group. That is, let's assume that there is always *someone* in the control group for a given combination of gender and age, but there isn't always for the treatment group. Then we can calculate the ATT. Because as you see in this table, for those two strata, 11-year-olds and 13-year-olds, there are both treatment and control group values for the calculation. So long as there exist controls for a given treatment strata, we can calculate the ATT. The equation to do so can be compactly written as:

$$\widehat{\delta}_{ATT} = \sum_{k=1}^{K} \left(\overline{Y}^{1,k} - \overline{Y}^{0,k} \right) \times \left(\frac{N_T^k}{N_T} \right)$$

We've seen a problem that arises with subclassification—in a finite sample, subclassification becomes less feasible as the number of covariates grows, because as K grows, the data becomes sparse. This is most likely caused by our sample being too small relative to the size of our covariate matrix. We will at some point be missing values, in other words, for those K categories. Imagine if we tried to add a third strata, say, race (black and white). Then we'd have two age categories, two gender categories, and two race categories, giving us eight possibilities. In this small sample, we probably will end up with many cells having missing information. This is called the *curse of dimensionality*. If sparseness occurs, it means many cells may contain either only treatment units or only control units, but not both. If that happens, we can't use subclassification, because we do not have common support. And therefore we are left searching for an alternative method to satisfy the backdoor criterion.

Exact Matching

Subclassification uses the difference between treatment and control group units and achieves covariate balance by using the K probability weights to weight the averages. It's a simple method, but it has the aforementioned problem of the curse of dimensionality. And probably, that's going to be an issue in any research you undertake because it may not be merely one variable you're worried about but several—in which case, you'll already be running into the curse. But the thing to emphasize here is that the subclassification method is using the raw data, but weighting it so as to achieve balance. We are weighting the differences, and then summing over those weighted differences.

But there are alternative approaches. For instance, what if we estimated $\widehat{\delta}_{ATT}$ by *imputing* the missing potential outcomes by conditioning on the confounding, observed covariate? Specifically, what if we filled in the missing potential outcome for each treatment unit using a control group unit that was "closest" to the treatment group unit for some X confounder? This would give us estimates of all the counterfactuals from which we could simply take the average over the differences. As we will show, this will also achieve covariate balance. This method is called *matching*.

There are two broad types of matching that we will consider: exact matching and approximate matching. We will first start by describing exact matching. Much of what I am going to be discussing is based on Abadie and Imbens [2006].

A simple matching estimator is the following:

$$\widehat{\delta}_{ATT} = \frac{1}{N_T} \sum_{D_i=1} (Y_i - Y_{j(i)})$$

where $Y_{j(i)}$ is the jth unit matched to the ith unit based on the jth being "closest to" the ith unit for some X covariate. For instance, let's say that a unit in the treatment group has a covariate with a value of 2 and we find another unit in the control group (exactly one unit) with a covariate value of 2. Then we will impute the treatment unit's missing counterfactual with the matched unit's, and take a difference.

But, what if there's more than one variable "closest to" the ith unit? For instance, say that the same ith unit has a covariate value of 2 and we find two j units with a value of 2. What can we then do? Well, one option is to simply take the average of those two units' Y outcome value. But what if we found 3 close units? What if we found 4? And so on. However many matches M that we find, we would assign the average outcome $\left(\frac{1}{M}\right)$ as the counterfactual for the treatment group unit.

Notationally, we can describe this estimator as

$$\widehat{\delta}_{ATT} = \frac{1}{N_T} \sum_{D_i=1} \left(Y_i - \left[\frac{1}{M} \sum_{m=1}^{M} Y_{jm(1)} \right] \right)$$

This estimator really isn't too different from the one just before it; the main difference is that this one averages over several close matches as opposed to just picking one. This approach works well when we can find a number of good matches for each treatment group unit. We usually define M to be small, like $M = 2$. If M is greater than 2, then we may simply randomly select two units to average outcomes over.

Those were both ATT estimators. You can tell that these are $\widehat{\delta}_{ATT}$ estimators because of the summing over the treatment group.[6] But we can also estimate the ATE. But note, when estimating the ATE, we are filling in both missing control group units like before and missing treatment group units. If observation i is treated, in other words, then we need to fill in the missing Y_i^0 using the control matches, and if the observation i is a control group unit, then we need to fill in the missing Y_i^1 using the treatment group matches. The estimator is below. It looks scarier than it really is. It's actually a very compact, nicely-written-out estimator equation.

$$\widehat{\delta}_{ATE} = \frac{1}{N} \sum_{i=1}^{N} (2D_i - 1) \left[Y_i - \left(\frac{1}{M} \sum_{m=1}^{M} Y_{jm(i)} \right) \right]$$

The $2D_i - 1$ is the nice little trick. When $D_i = 1$, then that leading term becomes a 1.[7] And when $D_i = 0$, then that leading term becomes a negative 1, and the outcomes reverse order so that the treatment observation can be imputed. Nice little mathematical form!

Let's see this work in action by working with an example. Table 28 shows two samples: a list of participants in a job trainings program and a list of non-participants, or non-trainees. The left-hand group is the treatment group and the right-hand group is the control group. The matching algorithm that we defined earlier will create a third group called the *matched sample*, consisting of each treatment

---

6 Notice the $D_i = 1$ in the subscript of the summation operator.

7 $2 \times 1 - 1 = 1$.

Table 28. Training example with exact matching.

Unit	Trainees Age	Earnings	Unit	Non-Trainees Age	Earnings
1	18	9500	1	20	8500
2	29	12250	2	27	10075
3	24	11000	3	21	8725
4	27	11750	4	39	12775
5	33	13250	5	38	12550
6	22	10500	6	29	10525
7	19	9750	7	39	12775
8	20	10000	8	33	11425
9	21	10250	9	24	9400
10	30	12500	10	30	10750
			11	33	11425
			12	36	12100
			13	22	8950
			14	18	8050
			15	43	13675
			16	39	12775
			17	19	8275
			18	30	9000
			19	51	15475
			20	48	14800
Mean	24.3	$11,075		31.95	$11,101.25

group unit's matched counterfactual. Here we will match on the age of the participant.

Before we do this, though, I want to show you how the ages of the trainees differ on average from the ages of the non-trainees. We can see that in Table 28—the average age of the participants is 24.3 years, and the average age of the non-participants is 31.95 years. Thus, the people in the control group are older, and since wages typically rise with age, we may suspect that part of the reason their average earnings are higher ($11,075 vs. $11,101) is because the control group is older. We say that the two groups are not *exchangeable* because the covariate is not *balanced*. Let's look at the age distribution. To illustrate this, we need to download the data first. We will create two histograms—the distribution of age for treatment and non-trainee group—as well as summarize earnings for each group. That information is also displayed in Figure 16.

STATA

training_example.do

```
1  use
   ↪ https://github.com/scunning1975/mixtape/raw/master/training_example.dta,
   ↪ clear
2  histogram age_treat, bin(10) frequency
3  histogram age_control, bin(10) frequency
4  su age_treat age_control
5  su earnings_treat earnings_control
6
7  histogram age_treat, bin(10) frequency
8  histogram age_matched, bin(10) frequency
9  su age_treat age_control
10 su earnings_matched earnings_matched
```

R

training_example.R

```
1  library(tidyverse)
2  library(haven)
3
4  read_data <- function(df)
5  {
6    full_path <- paste("https://raw.github.com/scunning1975/mixtape/master/",
7              df, sep = "")
8    df <- read_dta(full_path)
9    return(df)
10 }
11
12 training_example <- read_data("training_example.dta") %>%
13   slice(1:20)
14
15 ggplot(training_example, aes(x=age_treat)) +
16   stat_bin(bins = 10, na.rm = TRUE)
17
18 ggplot(training_example, aes(x=age_control)) +
19   geom_histogram(bins = 10, na.rm = TRUE)
```

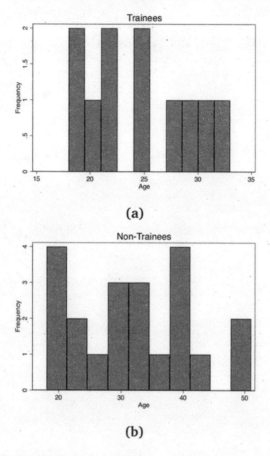

Figure 16. Covariate distribution by job trainings and control.

As you can see from Figure 16, these two populations not only have different means (Table 28); the entire distribution of age across the samples is different. So let's use our matching algorithm and create the missing counterfactuals for each treatment group unit. This method, since it only imputes the missing units for each treatment unit, will yield an estimate of the $\widehat{\delta}_{ATT}$.

Now let's move to creating the matched sample. As this is exact matching, the distance traveled to the nearest neighbor will be zero integers. This won't always be the case, but note that as the control group sample size grows, the likelihood that we find a unit with the

Table 29. Training example with exact matching (including matched sample).

	Trainees			Non-Trainees			Matched Sample	
Unit	Age	Earnings	Unit	Age	Earnings	Unit	Age	Earnings
1	18	9500	1	20	8500	14	18	8050
2	29	12250	2	27	10075	6	29	10525
3	24	11000	3	21	8725	9	24	9400
4	27	11750	4	39	12775	8	27	10075
5	33	13250	5	38	12550	11	33	11425
6	22	10500	6	29	10525	13	22	8950
7	19	9750	7	39	12775	17	19	8275
8	20	10000	8	33	11425	1	20	8500
9	21	10250	9	24	9400	3	21	8725
10	30	12500	10	30	10750	10,18	30	9875
			11	33	11425			
			12	36	12100			
			13	22	8950			
			14	18	8050			
			15	43	13675			
			16	39	12775			
			17	19	8275			
			18	30	9000			
			19	51	15475			
			20	48	14800			
Mean	24.3	$11,075		31.95	$11,101.25		24.3	$9,380

same covariate value as one in the treatment group grows. I've created a data set like this. The first treatment unit has an age of 18. Searching down through the non-trainees, we find exactly one person with an age of 18, and that's unit 14. So we move the age and earnings information to the new matched sample columns.

We continue doing that for all units, always moving the control group unit with the closest value on X to fill in the missing counterfactual for each treatment unit. If we run into a situation where there's more than one control group unit "close," then we simply average over them. For instance, there are two units in the non-trainees group with an age of 30, and that's 10 and 18. So we averaged their earnings and matched that average earnings to unit 10. This is filled out in Table 29.

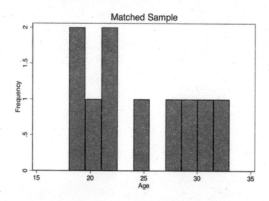

Figure 17. Covariate distribution by job trainings and matched sample.

Now we see that the mean age is the same for both groups. We can also check the overall age distribution (Figure 17). As you can see, the two groups are *exactly balanced* on age. We might say the two groups are *exchangeable*. And the difference in earnings between those in the treatment group and those in the control group is $1,695. That is, we estimate that the causal effect of the program was $1,695 in higher earnings.

Let's summarize what we've learned. We've been using a lot of different terms, drawn from different authors and different statistical traditions, so I'd like to map them onto one another. The two groups were different in ways that were likely a direction function of potential outcomes. This means that the independence assumption was violated. Assuming that treatment assignment was conditionally random, then matching on X created an exchangeable set of observations—the matched sample—and what characterized this matched sample was *balance*.

Approximate Matching

The previous example of matching was relatively simple—find a unit or collection of units that have the same value of some covariate X and substitute their outcomes as some unit *j*'s counterfactuals. Once you've done that, average the differences for an estimate of the ATE.

But what if you couldn't find another unit with that exact same value? Then you're in the world of approximate matching.

Nearest neighbor covariate matching. One of the instances where exact matching can break down is when the number of covariates, K, grows large. And when we have to match on more than one variable but are not using the sub-classification approach, then one of the first things we confront is the concept of *distance*. What does it mean for one unit's covariate to be "close" to someone else's? Furthermore, what does it mean when there are multiple covariates with measurements in multiple dimensions?

Matching on a single covariate is straightforward because distance is measured in terms of the covariate's own values. For instance, distance in age is simply how close in years or months or days one person is to another person. But what if we have several covariates needed for matching? Say, age and log income. A 1-point change in age is very different from a 1-point change in log income, not to mention that we are now measuring distance in two, not one, dimensions. When the number of matching covariates is more than one, we need a new definition of distance to measure closeness. We begin with the simplest measure of distance, the *Euclidean distance*:

$$||X_i - X_j|| = \sqrt{(X_i - X_j)'(X_i - X_j)}$$

$$= \sqrt{\sum_{n=1}^{k}(X_{ni} - X_{nj})^2}$$

The problem with this measure of distance is that the distance measure itself depends on the scale of the variables themselves. For this reason, researchers typically will use some modification of the Euclidean distance, such as the *normalized Euclidean distance*, or they'll use a wholly different alternative distance. The normalized Euclidean distance is a commonly used distance, and what makes it different is that the distance of each variable is scaled by the variable's variance. The distance is measured as:

$$||X_i - X_j|| = \sqrt{(X_i - X_j)'\widehat{V}^{-1}(X_i - X_j)}$$

where

$$\widehat{V}^{-1} = \begin{pmatrix} \widehat{\sigma}_1^2 & 0 & \cdots & 0 \\ 0 & \widehat{\sigma}_2^2 & \cdots & 0 \\ \vdots & \vdots & \ddots & \vdots \\ 0 & 0 & \cdots & \widehat{\sigma}_k^2 \end{pmatrix}$$

Notice that the normalized Euclidean distance is equal to:

$$||X_i - X_j|| = \sqrt{\sum_{n=1}^{k} \frac{(X_{ni} - X_{nj})}{\widehat{\sigma}_n^2}}$$

Thus, if there are changes in the scale of X, these changes also affect its variance, and so the normalized Euclidean distance does not change.

Finally, there is the *Mahalanobis* distance, which like the normalized Euclidean distance measure, is a scale-invariant distance metric. It is:

$$||X_i - X_j|| = \sqrt{(X_i - X_j)'\widehat{\Sigma}_X^{-1}(X_i - X_j)}$$

where $\widehat{\Sigma}_X$ is the sample variance-covariance matrix of X.

Basically, more than one covariate creates a lot of headaches. Not only does it create the curse-of-dimensionality problem; it also makes measuring distance harder. All of this creates some challenges for finding a good match in the data. As you can see in each of these distance formulas, there are sometimes going to be matching discrepancies. Sometimes $X_i \neq X_j$. What does this mean? It means that some unit i has been matched with some unit j on the basis of a similar covariate value of $X = x$. Maybe unit i has an age of 25, but unit j has an age of 26. Their difference is 1. Sometimes the discrepancies are small, sometimes zero, sometimes large. But, as they move away from zero, they become more problematic for our estimation and introduce bias.

How severe is this bias? First, the good news. What we know is that the matching discrepancies tend to converge to zero as the sample size increases—which is one of the main reasons that approximate matching is so data greedy. It demands a large sample size for the matching discrepancies to be trivially small. But what if there are many covariates? The more covariates, the longer it takes for that convergence to zero to occur. Basically, if it's hard to find good matches with an X that has a large dimension, then you will need a lot of observations as a result. *The larger the dimension, the greater likelihood of matching discrepancies, and the more data you need.* So you can take that to the bank—most likely, your matching problem requires a large data set in order to minimize the matching discrepancies.

Bias correction. Speaking of matching discrepancies, what sorts of options are available to us, putting aside seeking a large data set with lots of controls? Well, enter stage left, Abadie and Imbens [2011], who introduced bias-correction techniques with matching estimators when there are matching discrepancies in finite samples. So let's look at that more closely, as you'll likely need this in your work.

Everything we're getting at is suggesting that matching is biased because of these poor matching discrepancies. So let's derive this bias. First, we write out the sample ATT estimate, and then we subtract out the true ATT. So:

$$\widehat{\delta}_{ATT} = \frac{1}{N_T} \sum_{D_i=1} (Y_i - Y_{j(i)})$$

where each i and $j(i)$ units are matched, $X_i \approx X_{j(i)}$ and $D_{j(i)} = 0$. Next we define the conditional expection outcomes

$$\mu^0(x) = E[Y \mid X = x, D = 0] = E[Y^0 \mid X = x]$$
$$\mu^1(x) = E[Y \mid X = x, D = 1] = E[Y^1 \mid X = x]$$

Notice, these are just the expected conditional outcome functions based on the switching equation for both control and treatment groups.

As always, we write out the observed value as a function of expected conditional outcomes and some stochastic element:

$$Y_i = \mu^{D_i}(X_i) + \varepsilon_i$$

Now rewrite the ATT estimator using the above μ terms:

$$\widehat{\delta}_{ATT} = \frac{1}{N_T} \sum_{D_i=1} \left(\mu^1(X_i) + \varepsilon_i \right) - \left(\mu^0(X_{j(i)}) + \varepsilon_{j(i)} \right)$$

$$= \frac{1}{N_T} \sum_{D_i=1} \left(\mu^1(X_i) - \mu^0(X_{j(i)}) \right) + \frac{1}{N_T} \sum_{D_i=1} \left(\varepsilon_i - \varepsilon_{j(i)} \right)$$

Notice, the first line is just the ATT with the stochastic element included from the previous line. And the second line rearranges it so that we get two terms: the estimated ATT plus the average difference in the stochastic terms for the matched sample.

Now we compare this estimator with the true value of ATT.

$$\widehat{\delta}_{ATT} - \delta_{ATT} = \frac{1}{N_T} \sum_{D_i=1} \left(\mu^1(X_i) - \mu^0(X_{j(i)}) \right) - \delta_{ATT} + \frac{1}{N_T} \sum_{D_i=1} \left(\varepsilon_i - \varepsilon_{j(i)} \right)$$

which, with some simple algebraic manipulation is:

$$\widehat{\delta}_{ATT} - \delta_{ATT} = \frac{1}{N_T} \sum_{D_i=1} \left(\mu^1(X_i) - \mu^0(X_i) - \delta_{ATT} \right)$$

$$+ \frac{1}{N_T} \sum_{D_i=1} \left(\varepsilon_i - \varepsilon_{j(i)} \right)$$

$$+ \frac{1}{N_T} \sum_{D_i=1} \left(\mu^0(X_i) - \mu^0(X_{j(i)}) \right).$$

Applying the central limit theorem and the difference, $\sqrt{N_T}(\widehat{\delta}_{ATT} - \delta_{ATT})$ converges to a normal distribution with zero mean. But:

$$E\left[\sqrt{N_T}(\widehat{\delta}_{ATT} - \delta_{ATT}) \right] = E\left[\sqrt{N_T}(\mu^0(X_i) - \mu^0(X_{j(i)})) \mid D = 1 \right].$$

Now consider the implications if the number of covariates is large. First, the difference between X_i and $X_{j(i)}$ converges to zero slowly. This therefore makes the difference $\mu^0(X_i) - \mu(X_{j(i)})$ converge to zero very slowly. Third, $E[\sqrt{N_T}(\mu^0(X_i) - \mu^0(X_{j(i)})) \mid D = 1]$ may not converge to zero. And fourth, $E[\sqrt{N_T}(\widehat{\delta}_{ATT} - \delta_{ATT})]$ may not converge to zero.

As you can see, the bias of the matching estimator can be severe depending on the magnitude of these matching discrepancies. However, one piece of good news is that these discrepancies are observed. We can see the degree to which each unit's matched sample has

Table 30. Another matching example (this time to illustrate bias correction).

Unit	Y^1	Y^0	D	X
1	5		1	11
2	2		1	7
3	10		1	5
4	6		1	3
5		4	0	10
6		0	0	8
7		5	0	4
8		1	0	1

severe mismatch on the covariates themselves. Second, we can always make the matching discrepancy small by using a large donor pool of untreated units to select our matches, because recall, the likelihood of finding a good match grows as a function of the sample size, and so if we are content to estimating the ATT, then increasing the size of the donor pool can get us out of this mess. But let's say we can't do that and the matching discrepancies are large. Then we can apply bias-correction methods to minimize the size of the bias. So let's see what the bias-correction method looks like. This is based on Abadie and Imbens [2011].

Note that the total bias is made up of the bias associated with each individual unit i. Thus, each treated observation contributes $\mu^0(X_i) - \mu^0(X_{j(i)})$ to the overall bias. The bias-corrected matching is the following estimator:

$$\widehat{\delta}_{ATT}^{BC} = \frac{1}{N_T} \sum_{D_i=1} \left[(Y_i - Y_{j(i)}) - \left(\widehat{\mu}^0(X_i) - \widehat{\mu}^0(X_{j(i)}) \right) \right]$$

where $\widehat{\mu}^0(X)$ is an estimate of $E[Y \mid X = x, D = 0]$ using, for example, OLS. Again, I find it always helpful if we take a crack at these estimators with concrete data. Table 30 contains more make-believe data for eight units, four of whom are treated and the rest of whom are functioning as controls. According to the switching equation, we only observe the actual outcomes associated with the potential outcomes under treatment or control, which means we're missing the control values for our treatment group.

Table 31. Nearest-neighbor matched sample.

Unit	Y^1	Y^0	D	X
1	5	**4**	1	11
2	2	**0**	1	7
3	10	**5**	1	5
4	6	**1**	1	3
5		4	0	10
6		0	0	8
7		5	0	4
8		1	0	1

Notice in this example that we cannot implement exact matching because none of the treatment group units has an exact match in the control group. It's worth emphasizing that this is a consequence of finite samples; the likelihood of finding an exact match grows when the sample size of the control group grows faster than that of the treatment group. Instead, we use nearest-neighbor matching, which is simply going to match each treatment unit to the control group unit whose covariate value is *nearest* to that of the treatment group unit itself. But, when we do this kind of matching, we necessarily create *matching discrepancies*, which is simply another way of saying that the covariates are not perfectly matched for every unit. Nonetheless, the nearest-neighbor "algorithm" creates Table 31.

Recall that

$$\widehat{\delta}_{ATT} - \frac{5-4}{4} + \frac{2-0}{4} + \frac{10-5}{4} + \frac{6-1}{4} = 3.25$$

With the bias correction, we need to estimate $\widehat{\mu}^0(X)$. We'll use OLS. It should be clearer what $\widehat{\mu}^0(X)$ is. It is is the fitted values from a regression of Y on X. Let's illustrate this using the data set shown in Table 31.

STATA
training_bias_reduction.do

```
1  use
   ↪ https://github.com/scunning1975/mixtape/raw/master/training_bias_reduction.dta,
   ↪ clear
2  reg Y X
3  gen muhat = _b[_cons] + _b[X]*X
4  list
```

```
R
training_bias_reduction.R
1   library(tidyverse)
2   library(haven)
3
4   read_data <- function(df)
5   {
6    full_path <- paste("https://raw.github.com/scunning1975/mixtape/master/",
7                df, sep = "")
8    df <- read_dta(full_path)
9    return(df)
10  }
11
12  training_bias_reduction <- read_data("training_bias_reduction.dta") %>%
13    mutate(
14      Y1 = case_when(Unit %in% c(1,2,3,4) ~ Y),
15      Y0 = c(4,0,5,1,4,0,5,1))
16
17  train_reg <- lm(Y ~ X, training_bias_reduction)
18
19  training_bias_reduction <- training_bias_reduction %>%
20    mutate(u_hat0 = predict(train_reg))
```

When we regress Y onto X and D, we get the following estimated coefficients:

$$\widehat{\mu}^0(X) = \widehat{\beta}_0 + \widehat{\beta}_1 X$$
$$= 4.42 - 0.049X$$

This gives us the outcomes, treatment status, and predicted values in Table 32.

And then this would be done for the other three simple differences, each of which is added to a bias-correction term based on the fitted values from the covariate values.

Now, care must be given when using the fitted values for bias correction, so let me walk you through it. You are still going to be taking the simple differences (e.g., 5 − 4 for row 1), but now you will also subtract out the fitted values associated with each observation's unique covariate. So for instance, in row 1, the outcome 5 has a covariate of

Table 32. Nearest-neighbor matched sample with fitted values for bias correction.

Unit	Y^1	Y^0	Y	D	X	$\widehat{\mu}^0(X)$
1	5	4	5	1	11	3.89
2	2	0	2	1	7	4.08
3	10	5	10	1	5	4.18
4	6	1	6	1	3	4.28
5		4	4	0	10	3.94
6		0	0	0	8	4.03
7		5	5	0	4	4.23
8		1	1	0	1	4.37

11, which gives it a fitted value of 3.89, but the counterfactual has a value of 10, which gives it a predicted value of 3.94. So therefore we would use the following bias correction:

$$\widehat{\delta}_{ATT}^{BC} = \frac{5 - 4 - (3.89 - 3.94)}{4} + \dots$$

Now that we see how a specific fitted value is calculated and how it contributes to the calculation of the ATT, let's look at the entire calculation now.

$$\widehat{\delta}_{ATT}^{BC} = \frac{(5-4) - \left(\widehat{\mu}^0(11) - \widehat{\mu}^0(10)\right)}{4} + \frac{(2-0) - \left(\widehat{\mu}^0(7) - \widehat{\mu}^0(8)\right)}{4}$$
$$+ \frac{(10-5) - \left(\widehat{\mu}^0(5) - \widehat{\mu}^0(4)\right)}{4} + \frac{(6-1) - \left(\widehat{\mu}^0(3) - \widehat{\mu}^0(1)\right)}{4}$$
$$= 3.28$$

which is slightly higher than the unadjusted ATE of 3.25. Note that this bias-correction adjustment becomes more significant as the matching discrepancies themselves become more common. But, if the matching discrepancies are not very common in the first place, then by definition, bias adjustment doesn't change the estimated parameter very much.

Bias arises because of the effect of large matching discrepancies. To minimize these discrepancies, we need a small number of M (e.g., $M = 1$). Larger values of M produce large matching discrepancies. Second, we need matching with replacement. Because matching with replacement can use untreated units as a match more than once, matching with replacement produces smaller discrepancies. And finally, try to match covariates with a large effect on $\mu^0(.)$.

The matching estimators have a normal distribution in large samples provided that the bias is small. For matching without replacement, the usual variance estimator is valid. That is:

$$\widehat{\sigma}^2_{ATT} = \frac{1}{N_T} \sum_{D_i=1} \left(Y_i - \frac{1}{M} \sum_{m=1}^{M} Y_{jm(i)} - \widehat{\delta}_{ATT} \right)^2$$

For matching with replacement:

$$\widehat{\sigma}^2_{ATT} = \frac{1}{N_T} \sum_{D_i=1} \left(Y_i - \frac{1}{M} \sum_{m=1}^{M} Y_{jm(i)} - \widehat{\delta}_{ATT} \right)^2$$
$$+ \frac{1}{N_T} \sum_{D_i=0} \left(\frac{K_i(K_i-1)}{M^2} \right) \widehat{var}(\varepsilon \mid X_i, D_i = 0)$$

where K_i is the number of times that observation i is used as a match. Then $\widehat{var}(Y_i \mid X_i, D_i = 0)$ can be estimated by matching. For example, take two observations with $D_i = D_j = 0$ and $X_i \approx X_j$:

$$\widehat{var}(Y_i \mid X_i, D_i = 0) = \frac{(Y_i - Y_j)^2}{2}$$

is an unbiased estimator of $\widehat{var}(\varepsilon_i \mid X_i, D_i = 0)$. The bootstrap, though, doesn't create valid standard errors [Abadie and Imbens, 2008].

Propensity score methods. There are several ways of achieving the conditioning strategy implied by the backdoor criterion, and we've discussed several. But one popular one was developed by Donald Rubin in the mid-1970s to early 1980s called the propensity score method [Rosenbaum and Rubin, 1983; Rubin, 1977]. The propensity score is similar in many respects to both nearest-neighbor covariate matching by Abadie and Imbens [2006] and subclassification. It's a very popular method, particularly in the medical sciences, of addressing selection on observables, and it has gained some use among economists as well [Dehejia and Wahba, 2002].

Before we dig into it, though, a couple of words to help manage expectations. Despite some early excitement caused by Dehejia and Wahba [2002], subsequent enthusiasm was more tempered [King and Nielsen, 2019; Smith and Todd, 2001, 2005]. As such, propensity score matching has not seen as wide adoption among economists as in other nonexperimental methods like regression discontinuity or difference-in-differences. The most common reason given for this is that economists are oftentimes skeptical that CIA can be achieved in any data set—almost as an article of faith. This is because for many applications, economists as a group are usually more concerned about selection on unobservables than they are selection on observables, and as such, they reach for matching methods less often. But I am agnostic as to whether CIA holds or doesn't hold in your particular application. There's no theoretical reason to dismiss a procedure designed to estimate causal effects on some ad hoc principle one holds because of a hunch. Only prior knowledge and deep familiarity with the institutional details of your application can tell you what the appropriate identification strategy is, and insofar as the backdoor criterion can be met, then matching methods may be perfectly appropriate. And if it cannot, then matching is inappropriate. But then, so is a naïve multivariate regression in such cases.

We've mentioned that propensity score matching is an application used when a conditioning strategy can satisfy the backdoor criterion. But how exactly is it implemented? Propensity score matching takes those necessary covariates, estimates a maximum likelihood model of the conditional probability of treatment (usually a logit or probit so as to ensure that the fitted values are bounded between 0 and 1), and uses the predicted values from that estimation to collapse those covariates into a single scalar called the *propensity score*. All comparisons between the treatment and control group are then based on that value.

There is some subtlety to the propensity score in practice, though. Consider this scenario: two units, A and B, are assigned to treatment and control, respectively. But their propensity score is 0.6. Thus, they had the same 60% conditional probability of being assigned to treatment, but by random chance, A was assigned to treatment and B was assigned to control. The idea with propensity score methods is to compare units who, based on observables, had very similar probabilities of being placed into the treatment group even though those units differed with regard to actual treatment assignment. If conditional on X, two units have the same probability of being treated, then we say they have similar *propensity scores*, and all remaining variation in treatment assignment is due to chance. And insofar as the two units A and B have the same propensity score of 0.6, but one is the treatment group and one is not, and the *conditional independence assumption* credibly holds in the data, then differences between their observed outcomes are attributable to the treatment.

Implicit in that example, though, we see another assumption needed for this procedure, and that's the *common support* assumption. Common support simply requires that there be units in the treatment and control group across the estimated propensity score. We had common support for 0.6 because there was a unit in the treatment group (A) and one in the control group (B) for 0.6. In ways that are connected to this, the propensity score can be used to check for covariate balance between the treatment group and control group such that the two groups become observationally equivalent. But before walking through an example using real data, let's review some papers that use it.[8]

Example: The NSW job training program. The National Supported Work Demonstration (NSW) job-training program was operated by the Manpower Demonstration Research Corp (MRDC) in the mid-1970s. The NSW was a temporary employment program designed to help disadvantaged workers lacking basic job skills move into the labor market by giving them work experience and counseling in a sheltered environment. It was also unique in that it randomly assigned qualified applicants to training positions. The treatment group received all the benefits of the NSW program. The controls were basically left to fend for themselves. The program admitted women receiving Aid to Families

8 I cannot emphasize this enough—this method, like regression more generally, only has value for your project if you can satisfy the backdoor criterion by conditioning on X. If you cannot satisfy the backdoor criterion in your data, then the propensity score does not assist you in identifying a causal effect. At best, it helps you better understand issues related to balance on observables (but not unobservables). It is absolutely critical that your DAG be, in other words, credible, defensible, and accurate, as you depend on those theoretical relationships to design the appropriate identification strategy.

with Dependent Children, recovering addicts, released offenders, and men and women of both sexes who had not completed high school.

Treatment group members were guaranteed a job for nine to eighteen months depending on the target group and site. They were then divided into crews of three to five participants who worked together and met frequently with an NSW counselor to discuss grievances with the program and performance. Finally, they were paid for their work. NSW offered the trainees lower wages than they would've received on a regular job, but allowed for earnings to increase for satisfactory performance and attendance. After participants' terms expired, they were forced to find regular employment. The kinds of jobs varied within sites—some were gas-station attendants, some worked at a printer shop—and men and women frequently performed different kinds of work.

The MDRC collected earnings and demographic information from both the treatment and the control group at baseline as well as every nine months thereafter. MDRC also conducted up to four post-baseline interviews. There were different sample sizes from study to study, which can be confusing.

NSW was a randomized job-training program; therefore, the independence assumption was satisfied. So calculating average treatment effects was straightforward—it's the simple difference in means estimator that we discussed in the potential outcomes chapter.[9]

$$\frac{1}{N_T}\sum_{D_i=1} Y_i - \frac{1}{N_C}\sum_{D_i=0} Y_i \approx E[Y^1 - Y^0] \approx ATE$$

The good news for MDRC, and the treatment group, was that the treatment benefited the workers.[10] Treatment group participants' real earnings post-treatment in 1978 were more than earnings of the control group by approximately $900 [Lalonde, 1986] to $1,800 [Dehejia and Wahba, 2002], depending on the sample the researcher used.

---

9 Remember, randomization means that the treatment was independent of the potential outcomes, so simple difference in means identifies the average treatment effect.

10 Lalonde [1986] lists several studies that discuss the findings from the program.

Lalonde [1986] is an interesting study both because he is evaluating the NSW program and because he is evaluating commonly used econometric methods from that time. He evaluated the econometric estimators' performance by trading out the experimental control group data with data on the non-experimental control group drawn from the population of US citizens. He used three samples of the Current Population Survey (CPS) and three samples of the Panel Survey of Income Dynamics (PSID) for this non-experimental control group data, but I will use just one for each. Non-experimental data is, after all, the typical situation an economist finds herself in. But the difference with the NSW is that it was a randomized experiment, and therefore we know the average treatment effect. Since we know the average treatment effect, we can see how well a variety of econometric models perform. If the NSW program increased earnings by approximately $900, then we should find that if the other econometrics estimators does a good job, right?

Lalonde [1986] reviewed a number of popular econometric methods used by his contemporaries with both the PSID and the CPS samples as nonexperimental comparison groups, and his results were consistently *horrible*. Not only were his estimates usually very different in magnitude, but his results were almost always the wrong sign! This paper, and its pessimistic conclusion, was influential in policy circles and led to a greater push for more experimental evaluations.[11] We can see these results in the following tables from Lalonde [1986]. Table 33 shows the effect of the treatment when comparing the treatment group to the experimental control group. The baseline difference in real earnings between the two groups was negligible. The treatment group made $39 more than the control group in the pre-treatment period without controls and $21 less in the multivariate regression model, but neither is statistically significant. But the post-treatment difference in average earnings was between $798 and $886.[12]

Table 33 also shows the results he got when he used the non-experimental data as the comparison group. Here I report his results when using one sample from the PSID and one from the CPS, although

11 It's since been cited a little more than 1,700 times.

12 Lalonde reports a couple different diff-in-diff models, but for simplicity, I will only report one.

Table 33. Earnings comparisons and estimated training effects for the NSW male participants using comparison groups from the PSID and the CPS-SSA.

Name of comparison group	NSW Treatment minus Control Earnings				Difference-in-differences
	Pre-treatment		Post-treatment		
	Unadj.	Adj.	Unadj.	Adj.	
Experimental controls	$ 39	$ −21	$ 886	$ 798	$ 856
	(383)	(378)	(476)	(472)	(558)
PSID-1	−$15,997	−$7,624	−$15,578	−$8,067	−$749
	(795)	(851)	(913)	(990)	(692)
CPS-SSA-1	−$10,585	−$4,654	−$8,870	−$4,416	$195
	(539)	(509)	(562)	(557)	(441)

Note: Each column represents an estimated treatment effect per econometric measure and for different comparison groups. The dependent variable is earnings in 1978. Based on experimental treatment and controls, the estimated impact of trainings is $886. Standard errors are in parentheses. Exogenous covariates used in the regression adjusted equations are age, age squared, years of schooling, high school completion status, and race.

in his original paper he used three of each. In nearly every point estimate, the effect is negative. The one exception is the difference-in-differences model which is positive, small, and insignificant.

So why is there such a stark difference when we move from the NSW control group to either the PSID or CPS? The reason is because of selection bias:

$$E[Y^0 \mid D = 1] \neq E[Y^0 \mid D = 0]$$

In other words, it's highly likely that the real earnings of NSW participants would have been much lower than the non-experimental control group's earnings. As you recall from our decomposition of the simple difference in means estimator, the second form of bias is selection bias, and if $E[Y^0 \mid D = 1] < E[Y^0 \mid D = 0]$, this will bias the estimate of the ATE downward (e.g., estimates that show a negative effect).

But as I will show shortly, a violation of independence also implies that covariates will be unbalanced across the propensity score—something we call the *balancing property*. Table 34 illustrates this showing the mean values for each covariate for the treatment and

Table 34. Completed matching example with single covariate.

Covariate	All Mean	S.D.	CPS Controls $N_c = 15,992$ Mean	NSW Trainees $N_t = 297$ Mean	T-static	Diff.
Black	0.09	0.28	0.07	0.80	47.04	−0.73
Hispanic	0.07	0.26	0.07	0.94	1.47	−0.02
Age	33.07	11.04	33.2	24.63	13.37	8.6
Married	0.70	0.46	0.71	0.17	20.54	0.54
No degree	0.30	0.46	0.30	0.73	16.27	−0.43
Education	12.0	2.86	12.03	10.38	9.85	1.65
1975 Earnings	13.51	9.31	13.65	3.1	19.63	10.6
1975 Unemp.	0.11	0.32	0.11	0.37	14.29	−0.26

control groups, where the control is the 15,992 observations from the CPS. As you can see, the treatment group appears to be very different on average from the control group CPS sample along nearly every covariate listed. The NSW participants are more black, more Hispanic, younger, less likely to be married, more likely to have no degree and less schooling, more likely to be unemployed in 1975, and more likely to have considerably lower earnings in 1975. In short, the two groups are not *exchangeable* on observables (and likely not exchangeable on unobservables either).

The first paper to reevaluate Lalonde [1986] using propensity score methods was Dehejia and Wahba [1999]. Their interest was twofold. First, they wanted to examine whether propensity score matching could be an improvement in estimating treatment effects using non-experimental data. And second, they wanted to show the diagnostic value of propensity score matching. The authors used the same non-experimental control group data sets from the CPS and PSID as Lalonde [1986] did.

Let's walk through this, and what they learned from each of these steps. First, the authors estimated the propensity score using maximum likelihood modeling. Once they had the estimated propensity score, they compared treatment units to control units within intervals of the propensity score itself. This process of checking whether there are units in both treatment and control for intervals of the propensity score is called checking for common support.

One easy way to check for common support is to plot the number of treatment and control group observations separately across the propensity score with a histogram. Dehejia and Wahba [1999] did this using both the PSID and CPS samples and found that the overlap was nearly nonexistent, but here I'll focus on their CPS sample. The overlap was so bad that they opted to drop 12,611 observations in the control group because their propensity scores were outside the treatment group range. Also, a large number of observations have low propensity scores, evidenced by the fact that the first bin contains 2,969 comparison units. Once this "trimming" was done, the overlap improved, though still wasn't great.

We learn some things from this kind of diagnostic, though. We learn, for one, that the selection bias on observables is probably extreme if for no other reason than the fact that there are so few units in both treatment and control for given values of the propensity score. When there is considerable bunching at either end of the propensity score distribution, it suggests you have units who differ remarkably on observables with respect to the treatment variable itself. Trimming around those extreme values has been a way of addressing this when employing traditional propensity score adjustment techniques.

With estimated propensity score in hand, Dehejia and Wahba [1999] estimated the treatment effect on real earnings 1978 using the experimental treatment group compared with the non-experimental control group. The treatment effect here differs from what we found in Lalonde because Dehejia and Wahba [1999] used a slightly different sample. Still, using their sample, they find that the NSW program caused earnings to increase between $1,672 and $1,794 depending on whether exogenous covariates were included in a regression. Both of these estimates are highly significant.

The first two columns labeled "unadjusted" and "adjusted" represent OLS regressions with and without controls. Without controls, both PSID and CPS estimates are extremely negative and precise. This, again, is because the selection bias is so severe with respect to the NSW program. When controls are included, effects become positive

Table 35. Estimated training effects using propensity scores.

| Comparison group | NSW T-C Earnings | | Propensity score adjusted | | | | |
| | | | | Stratification | | Matching | |
	Unadj.	Adj.	Quadratic score	Unadj.	Adj.	Unadj.	Adj.
Experimental controls	1,794 (633)	1,672 (638)					
PSID-1	−15,205 (1154)	731 (886)	294 (1389)	1,608 (1571)	1,494 (1581)	1,691 (2209)	1,473 (809)
CPS-1	−8498 (712)	972 (550)	1,117 (747)	1,713 (1115)	1,774 (1152)	1,582 (1069)	1,616 (751)

Note: Adjusted column 2 is OLS regressed onto treatment indicator, age and age squared, education, no degree, black hispanic, real earnings 1974 and 1975. Quadratic score in column 3 is OLS regressed onto a quadratic on the propensity score and a treatment indicator. Last column labeled "adjusted" is weighted least squares.

and imprecise for the PSID sample though almost significant at 5% for CPS. But each effect size is only about half the size of the true effect.

Table 35 shows the results using propensity score weighting or matching.[13] As can be seen, the results are a considerable improvement over Lalonde [1986]. I won't review every treatment effect the authors calculated, but I will note that they are all positive and similar in magnitude to what they found in columns 1 and 2 using only the experimental data.

Finally, the authors examined the balance between the covariates in the treatment group (NSW) and the various non-experimental (matched) samples in Table 36. In the next section, I explain why we expect covariate values to balance along the propensity score for the treatment and control group after trimming the outlier propensity score units from the data. Table 36 shows the sample means of characteristics in the matched control sample versus the experimental NSW sample (first row). Trimming on the propensity score, in effect, helped

---

13 Let's hold off digging into exactly how they used the propensity score to generate these estimates.

Table 36. Sample means of characteristics for matched control samples.

Matched Sample	N	Age	Education	Black	Hispanic	No degree	Married	RE74	RE75
NSW	185	25.81	10.335	0.84	0.06	0.71	0.19	2,096	1,532
PSID	56	26.39	10.62	0.86	0.02	0.55	0.15	1,794	1,126
		(2.56)	(0.63)	(0.13)	(0.06)	(0.13)	(0.13)	(0.12)	(1,406)
CPS	119	26.91	10.52	0.86	0.04	0.64	0.19	2,110	1,396
		(1.25)	(0.32)	(0.06)	(0.04)	(0.07)	(0.06)	(841)	(563)

Note: Standard error on the difference in means with NSW sample is given in parentheses. RE74 stands for real earnings in 1974.

balance the sample. Covariates are much closer in mean value to the NSW sample after trimming on the propensity score.

Propensity score is best explained using actual data. We will use data from Dehejia and Wahba [2002] for the following exercises. But before using the propensity score methods for estimating treatment effects, let's calculate the average treatment effect from the actual experiment. Using the following code, we calculate that the NSW job-training program caused real earnings in 1978 to increase by $1,794.343.

STATA
nsw_experimental.do

```
1   use https://github.com/scunning1975/mixtape/raw/master/nsw_mixtape.dta,
    ↪   clear
2   su re78 if treat
3   gen y1 = r(mean)
4   su re78 if treat==0
5   gen y0 = r(mean)
6   gen ate = y1-y0
7   su ate
8   di 6349.144 - 4554.801
9   * ATE is 1794.34
10  drop if treat==0
11  drop y1 y0 ate
12  compress
```

R
nsw_experimental.R

```r
1   library(tidyverse)
2   library(haven)
3
4   read_data <- function(df)
5   {
6   full_path <- paste("https://raw.github.com/scunning1975/mixtape/master/",
7               df, sep = "")
8    df <- read_dta(full_path)
9    return(df)
10  }
11
12  nsw_dw <- read_data("nsw_mixtape.dta")
13
14  nsw_dw %>%
15   filter(treat == 1) %>%
16    summary(re78)
17
18  mean1 <- nsw_dw %>%
19   filter(treat == 1) %>%
20   pull(re78) %>%
21   mean()
22
23  nsw_dw$y1 <- mean1
24
25  nsw_dw %>%
26   filter(treat == 0) %>%
27    summary(re78)
28
29  mean0 <- nsw_dw %>%
30   filter(treat == 0) %>%
31   pull(re78) %>%
32   mean()
33
34  nsw_dw$y0 <- mean0
35
36  ate <- unique(nsw_dw$y1 - nsw_dw$y0)
37
38  nsw_dw <- nsw_dw %>%
39   filter(treat == 1) %>%
40   select(-y1, -y0)
```

Next we want to go through several examples in which we estimate the average treatment effect or some if its variants such as the average treatment effect on the treatment group or the average treatment effect on the untreated group. But here, rather than using the experimental control group from the original randomized experiment, we will use the non-experimental control group from the Current Population Survey. It is very important to stress that while the treatment group is an experimental group, the control group now consists of a random sample of Americans from that time period. Thus, the control group suffers from extreme selection bias since most Americans would not function as counterfactuals for the distressed group of workers who selected into the NSW program. In the following, we will append the CPS data to the experimental data and estimate the propensity score using logit so as to be consistent with Dehejia and Wahba [2002].

STATA

nsw_pscore.do

```
1   * Reload experimental group data
2   use https://github.com/scunning1975/mixtape/raw/master/nsw_mixtape.dta,
    ↪  clear
3   drop if treat==0
4
5   * Now merge in the CPS controls from footnote 2 of Table 2 (Dehejia and Wahba
    ↪  2002)
6   append using
    ↪  https://github.com/scunning1975/mixtape/raw/master/cps_mixtape.dta
7   gen agesq=age*age
8   gen agecube=age*age*age
9   gen edusq=educ*edu
10  gen u74 = 0 if re74!=.
11  replace u74 = 1 if re74==0
12  gen u75 = 0 if re75!=.
13  replace u75 = 1 if re75==0
14  gen interaction1 = educ*re74
15  gen re74sq=re74^2
16  gen re75sq=re75^2
17  gen interaction2 = u74*hisp
18
19  * Now estimate the propensity score
```

(continued)

STATA *(continued)*

```
20   logit treat age agesq agecube educ edusq marr nodegree black hisp re74 re75
     ↪  u74 u75 interaction1
21   predict pscore
22
23   * Checking mean propensity scores for treatment and control groups
24   su pscore if treat==1, detail
25   su pscore if treat==0, detail
26
27   * Now look at the propensity score distribution for treatment and control groups
28   histogram pscore, by(treat) binrescale
```

R

nsw_pscore.R

```
1    library(tidyverse)
2    library(haven)
3
4    read_data <- function(df)
5    {
6      full_path <- paste("https://raw.github.com/scunning1975/mixtape/master/",
7                  df, sep = "")
8      df <- read_dta(full_path)
9      return(df)
10   }
11
12   nsw_dw_cpscontrol <- read_data("cps_mixtape.dta") %>%
13     bind_rows(nsw_dw) %>%
14     mutate(agesq = age^2,
15         agecube = age^3,
16         educsq = educ*educ,
17         u74 = case_when(re74 == 0 ~ 1, TRUE ~ 0),
18         u75 = case_when(re75 == 0 ~ 1, TRUE ~ 0),
19         interaction1 = educ*re74,
20         re74sq = re74^2,
21         re75sq = re75^2,
22         interaction2 = u74*hisp)
23
24   # estimating
25   logit_nsw <- glm(treat ~ age + agesq + agecube + educ + educsq +
26             marr + nodegree + black + hisp + re74 + re75 + u74 +
27             u75 + interaction1, family = binomial(link = "logit"),
```

(continued)

R *(continued)*

```
28              data = nsw_dw_cpscontrol)
29
30  nsw_dw_cpscontrol <- nsw_dw_cpscontrol %>%
31    mutate(pscore = logit_nsw$fitted.values)
32
33  # mean pscore
34  pscore_control <- nsw_dw_cpscontrol %>%
35    filter(treat == 0) %>%
36    pull(pscore) %>%
37    mean()
38
39  pscore_treated <- nsw_dw_cpscontrol %>%
40    filter(treat == 1) %>%
41    pull(pscore) %>%
42    mean()
43
44  # histogram
45  nsw_dw_cpscontrol %>%
46    filter(treat == 0) %>%
47    ggplot() +
48    geom_histogram(aes(x = pscore))
49
50  nsw_dw_cpscontrol %>%
51    filter(treat == 1) %>%
52    ggplot() +
53    geom_histogram(aes(x = pscore))
54
```

The propensity score is the fitted values of the logit model. Put differently, we used the estimated coefficients from that logit regression to estimate the conditional probability of treatment, assuming that probabilities are based on the cumulative logistic distribution:

$$\Pr\left(D = 1 \mid X\right) = F(\beta_0 + \gamma\, \text{Treat} + \alpha X)$$

where $F() = \dfrac{e}{(1+e)}$ and X is the exogenous covariates we are including in the model.

As I said earlier, the propensity score used the fitted values from the maximum likelihood regression to calculate each unit's conditional probability of treatment *regardless of actual treatment status*. The propensity score is just the predicted conditional probability of treatment or fitted value for each unit. It is advisable to use maximum likelihood when estimating the propensity score so that the fitted values are in the range $[0,1]$. We could use a linear probability model, but linear probability models routinely create fitted values below 0 and above 1, which are not true probabilities since $0 \le p \le 1$.

The definition of the propensity score is the selection probability conditional on the confounding variables; $p(X) = \Pr(D = 1 \mid X)$. Recall that we said there are two identifying assumptions for propensity score methods. The first assumption is CIA. That is, $(Y^0, Y^1) \perp\!\!\!\perp D \mid X$. It is not testable, because the assumption is based on unobservable potential outcomes. The second assumption is called the *common support* assumption. That is, $0 < \Pr(D = 1 \mid X) < 1$. This simply means that for any probability, there must be units in both the treatment group *and* the control group. The conditional independence assumption simply means that the backdoor criterion is met in the data by conditioning on a vector X. Or, put another way, conditional on X, the assignment of units to the treatment is *as good as random*.[14]

Common support is required to calculate any particular kind of defined average treatment effect, and without it, you will just get some kind of weird weighted average treatment effect for only those regions that do have common support. The reason it is "weird" is that average treatment effect doesn't correspond to any of the interesting treatment effects the policymaker needed. Common support requires that for each value of X, there is a positive probability of being both treated and untreated, or $0 < \Pr(D_i = 1 \mid X_i) < 1$. This implies that the probability of receiving treatment for every value of the vector X is strictly within the unit interval. Common support ensures there is sufficient overlap in the characteristics of treated and untreated units to find adequate matches. Unlike CIA, the common support requirement is testable by simply plotting histograms

---

14 CIA is expressed in different ways according to the econometric or statistical tradition. Rosenbaum and Rubin [1983] called it the ignorable treatment assignment, or unconfoundedness. Barnow et al. [1981] and Dale and Krueger [2002] called it *selection on observables*. In the traditional econometric pedagogy, as we discussed earlier, it's called the zero conditional mean assumption.

Table 37. Distribution of propensity score for treatment group.

Percentiles	Treatment Group Values	Smallest
1%	0.0011757	0.0010614
5%	0.0072641	0.0011757
10%	0.0260147	0.0018463
25%	0.1322174	0.0020981
50%	0.4001992	

Percentiles	Values	Largest
75%	0.6706164	0.935645
90%	0.8866026	0.93718
95%	0.9021386	0.9374608
99%	0.9374608	0.9384554

Table 38. Distribution of propensity score for CPS control group.

Percentiles	CPS Control Group Values	Smallest
1%	5.90e-07	1.18e-09
5%	1.72e-06	4.07e-09
10%	3.58e-06	4.24e-09
25%	0.0000193	1.55e-08
50%	0.0001187	
50%	.0003544	

Percentiles	Values	Largest
75%	0.0009635	0.8786677
90%	0.0066319	0.8893389
95%	0.0163109	0.9099022
99%	0.1551548	0.9239787

or summarizing the data. Here we do that two ways: by looking at the summary statistics and by looking at a histogram. Let's start with looking at a distribution in table form before looking at the histogram.

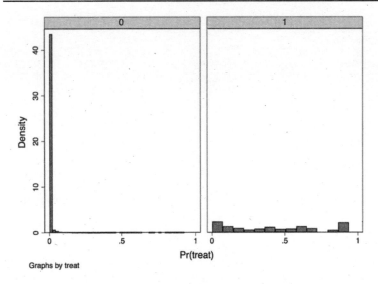

Graphs by treat

Figure 18. Histogram of propensity score by treatment status.

The mean value of the propensity score for the treatment group is 0.43, and the mean for the CPS control group is 0.007. The 50th percentile for the treatment group is 0.4, but the control group doesn't reach that high a number until the 99th percentile. Let's look at the distribution of the propensity score for the two groups using a histogram now.

These two simple diagnostic tests show what is going to be a problem later when we use inverse probability weighting. The probability of treatment is spread out across the units in the treatment group, but there is a very large mass of nearly zero propensity scores in the CPS. How do we interpret this? What this means is that the characteristics of individuals in the treatment group are rare in the CPS sample. This is not surprising given the strong negative selection into treatment. These individuals are younger, less likely to be married, and more likely to be uneducated and a minority. The lesson is, if the two groups are significantly different on background characteristics, then the propensity scores will have grossly different distributions by treatment status. We will discuss this in greater detail later.

For now, let's look at the treatment parameter under both assumptions.

$$E[\delta_i(X_i)] = E[Y_i^1 - Y_i^0 \mid X_i = x]$$
$$= E[Y_i^1 \mid X_i = x] - E[Y_i^0 \mid X_i = x]$$

The conditional independence assumption allows us to make the following substitution,

$$E[Y_i^1 \mid D_i = 1, X_i = x] = E[Y_i \mid D_i = 1, X_i = x]$$

and same for the other term. Common support means we can estimate both terms. Therefore, under both assumptions:

$$\delta = E[\delta(X_i)]$$

From these assumptions we get the *propensity score theorem*, which states that under CIA

$$(Y^1, Y^0) \perp\!\!\!\perp D \mid X$$

This then yields

$$(Y^1, Y^0) \perp\!\!\!\perp D \mid p(X)$$

where $p(X) = \Pr(D = 1 \mid X)$, the propensity score. In English, this means that in order to achieve independence, assuming CIA, all we have to do is condition on the propensity score. Conditioning on the propensity score is enough to have independence between the treatment and the potential outcomes.

This is an extremely valuable theorem because stratifying on X tends to run into the sparseness-related problems (i.e., empty cells) in finite samples for even a moderate number of covariates. But the propensity scores are just a scalar. So stratifying across a probability is going to reduce that dimensionality problem.

The proof of the propensity score theorem is fairly straightforward, as it's just an application of the law of iterated expectations with nested conditioning.[15] If we can show that the probability an individual receives treatment conditional on potential outcomes and the propensity score is not a function of potential outcomes, then we will have

---

15 See Angrist and Pischke [2009], 80–81.

proved that there is independence between the potential outcomes and the treatment conditional on X. Before diving into the proof, first recognize that

$$\Pr\left(D=1\mid Y^0,Y^1,p(X)\right)=E[D\mid Y^0,Y^1,p(X)]$$

because

$$E[D\mid Y^0,Y^1,p(X)]=1\times\Pr\left(D=1\mid Y^0,Y^1,p(X)\right)$$
$$+0\times\Pr\left(D=0\mid Y^0,Y^1,p(X)\right)$$

and the second term cancels out because it's multiplied by zero. The formal proof is as follows:

$$\Pr\left(D=1\mid Y^1,Y^0,p(X)\right)=\underbrace{E[D\mid Y^1,Y^0,p(X)]}_{\text{See previous equation}}$$

$$=\underbrace{E\Big[E[D\mid Y^1,Y^0,p(X),X]\mid Y^1,Y^0,p(X)\Big]}_{\text{by LIE}}$$

$$=\underbrace{E\Big[E[D\mid Y^1,Y^0,X]\mid Y^1,Y^0,p(X)\Big]}_{\text{Given }X\text{, we know }p(X)}$$

$$=\underbrace{E\Big[E[D\mid X]\mid Y^1,Y^0,p(X)\Big]}_{\text{by conditional independence}}$$

$$=\underbrace{E\Big[p(X)\mid Y^1,Y^0,p(X)\Big]}_{\text{propensity score definition}}$$

$$=p(X)$$

Using a similar argument, we obtain:

$$\Pr\left(D=1\mid p(X)\right)=\underbrace{E[D\mid p(X)]}_{\text{Previous argument}}$$

$$=\underbrace{E\Big[E[D\mid X]\mid p(X)\Big]}_{\text{LIE}}$$

$$= \underbrace{E[p(X) \mid p(X) \mid]}_{\text{definition}}$$

$$= p(X)$$

and $\Pr(D = 1 \mid Y^1, Y^0, p(X)) = \Pr(D = 1 \mid p(X))$ by CIA.

Like the omitted variable bias formula for regression, the propensity score theorem says that you need only control for covariates that determine the likelihood a unit receives the treatment. But it also says something more than that. It technically says that the *only* covariate you need to condition on is the propensity score. All of the information from the X matrix has been collapsed into a single number: the propensity score.

A corollary of the propensity score theorem, therefore, states that given CIA, we can estimate average treatment effects by weighting appropriately the simple difference in means.[16]

Because the propensity score is a function of X, we know

$$\Pr(D = 1 \mid X, p(X)) = \Pr(D = 1 \mid X)$$

$$= p(X)$$

Therefore, conditional on the propensity score, the probability that $D = 1$ does not depend on X any longer. That is, D and X are independent of one another conditional on the propensity score, or

$$D \perp\!\!\!\perp \mid p(X)$$

So from this we also obtain the *balancing property* of the propensity score:

$$\Pr(X \mid \mid D = 1, p(X)) = \Pr(X \mid D = 0, p(X))$$

which states that conditional on the propensity score, the distribution of the covariates is the same for treatment as it is for control group units. See this in the following DAG:

---

16 This all works if we match on the propensity score and then calculate differences in means. Direct propensity score matching works in the same way as the covariate matching we discussed earlier (e.g., nearest-neighbor matching), except that we match on the *score* instead of the *covariates* directly.

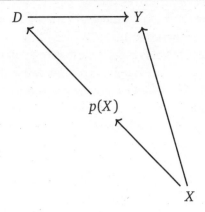

Notice that there exist two paths between X and D. There's the direct path of $X \rightarrow p(X) \rightarrow D$, and there's the backdoor path $X \rightarrow Y \leftarrow D$. The backdoor path is blocked by a collider, so there is no systematic correlation between X and D through it. But there is systematic correlation between X and D through the first directed path. But, when we condition on $p(X)$, the propensity score, notice that D and X are statistically *independent*. This implies that $D \perp\!\!\!\perp X \mid p(X)$, which implies

$$\Pr\left(X \mid D = 1, \widehat{p}(X)\right) = \Pr\left(X \mid D = 0, \widehat{p}(X)\right)$$

This is something we can directly test, but note the implication: conditional on the propensity score, treatment and control should on average be the same with respect to X. In other words, the propensity score theorem implies *balanced* observable covariates.[17]

Weighting on the propensity score. There are several ways researchers can estimate average treatment effects using an estimated propensity score. Busso et al. [2014] examined the properties of various approaches and found that inverse probability weighting was competitive in several simulations. As there are different ways in which the weights are incorporated into a weighting design, I discuss a few canonical versions of the method of inverse probability weighting and associated methods for inference. This is an expansive area in causal

---

17 Just because something is exchangeable on observables does not make it exchangeable on unobservables. The propensity score theorem does *not* imply balanced unobserved covariates. See Brooks and Ohsfeldt [2013].

inference econometrics, so consider this merely an overview of and introduction to the main concepts.

Assuming that CIA holds in our data, then one way we can estimate treatment effects is to use a weighting procedure in which each individual's propensity score is a weight of that individual's outcome [Imbens, 2000]. When aggregated, this has the potential to identify some average treatment effect. This estimator is based on earlier work in survey methodology first proposed by Horvitz and Thompson [1952]. The weight enters the expression differently depending on each unit's treatment status and takes on two different forms depending on whether the target parameter is the ATE or the ATT (or the ATU, which is not shown here):

$$\delta_{ATE} = E[Y^1 - Y^0]$$

$$= E\left[Y \cdot \frac{D - p(X)}{p(X) \cdot (1 - p(X))} \right] \tag{5.4}$$

$$\delta_{ATT} = E[Y^1 - Y^0 \mid D = 1]$$

$$= \frac{1}{\Pr(D=1)} \cdot E\left[Y \cdot \frac{D - p(X)}{1 - p(X)} \right] \tag{5.5}$$

A proof for ATE is provided:

$$E\left[Y \frac{D - p(X)}{p(X)(1 - p(X))} \Big| X \right] = E\left[\frac{Y}{p(X)} \Big| X, D = 1 \right] p(X)$$

$$+ E\left[\frac{-Y}{1 - p(X)} \Big| X, D = 0 \right] (1 - p(X))$$

$$= E[Y \mid X, D = 1] - E[Y \mid X, D = 0] \tag{5.6}$$

and the results follow from integrating over $P(X)$ and $P(X \mid D = 1)$.

The sample versions of both ATE and ATT are obtained by a two-step estimation procedure. In the first step, the researcher estimates the propensity score using logit or probit. In the second step, the researcher uses the estimated score to produce sample versions of one of the average treatment effect estimators shown above. Those sample versions can be written as follows:

$$\widehat{\delta}_{ATE} = \frac{1}{N} \sum_{i=1}^{N} Y_i \cdot \frac{D_i - \widehat{p}(X_i)}{\widehat{p}(X_i) \cdot (1 - \widehat{p}(X_i))} \tag{5.7}$$

$$\widehat{\delta}_{ATT} = \frac{1}{N_T} \sum_{i=1}^{N} Y_i \cdot \frac{D_i - \widehat{p}(X_i)}{1 - \widehat{p}(X_i)} \qquad (5.8)$$

We have a few options for estimating the variance of this estimator, but one is simply to use bootstrapping. First created by Efron [1979], bootstrapping is a procedure used to estimate the variance of an estimator. In the context of inverse probability weighting, we would repeatedly draw ("with replacement") a random sample of our original data and then use that smaller sample to calculate the sample analogs of the ATE or ATT. More specifically, using the smaller "bootstrapped" data, we would first estimate the propensity score and then use the estimated propensity score to calculate sample analogs of the ATE or ATT over and over to obtain a distribution of treatment effects corresponding to different cuts of the data itself.[18] If we do this 1,000 or 10,000 times, we get a distribution of parameter estimates from which we can calculate the standard deviation. This standard deviation becomes like a standard error and gives us a measure of the dispersion of the parameter estimate under uncertainty regarding the sample itself.[19] Adudumilli [2018] and Bodory et al. [2020] discuss the performance of various bootstrapping procedures, such as the standard bootstrap or the wild bootstrap. I encourage you to read these

---

18 Bootstrapping and randomization inference are mechanically similar. Each randomizes *something* over and over, and under each randomization, reestimates treatment effects to obtain a distribution of treatment effects. But that is where the similarity ends. Bootstrapping is a method for computing the variance in an estimator where we take the treatment assignment as given. The uncertainty in bootstrapping stems from the sample, not the treatment assignment. And thus with each bootstrapped sample, we use fewer observations than exist in our real sample. That is not the source of uncertainty in randomization inference, though. In randomization inference, as you recall from the earlier chapter, the uncertainty in question regards the treatment assignment, not the sample. And thus in randomization inference, we randomly assign the treatment in order to reject or fail to reject Fisher's sharp null of no individual treatment effects.

19 Abadie and Imbens [2008] show that the bootstrap fails for *matching*, but inverse probability weighting is not matching. This may seem like a subtle point, but in my experience many people conflate propensity score based matching with other methods that use the propensity score, calling all of them "matching." But inverse probability weighting is *not* a matching procedure. Rather, it is a weighting procedure whose properties differ from that of using imputation and generally the bootstrap is fine.

papers more closely when choosing which bootstrap is suitable for your question.

The sensitivity of inverse probability weighting to extreme values of the propensity score has led some researchers to propose an alternative that can handle extremes a bit better. Hirano and Imbens [2001] propose an inverse probability weighting estimator of the average treatment effect that assigns weights normalized by the sum of propensity scores for treated and control groups as opposed to equal weights of $\frac{1}{N}$ to each observation. This procedure is sometimes associated with Hájek [1971]. Millimet and Tchernis [2009] refer to this estimator as the normalized estimator. Its weights sum to one within each group, which tends to make it more stable. The expression of this normalized estimator is shown here:

$$\widehat{\delta}_{ATT} = \left[\sum_{i=1}^{N} \frac{Y_i D_i}{\widehat{p}}\right] / \left[\sum_{i=1}^{N} \frac{D_i}{\widehat{p}}\right] - \left[\sum_{i=1}^{N} \frac{Y_i(1-D_i)}{(1-\widehat{p})}\right] / \left[\sum_{i=1}^{N} \frac{(1-D_i)}{(1-\widehat{p})}\right] \quad (5.9)$$

Most software packages have programs that will estimate the sample analog of these inverse probability weighted parameters that use the second method with normalized weights. For instance, Stata's -teffects- and R's -ipw- can both be used. These packages will also generate standard errors. But I'd like to manually calculate these point estimates so that you can see more clearly exactly how to use the propensity score to construct either non-normalized or normalized weights and then estimate ATT.

STATA
ipw.do

```
1    * Manual with non-normalized weights using all the data
2    gen d1=treat/pscore
3    gen d0=(1-treat)/(1-pscore)
4    egen s1=sum(d1)
5    egen s0=sum(d0)
6
```

(continued)

STATA *(continued)*

```
7    gen y1=treat*re78/pscore
8    gen y0=(1-treat)*re78/(1-pscore)
9    gen ht=y1-y0
10
11   * Manual with normalized weights
12   replace y1=(treat*re78/pscore)/(s1/_N)
13   replace y0=((1-treat)*re78/(1-pscore))/(s0/_N)
14   gen norm=y1-y0
15   su ht norm
16
17   * ATT under non-normalized weights is -$11,876
18   * ATT under normalized weights is -$7,238
19
20   drop d1 d0 s1 s0 y1 y0 ht norm
21
22   * Trimming the propensity score
23   drop if pscore <= 0.1
24   drop if pscore >= 0.9
25
26   * Manual with non-normalized weights using trimmed data
27   gen d1=treat/pscore
28   gen d0=(1-treat)/(1-pscore)
29   egen s1=sum(d1)
30   egen s0=sum(d0)
31
32   gen y1=treat*re78/pscore
33   gen y0=(1-treat)*re78/(1-pscore)
34   gen ht=y1-y0
35
36   * Manual with normalized weights using trimmed data
37   replace y1=(treat*re78/pscore)/(s1/_N)
38   replace y0=((1-treat)*re78/(1-pscore))/(s0/_N)
39   gen norm=y1-y0
40   su ht norm
41
42   * ATT under non-normalized weights is $2,006
43   * ATT under normalized weights is $1,806
```

R

ipw.R

```r
1   library(tidyverse)
2   library(haven)
3
4   #continuation
5   N <- nrow(nsw_dw_cpscontrol)
6   #- Manual with non-normalized weights using all data
7   nsw_dw_cpscontrol <- nsw_dw_cpscontrol %>%
8     mutate(d1 = treat/pscore,
9            d0 = (1-treat)/(1-pscore))
10
11  s1 <- sum(nsw_dw_cpscontrol$d1)
12  s0 <- sum(nsw_dw_cpscontrol$d0)
13
14
15  nsw_dw_cpscontrol <- nsw_dw_cpscontrol %>%
16    mutate(y1 = treat * re78/pscore,
17           y0 = (1-treat) * re78/(1-pscore),
18           ht = y1 - y0)
19
20  #- Manual with normalized weights
21  nsw_dw_cpscontrol <- nsw_dw_cpscontrol %>%
22    mutate(y1 = (treat*re78/pscore)/(s1/N),
23           y0 = ((1-treat)*re78/(1-pscore))/(s0/N),
24           norm = y1 - y0)
25
26  nsw_dw_cpscontrol %>%
27    pull(ht) %>%
28    mean()
29
30  nsw_dw_cpscontrol %>%
31    pull(norm) %>%
32    mean()
33
34  #-- trimming propensity score
35  nsw_dw_cpscontrol <- nsw_dw_cpscontrol %>%
36    select(-d1, -d0, -y1, -y0, -ht, -norm) %>%
37    filter(!(pscore >= 0.9)) %>%
38    filter(!(pscore <= 0.1))
39
40  N <- nrow(nsw_dw_cpscontrol)
```

(continued)

R *(continued)*

```
41
42   #- Manual with non-normalized weights using trimmed data
43   nsw_dw_cpscontrol <- nsw_dw_cpscontrol %>%
44     mutate(d1 = treat/pscore,
45         d0 = (1-treat)/(1-pscore))
46
47   s1 <- sum(nsw_dw_cpscontrol$d1)
48   s0 <- sum(nsw_dw_cpscontrol$d0)
49
50   nsw_dw_cpscontrol <- nsw_dw_cpscontrol %>%
51     mutate(y1 = treat * re78/pscore,
52         y0 = (1-treat) * re78/(1-pscore),
53         ht = y1 - y0)
54
55   #- Manual with normalized weights with trimmed data
56   nsw_dw_cpscontrol <- nsw_dw_cpscontrol %>%
57     mutate(y1 = (treat*re78/pscore)/(s1/N),
58         y0 = ((1-treat)*re78/(1-pscore))/(s0/N),
59         norm = y1 - y0)
60
61   nsw_dw_cpscontrol %>%
62     pull(ht) %>%
63     mean()
64
65   nsw_dw_cpscontrol %>%
66     pull(norm) %>%
67     mean()
```

When we estimate the treatment effect using inverse probability weighting using the non-normalized weighting procedure described earlier, we find an estimated ATT of −$11,876. Using the normalization of the weights, we get −$7,238. Why is this so much different than what we get using the experimental data?

Recall what inverse probability weighting is doing. It is weighting treatment and control units according to $\hat{p}(X)$, which is causing units with very small values of the propensity score to blow up and become unusually influential in the calculation of ATT. Thus, we will need to trim the data. Here we will do a very small trim to eliminate the mass of values at the far-left tail. Crump et al. [2009] develop a principled method for addressing a lack of overlap. A good rule of thumb, they note, is to keep only observations on the interval [0.1,0.9], which was performed at the end of the program.

Now let's repeat the analysis having trimmed the propensity score, keeping only values whose scores are between 0.1 and 0.9. Now we find $2,006 using the non-normalized weights and $1,806 using the normalized weights. This is very similar to what we know is the true causal effect using the experimental data, which was $1,794. And we can see that the normalized weights are even closer. We still need to calculate standard errors, such as based on a bootstrapping method, but I leave it to you investigate that more carefully by reading Adudumilli [2018] and Bodory et al. [2020], who, as I mentioned, discuss the performance of various bootstrapping procedures such as the standard bootstrap and the wild bootstrap.

Nearest-neighbor matching. An alternative, very popular approach to inverse probability weighting is matching on the propensity score. This is often done by finding a couple of units with comparable propensity scores from the control unit donor pool within some ad hoc chosen radius distance of the treated unit's own propensity score. The researcher then averages the outcomes and then assigns that average as an imputation to the original treated unit as a proxy for the potential outcome under counterfactual control. Then effort is made to enforce common support through trimming.

But this method has been criticized by King and Nielsen [2019]. The King and Nielsen [2019] critique is not of the propensity score itself. For instance, the critique does not apply to stratification based on the propensity score [Rosenbaum and Rubin, 1983], regression adjustment or inverse probability weighting. The problem is only focused on nearest-neighbor matching and is related to the forced balance through trimming as well as myriad other common research choices made in the course of the project that together ultimately amplify bias. King and Nielsen write: "The more balanced the data, or the more balance it becomes by [trimming] some of the observations through matching, the more likely propensity score matching will degrade inferences" [2019, 1].

Nevertheless, nearest-neighbor matching, along with inverse probability weighting, is perhaps the most common method for estimating a propensity score model. Nearest-neighbor matching using the propensity score pairs each treatment unit i with one or more comparable control group units j, where comparability is measured in terms of distance to the nearest propensity score. This control group unit's outcome is then plugged into a matched sample. Once we have the matched sample, we can calculate the ATT as

$$\widehat{ATT} = \frac{1}{N_T}(Y_i - Y_{i(j)})$$

where $Y_{i(j)}$ is the matched control group unit to i. We will focus on the ATT because of the problems with overlap that we discussed earlier.

STATA
teffects_nn.do

```
1   teffects psmatch (re78) (treat age agesq agecube educ edusq marr nodegree
    ↪   black hisp re74 re75 u74 u75 interaction1, logit), atet gen(pstub_cps) nn(5)
```

R
teffects_nn.R

```
1    library(MatchIt)
2    library(Zelig)
3
4    m_out <- matchit(treat ~ age + agesq + agecube + educ +
5                educsq + marr + nodegree +
6                black + hisp + re74 + re75 + u74 + u75 + interaction1,
7                data = nsw_dw_cpscontrol, method = "nearest",
8                distance = "logit", ratio =5)
9
10   m_data <- match.data(m_out)
11
12   z_out <- zelig(re78 ~ treat + age + agesq + agecube + educ +
13                educsq + marr + nodegree +
14                black + hisp + re74 + re75 + u74 + u75 + interaction1,
15                model = "ls", data = m_data)
16
17   x_out <- setx(z_out, treat = 0)
18   x1_out <- setx(z_out, treat = 1)
19
20   s_out <- sim(z_out, x = x_out, x1 = x1_out)
21
22   summary(s_out)
```

I chose to match using five nearest neighbors. Nearest neighbors, in other words, will find the five nearest units in the control group, where "nearest" is measured as closest on the propensity score itself. Unlike covariate matching, distance here is straightforward because of the dimension reduction afforded by the propensity score. We then average actual outcome, and match that average outcome to each treatment unit. Once we have that, we subtract each unit's matched control from its treatment value, and then divide by N_T, the number of treatment units. When we do that in Stata, we get an ATT of \$1,725 with $p < 0.05$. Thus, it is both relatively precise and similar to what we find with the experiment itself.

Coarsened exact matching. There are two kinds of matching we've reviewed so far. Exact matching matches a treated unit to all of the control units with the same covariate value. But sometimes this is impossible, and therefore there are matching discrepancies. For instance, say that we are matching continuous age and continuous income. The probability we find another person with the exact same value of both is very small, if not zero. This leads therefore to mismatching on the covariates, which introduces bias.

The second kind of matching we've discussed are approximate matching methods, which specify a metric to find control units that are "close" to the treated unit. This requires a distance metric, such as Euclidean, Mahalanobis, or the propensity score. All of these can be implemented in Stata or R.

Iacus et al. [2012] introduced a kind of exact matching called coarsened exact matching (CEM). The idea is very simple. It's based on the notion that sometimes it's possible to do exact matching once we coarsen the data enough. If we coarsen the data, meaning we create categorical variables (e.g., 0- to 10-year-olds, 11- to 20-year olds), then oftentimes we can find exact matches. Once we find those matches, we calculate weights on the basis of where a person fits in some strata, and those weights are used in a simple weighted regression.

First, we begin with covariates X and make a copy called $X*$. Next we coarsen $X*$ according to user-defined cutpoints or CEM's automatic binning algorithm. For instance, schooling becomes less than high school, high school only, some college, college graduate, post college. Then we create one stratum per unique observation of $X*$ and place each observation in a stratum. Assign these strata to the original and uncoarsened data, X, and drop any observation whose stratum doesn't contain at least one treated and control unit. Then add weights for stratum size and analyze without matching.

But there are trade-offs. Larger bins mean more coarsening of the data, which results in fewer strata. Fewer strata result in more diverse observations within the same strata and thus higher covariate imbalance. CEM prunes both treatment and control group units, which changes the parameter of interest, but so long as you're transparent about this and up front, readers may be willing to give you the benefit of the doubt.[20] Just know, though, that you are not estimating the ATE or the ATT when you start trimming (just as you aren't doing so when you trim propensity scores).

The key benefit of CEM is that it is part of a class of matching methods called monotonic imbalance bounding (MIB). MIB methods bound the maximum imbalance in some feature of the empirical distributions by an ex ante decision by the user. In CEM, this ex ante choice is the coarsening decision. By choosing the coarsening beforehand, users can control the amount of imbalance in the matching solution. It's also very fast.

There are several ways of measuring imbalance, but here we focus on the $L1(f,g)$ measure, which is

$$L1(f,g) = \frac{1}{2} \sum_{l_1 \ldots l_k} \left| f_{l_1 \ldots l_k} - g_{l_1 \ldots l_k} \right|$$

where f and g record the relative frequencies for the treatment and control group units. Perfect global balance is indicated by $L1 = 0$. Larger values indicate larger imbalance between the groups, with a maximum of $L1 = 1$. Hence the "imbalance bounding" between 0 and 1.

Now let's get to the fun part: estimation. We will use the same job-training data we've been working with for this estimation.

---

20 They also may not. The methods are easy. It's convincing readers that's hard.

STATA
cem.do

```
1   ssc install cem
2
3   * Reload experimental group data
4   use https://github.com/scunning1975/mixtape/raw/master/nsw_mixtape.dta,
    ↪ clear
5   drop if treat==0
6
7   * Now merge in the CPS controls from footnote 2 of Table 2 (Dehejia and Wahba
    ↪ 2002)
8   append using
    ↪ https://github.com/scunning1975/mixtape/raw/master/cps_mixtape.dta
9   gen agesq=age*age
10  gen agecube=age*age*age
11  gen edusq=educ*edu
12  gen u74 = 0 if re74!=.
13  replace u74 = 1 if re74==0
14  gen u75 = 0 if re75!=.
15  replace u75 = 1 if re75==0
16  gen interaction1 = educ*re74
17  gen re74sq=re74^2
18  gen re75sq=re75^2
19  gen interaction2 = u74*hisp
20
21  cem age (10 20 30 40 60) age agesq agecube educ edusq marr nodegree black
    ↪ hisp re74 re75 u74 u75 interaction1, treatment(treat)
22  reg re78 treat [iweight=cem_weights], robust
```

R
cem.R

```
1   library(cem)
2   library(MatchIt)
3   library(Zelig)
4   library(tidyverse)
5   library(estimatr)
6
7
```

(continued)

	R *(continued)*
8	m_out <- matchit(treat ~ age + agesq + agecube + educ +
9	educsq + marr + nodegree +
10	black + hisp + re74 + re75 +
11	u74 + u75 + interaction1,
12	data = nsw_dw_cpscontrol,
13	method = "cem",
14	distance = "logit")
15	
16	m_data <- match.data(m_out)
17	
18	m_ate <- lm_robust(re78 ~ treat,
19	data = m_data,
20	weights = m_data$weights)

The estimated ATE is \$2,152, which is larger than our estimated experimental effect. But this ensured a high degree of balance on the covariates, as can be seen from the output from the cem command itself.

As can be seen from Table 39, the values of $L1$ are close to zero in most cases. The largest $L1$ gets is 0.12 for age squared.

Table 39. Balance in covariates after coarsened exact matching.

Covariate	$L1$	Mean	Min.	25%	50%	75%	Max.
age	.08918	.55337	1	1	0	1	0
agesq	.1155	21.351	33	35	0	49	0
agecube	.05263	626.9	817	919	0	1801	0
school	6.0e-16	$-2.3e-14$	0	0	0	0	0
schoolsq	5.4e-16	$-2.8e-13$	0	0	0	0	0
married	1.1e-16	$-1.1e-16$	0	0	0	0	0
nodegree	4.7e-16	$-3.3e-16$	0	0	0	0	0
black	4.7e-16	$-8.9e-16$	0	0	0	0	0
hispanic	7.1e-17	$-3.1e-17$	0	0	0	0	0
re74	.06096	42.399	0	0	0	0	−94.801
re75	.03756	−73.999	0	0	0	−222.85	−545.65
u74	1.9e-16	$-2.2e-16$	0	0	0	0	0
u75	2.5e-16	$-1.1e-16$	0	0	0	0	0
interaction1	.06535	425.68	0	0	0	0	−853.21

Conclusion. Matching methods are an important member of the causal inference arsenal. Propensity scores are an excellent tool to check the balance and overlap of covariates. It's an under-appreciated diagnostic, and one that you might miss if you only ran regressions. There are extensions for more than two treatments, like multinomial models, but I don't cover those here. The propensity score can make groups comparable, but only on the variables used to estimate the propensity score in the first place. It is an area that continues to advance to include covariate balancing [Imai and Ratkovic, 2013; Zhao, 2019; Zubizarreta, 2015] and doubly robust estimators [Band and Robins, 2005]. Consider this chapter more about the mechanics of matching when you have exact and approximate matching situations.

Learning about the propensity score is particularly valuable given that it appears to have a very long half-life. For instance, propensity scores make their way into other contemporary designs too, such as difference-in-differences [Sant'Anna and Zhao, 2018]. So investing in a basic understanding of these ideas and methods is likely worthwhile. You never know when the right project comes along for which these methods are the perfect solution, so there's no intelligent reason to write them off.

But remember, every matching solution to a causality problem requires a credible belief that the backdoor criterion can be achieved by conditioning on some matrix X, or what we've called CIA. This explicitly requires that there are no unobservable variables opening backdoor paths as confounders, which to many researchers requires a leap of faith so great they are unwilling to make it. In some respects, CIA is somewhat advanced because it requires deep institutional knowledge to say with confidence that no such unobserved confounder exists. The method is easy compared to such domain-specific knowledge. So if you have good reason to believe that there are important, unobservable variables, you will need another tool. But if you are willing to make such an assumption, then these methods and others could be useful for you in your projects.

Regression Discontinuity

> *Jump around!*
> *Jump around!*
> *Jump up, jump up, and get*
>   *down!*
> *Jump!*

> **House of Pain**

Huge Popularity of Regression Discontinuity

Waiting for life. Over the past twenty years, interest in the *regression-discontinuity design* (RDD) has increased (Figure 19). It was not always so popular, though. The method dates back about sixty years to Donald Campbell, an educational psychologist, who wrote several studies using it, beginning with Thistlehwaite and Campbell [1960].[1] In a wonderful article on the history of thought around RDD, Cook [2008] documents its social evolution. Despite Campbell's many efforts to advocate for its usefulness and understand its properties, RDD did not catch on beyond a few doctoral students and a handful of papers here and there. Eventually, Campbell too moved on from it.

---

1 Thistlehwaite and Campbell [1960] studied the effect of merit awards on future academic outcomes. Merit awards were given out to students based on a score, and anyone with a score above some cutoff received the merit award, whereas everyone below that cutoff did not. Knowing the treatment assignment allowed the authors to carefully estimate the causal effect of merit awards on future academic performance.

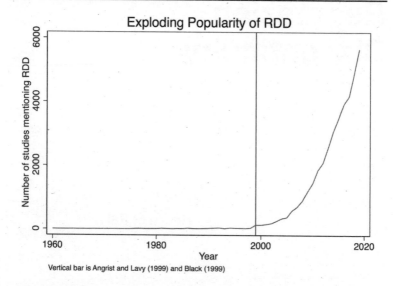

Figure 19. Regression discontinuity over time.

To see its growing popularity, let's look at counts of papers from Google Scholar by year that mentioned the phrase "regression discontinuity design" (see Figure 19).[2] Thistlehwaite and Campbell [1960] had no influence on the broader community of scholars using his design, confirming what Cook [2008] wrote. The first time RDD appears in the economics community is with an unpublished econometrics paper [Goldberger, 1972]. Starting in 1976, RDD finally gets annual double-digit usage for the first time, after which it begins to slowly tick upward. But for the most part, adoption was imperceptibly slow.

But then things change starting in 1999. That's the year when a couple of notable papers in the prestigious *Quarterly Journal of Economics* resurrected the method. These papers were Angrist and Lavy [1999] and Black [1999], followed by Hahn et al. [2001] two years later. Angrist and Lavy [1999] studied the effect of class size on pupil achievement using an unusual feature in Israeli public schools that created smaller classes when the number of students passed a particular

2 Hat tip to John Holbein for giving me these data.

threshold. Black [1999] used a kind of RDD approach when she creatively exploited discontinuities at the geographical level created by school district zoning to estimate people's willingness to pay for better schools. The year 1999 marks a watershed in the design's widespread adoption. A 2010 *Journal of Economic Literature* article by Lee and Lemieux, which has nearly 4,000 cites shows up in a year with nearly 1,500 new papers mentioning the method. By 2019, RDD output would be over 5,600. The design is today incredibly popular and shows no sign of slowing down.

But 1972 to 1999 is a long time without so much as a peep for what is now considered one of the most credible research designs with observational data, so what gives? Cook [2008] says that RDD was "waiting for life" during this time. The conditions for life in empirical microeconomics were likely the growing acceptance of the potential outcomes framework among microeconomists (i.e., the so-called credibility revolution led by Josh Angrist, David Card, Alan Krueger, Steven Levitt, and many others) as well as, and perhaps even more importantly, the increased availability of large digitized the administrative data sets, many of which often captured unusual administrative rules for treatment assignments. These unusual rules, combined with the administrative data sets' massive size, provided the much-needed necessary conditions for Campbell's original design to bloom into thousands of flowers.

Graphical representation of RDD. So what's the big deal? Why is RDD so special? The reason RDD is so appealing to many is because of its ability to convincingly eliminate selection bias. This appeal is partly due to the fact that its underlying identifying assumptions are viewed by many as easier to accept and evaluate. Rendering selection bias impotent, the procedure is capable of recovering average treatment effects for a given subpopulation of units. The method is based on a simple, intuitive idea. Consider the following DAG developed by Steiner et al. [2017] that illustrates this method very well.

In the first graph, X is a continuous variable assigning units to treatment D ($X \rightarrow D$). This assignment of units to treatment is based on a "cutoff" score c_0 such that any unit with a score above the cutoff gets placed into the treatment group, and units below do not. An example

(A) Data generating graph

(B) Limiting graph

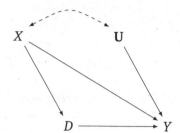
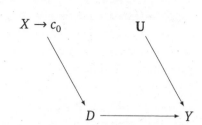

might be a charge of driving while intoxicated (or impaired; DWI). Individuals with a blood-alcohol content of 0.08 or higher are arrested and charged with DWI, whereas those with a blood-alcohol level below 0.08 are not [Hansen, 2015]. The assignment variable may itself independently affect the outcome via the $X \to Y$ path and may even be related to a set of variables U that independently determine Y. Notice for the moment that a unit's treatment status is *exclusively* determined by the assignment rule. Treatment is not determined by U.

This DAG clearly shows that the assignment variable X—or what is often called the "running variable"—is an observable confounder since it causes both D and Y. Furthermore, because the assignment variable assigns treatment on the basis of a cutoff, we are never able to observe units in both treatment and control for the same value of X. Calling back to our matching chapter, this means a situation such as this one does not satisfy the overlap condition needed to use matching methods, and therefore the backdoor criterion cannot be met.[3]

However, we can identify causal effects using RDD, which is illustrated in the limiting graph DAG. We can identify causal effects for those subjects whose score is in a close neighborhood around some cutoff c_0. Specifically, as we will show, the average causal effect for

---

3 Think about it for a moment. The backdoor criterion calculates differences in expected outcomes between treatment and control *for a given value of X*. But if the assignment variable only moves units into treatment when X passes some cutoff, then such calculations are impossible because there will not be units in treatment and control for any given value of X.

this subpopulation is identified as $X \to c_0$ in the limit. This is possible because the cutoff is the sole point where treatment and control subjects overlap in the limit.

There are a variety of explicit assumptions buried in this graph that must hold in order for the methods we will review later to recover any average causal effect. But the main one I discuss here is that the cutoff itself cannot be endogenous to some competing intervention occurring at precisely the same moment that the cutoff is triggering units into the D treatment category. This assumption is called *continuity*, and what it formally means is that the expected potential outcomes are continuous at the cutoff. If expected potential outcomes are continuous at the cutoff, then it necessarily rules out competing interventions occurring at the same time.

The continuity assumption is reflected graphically by the absence of an arrow from $X \to Y$ in the second graph because the cutoff c_0 has cut it off. At c_0, the assignment variable X no longer has a direct effect on Y. Understanding continuity should be one of your main goals in this chapter. It is my personal opinion that the null hypothesis should always be continuity and that any discontinuity necessarily implies some cause, because the tendency for things to change gradually is what we have come to expect in nature. Jumps are so unnatural that when we see them happen, they beg for explanation. Charles Darwin, in his *On the Origin of Species*, summarized this by saying *Natura non facit saltum*, or "nature does not make jumps." Or to use a favorite phrase of mine from growing up in Mississippi, if you see a turtle on a fencepost, you know he didn't get there by himself.

That's the heart and soul of RDD. We use our knowledge about selection into treatment in order to estimate average treatment effects. Since we know the probability of treatment assignment changes discontinuously at c_0, then our job is simply to compare people above and below c_0 to estimate a particular kind of average treatment effect called the *local average treatment effect*, or LATE [Imbens and Angrist, 1994]. Because we do not have overlap, or "common support," we must rely on extrapolation, which means we are comparing units with different values of the running variable. They only overlap in the limit as X approaches the cutoff from either direction. All methods used for RDD are ways of handling the bias from extrapolation as cleanly as possible.

A picture is worth a thousand words. As I've said before, and will say again and again—pictures of your main results, including your identification strategy, are absolutely essential to any study attempting to convince readers of a causal effect. And RDD is no different. In fact, pictures are the comparative advantage of RDD. RDD is, like many modern designs, a very visually intensive design. It and synthetic control are probably two of the most visually intensive designs you'll ever encounter, in fact. So to help make RDD concrete, let's first look at a couple of pictures. The following discussion derives from Hoekstra [2009].[4]

Labor economists had for decades been interested in estimating the causal effect of college on earnings. But Hoekstra wanted to crack open the black box of college's returns a little by checking whether there were heterogeneous returns to college. He does this by estimating the causal effect of attending the state flagship university on earnings. State flagship universities are often more selective than other public universities in the same state. In Texas, the top 7% of graduating high school students can select their university in state, and the modal first choice is University of Texas at Austin. These universities are often environments of higher research, with more resources and strongly positive peer effects. So it is natural to wonder whether there are heterogeneous returns across public universities.

The challenge in this type of question should be easy to see. Let's say that we were to compare individuals who attended the University of Florida to those who attended the University of South Florida. Insofar as there is positive selection into the state flagship school, we might expect individuals with higher observed and unobserved ability to sort into the state flagship school. And insofar as that ability increases one's marginal product, then we expect those individuals to earn more in the workforce regardless of whether they had in fact attended the state flagship. Such basic forms of selection bias confound our ability to estimate the causal effect of attending the state flagship on earnings. But Hoekstra [2009] had an ingenious strategy to disentangle the causal effect from the selection bias using an RDD. To illustrate, let's look at two pictures associated with this interesting study.

---

4 Mark Hoekstra is one of the more creative microeconomists I have met when it comes to devising compelling strategies for identifying causal effects in observational data, and this is one of my favorite papers by him.

Before talking about the picture, I want to say something about the data. Hoekstra has data on all applications to the state flagship university. To get these data, he would've had to build a relationship with the admissions office. This would have involved making introductions, holding meetings to explain his project, convincing administrators the project had value for them as well as him, and ultimately winning their approval to cooperatively share the data. This likely would've involved the school's general counsel, careful plans to de-identify the data, agreements on data storage, and many other assurances that students' names and identities were never released and could not be identified. There is a lot of trust and social capital that must be created to do projects like this, and this is the secret sauce in most RDDs—your acquisition of the data requires far more soft skills, such as friendship, respect, and the building of alliances, than you may be accustomed to. This isn't as straightforward as simply downloading the CPS from IPUMS; it's going to take genuine smiles, hustle, and luck. Given that these agencies have considerable discretion in whom they release data to, it is likely that certain groups will have more trouble than others in acquiring the data. So it is of utmost importance that you approach these individuals with humility, genuine curiosity, and most of all, scientific integrity. They ultimately are the ones who can give you the data if it is not public use, so don't be a jerk.[5]

But on to the picture. Figure 20 has a lot going on, and it's worth carefully unpacking each element for the reader. There are four distinct elements to this picture that I want to focus on. First, notice the horizontal axis. It ranges from a negative number to a positive number with a zero around the center of the picture. The caption reads "SAT Points Above or Below the Admission Cutoff." Hoekstra has "recentered" the university's admissions criteria by subtracting the admission cutoff from the students' actual score, which is something I discuss in more detail later in this chapter. The vertical line at zero marks the "cutoff," which was this university's minimum SAT score for admissions. It appears it was binding, but not deterministically, for there are some students who enrolled but did not have the minimum SAT

---

5 "Don't be a jerk" applies even to situations when you aren't seeking proprietary data.

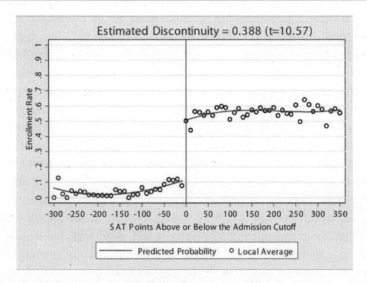

Figure 20. Attending the state flagship university as a function of recentered standard-ized test scores. Reprinted from Mark Hoekstra, "The Effect of Attending the Flagship State University on Earnings: A Discontinuity-Based Approach," *The Review of Economics and Statistics*, 91:4 (November, 2009), pp. 717–724. © 2009 by the President and Fellows of Harvard College and the Massachusetts Institute of Technology.

requirements. These individuals likely had other qualifications that compensated for their lower SAT scores. This recentered SAT score is in today's parlance called the "running variable."

Second, notice the dots. Hoekstra used hollow dots at regular intervals along the recentered SAT variable. These dots represent conditional mean enrollments per recentered SAT score. While his administrative data set contains thousands and thousands of observations, he only shows the conditional means along evenly spaced out bins of the recentered SAT score.

Third are the curvy lines fitting the data. Notice that the picture has *two* such lines—there is a curvy line fitted to the left of zero, and there is a separate line fit to the right. These lines are the least squares fitted values of the running variable, where the running variable was allowed to take on higher-order terms. By including higher-order terms in the regression itself, the fitted values are allowed to more flexibly track the central tendencies of the data itself. But the thing I really want to focus your attention on is that there are two lines, not one. He fit the lines separately to the left and right of the cutoff.

Finally, and probably the most vivid piece of information in this picture—the gigantic jump in the dots at zero on the recentered running variable. What is going on here? Well, I think you probably know, but let me spell it out. The probability of enrolling at the flagship state university jumps discontinuously when the student just barely hits the minimum SAT score required by the school. Let's say that the score was 1250. That means a student with 1240 had a lower chance of getting in than a student with 1250. Ten measly points and they have to go a different path.

Imagine two students—the first student got a 1240, and the second got a 1250. Are these two students really so different from one another? Well, sure: those two *individual* students are likely very different. But what if we had hundreds of students who made 1240 and hundreds more who made 1250. Don't you think those two groups are probably pretty similar to one another on observable and unobservable characteristics? After all, why would there be suddenly at 1250 a major difference in the characteristics of the students in a large sample? That's the question you should reflect on. If the university is arbitrarily picking a reasonable cutoff, are there reasons to believe they are also picking a cutoff where the natural ability of students jumps at that exact spot?

But I said Hoekstra is evaluating the effect of attending the state flagship university on future earnings. Here's where the study gets even more intriguing. States collect data on workers in a variety of ways. One is through unemployment insurance tax reports. Hoekstra's partner, the state flagship university, sent the university admissions data directly to a state office in which employers submit unemployment insurance tax reports. The university had social security numbers, so the matching of student to future worker worked quite well since a social security number uniquely identifies a worker. The social security numbers were used to match quarterly earnings records from 1998 through the second quarter of 2005 to the university records. He then estimated:

$$\ln(\text{Earnings}) = \psi_{\text{Year}} + \omega_{\text{Experience}} + \theta_{\text{Cohort}} + \varepsilon$$

where ψ is a vector of year dummies, ω is a dummy for years after high school that earnings were observed, and θ is a vector of dummies controlling for the cohort in which the student applied to the university (e.g., 1988). The residuals from this regression were then averaged for each applicant, with the resulting average residual earnings measure being used to implement a partialled out future earnings variable according to the Frisch-Waugh-Lovell theorem. Hoekstra then takes each students' residuals from the natural log of earnings regression and collapses them into conditional averages for bins along the recentered running variable. Let's look at that in Figure 21.

In this picture, we see many of the same elements we saw in Figure 20. For instance, we see the recentered running variable along the horizontal axis, the little hollow dots representing conditional means, the curvy lines which were fit left and right of the cutoff at zero, and a helpful vertical line at zero. But now we also have an interesting title: "Estimated Discontinuity = 0.095 (z = 3.01)." What is this exactly?

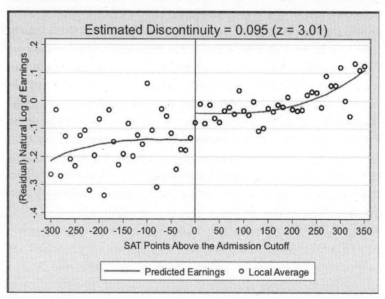

Figure 21. Future earnings as a function of recentered standardized test scores. Reprinted from Mark Hoekstra, "The Effect of Attending the Flagship State University on Earnings: A Discontinuity-Based Approach," *The Review of Economics and Statistics*, 91:4 (November, 2009), pp. 717–724. © 2009 by the President and Fellows of Harvard College and the Massachusetts Institute of Technology.

The visualization of a discontinuous jump at zero in earnings isn't as compelling as the prior figure, so Hoekstra conducts hypothesis tests to determine if the mean between the groups just below and just above are the same. He finds that they are not: those just above the cutoff earn 9.5% higher wages in the long term than do those just below. In his paper, he experiments with a variety of binning of the data (what he calls the "bandwidth"), and his estimates when he does so range from 7.4% to 11.1%.

Now let's think for a second about what Hoekstra is finding. Hoekstra is finding that at exactly the point where workers experienced a jump in the probability of enrolling at the state flagship university, there is, ten to fifteen years later, a separate jump in logged earnings of around 10%. Those individuals who just barely made it in to the state flagship university made around 10% more in long-term earnings than those individuals who just barely missed the cutoff.

This, again, is the heart and soul of the RDD. By exploiting institutional knowledge about how students were accepted (and subsequently enrolled) into the state flagship university, Hoekstra was able to craft an ingenious natural experiment. And insofar as the two groups of applicants right around the cutoff have comparable future earnings in a world where neither attended the state flagship university, then there is no selection bias confounding his comparison. And we see this result in powerful, yet simple graphs. This study was an early one to show that not only does college matter for long-term earnings, but the sort of college you attend—even among public universities—matters as well.

Data requirements for RDD. RDD is all about finding "jumps" in the probability of treatment as we move along some running variable X. So where do we find these jumps? Where do we find these *discontinuities*? The answer is that humans often embed jumps into rules. And sometimes, if we are lucky, someone gives us the data that allows us to use these rules for our study.

I am convinced that firms and government agencies are unknowingly sitting atop a mountain of potential RDD-based projects. Students looking for thesis and dissertation ideas might try to find them. I encourage you to find a topic you are interested in and begin building relationships with local employers and government administrators for whom that topic is a priority. Take them out for coffee, get to know them, learn about their job, and ask them how treatment assignment works. Pay close attention to precisely how individual units get assigned to the program. Is it random? Is it via a rule? Oftentimes they will describe a process whereby a running variable is used for treatment assignment, but they won't call it that. While I can't promise this will yield pay dirt, my hunch, based in part on experience, is that they will end up describing to you some running variable that when it exceeds a threshold, people switch into some intervention. Building alliances with local firms and agencies can pay when trying to find good research ideas.

The validity of an RDD doesn't require that the assignment rule be arbitrary. It only requires that it be known, precise and free of manipulation. The most effective RDD studies involve programs where X has a "hair trigger" that is not tightly related to the outcome being studied. Examples include the probability of being arrested for DWI jumping at greater than 0.08 blood-alcohol content [Hansen, 2015]; the probability of receiving health-care insurance jumping at age 65, [Card et al., 2008]; the probability of receiving medical attention jumping when birthweight falls below 1,500 grams [Almond et al., 2010; Barreca et al., 2011]; the probability of attending summer school when grades fall below some minimum level [Jacob and Lefgen, 2004], and as we just saw, the probability of attending the state flagship university jumping when the applicant's test scores exceed some minimum requirement [Hoekstra, 2009].

In all these kinds of studies, we need data. But specifically, we need a lot of data *around* the discontinuities, which itself implies that the data sets useful for RDD are likely very large. In fact, large sample sizes are characteristic features of the RDD. This is also because in the face of strong trends in the running variable, sample-size requirements get even larger. Researchers are typically using administrative data or settings such as birth records where there are many observations.

Estimation Using an RDD

The Sharp RD Design. There are generally accepted two kinds of RDD studies. There are designs where the probability of treatment goes

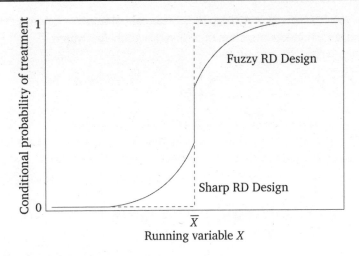

Figure 22. Sharp vs. Fuzzy RDD.

from 0 to 1 at the cutoff, or what is called a "sharp" design. And there are designs where the probability of treatment discontinuously increases at the cutoff. These are often called "fuzzy" designs. In all of these, though, there is some running variable X that, upon reaching a cutoff c_0, the likelihood of receiving some treatment flips. Let's look at the diagram in Figure 22, which illustrates the similarities and differences between the two designs.

Sharp RDD is where treatment is a deterministic function of the running variable X.[6] An example might be Medicare enrollment, which happens sharply at age 65, excluding disability situations. A fuzzy RDD represents a discontinuous "jump" in the probability of treatment when $X > c_0$. In these fuzzy designs, the cutoff is used as an instrumental variable for treatment, as did Angrist and Lavy [1999], whose instrument for class size with a class-size function they created from the rules used by Israeli schools to construct class sizes.

---

6 Van der Klaauw [2002] called the running variable the "selection variable." This is because Van der Klaauw [2002] is an early paper in the new literature, and the terminology hadn't yet been hammered out. But here they mean the same thing.

More formally, in a sharp RDD, treatment status is a deterministic and discontinuous function of a running variable X_i, where

$$D_i = \begin{cases} 1 \text{ if } & X_i \geq c_0 \\ 0 \text{ if } & X_i < c_0 \end{cases}$$

where c_0 is a known threshold or cutoff. If you know the value of X_i for unit i, then you know treatment assignment for unit i with certainty. But, if for every value of X you can perfectly predict the treatment assignment, then it necessarily means that there are no overlap along the running variable.

If we assume constant treatment effects, then in potential outcomes terms, we get

$$Y_i^0 = \alpha + \beta X_i$$
$$Y_i^1 = Y_i^0 + \delta$$

Using the switching equation, we get

$$Y_i = Y_i^0 + (Y_i^1 - Y_i^0)D_i$$
$$Y_i = \alpha + \beta X_i + \delta D_i + \varepsilon_i$$

where the treatment effect parameter, δ, is the discontinuity in the conditional expectation function:

$$\delta = \lim_{X_i \to X_0} E[Y_i^1 \mid X_i = X_0] - \lim_{X_0 \leftarrow X_i} E[Y_i^0 \mid X_i = X_0] \qquad (6.1)$$

$$= \lim_{X_i \to X_0} E[Y_i \mid X_i = X_0] - \lim_{X_0 \leftarrow X_i} E[Y_i \mid X_i = X_0] \qquad (6.2)$$

The sharp RDD estimation is interpreted as an average causal effect of the treatment as the running variable approaches the cutoff in the limit, for it is only in the limit that we have overlap. This average causal effect is the local average treatment effect (LATE). We discuss LATE in greater detail in the instrumental variables, but I will say one thing about it here. Since identification in an RDD is a limiting case, we are technically only identifying an average causal effect for those units at the cutoff. Insofar as those units have treatment effects that differ from units along the rest of the running variable, then we have only estimated an average treatment effect that is local to the range around the cutoff. We define this local average treatment effect as follows:

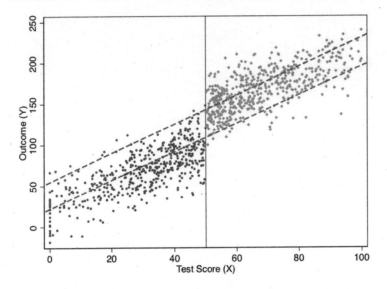

Figure 23. Simulated data representing observed data points along a running variable below and above some binding cutoff.

Note: Dashed lines are extrapolations.

$$\delta_{SRD} = E[Y_i^1 - Y_i^0 \mid X_i = c_0] \qquad (6.3)$$

Notice the role that *extrapolation* plays in estimating treatment effects with sharp RDD. If unit i is just below c_0, then $D_i = 0$. But if unit i is just above c_0, then the $D_i = 1$. But for any value of X_i, there are either units in the treatment group or the control group, but not both. Therefore, the RDD does not have common support, which is one of the reasons we rely on extrapolation for our estimation. See Figure 23.

Continuity assumption. The key identifying assumption in an RDD is called the continuity assumption. It states that $E[Y_i^0 \mid X = c_0]$ and $E[Y_i^1 \mid X = c_0]$ are continuous (smooth) functions of X even across the c_0 threshold. Absent the treatment, in other words, the expected potential outcomes wouldn't have jumped; they would've remained smooth functions of X. But think about what that means for a moment. If the expected potential outcomes are not jumping at c_0, then there necessarily are no competing interventions occurring at c_0. Continuity, in other words, explicitly rules out omitted variable bias at the cutoff itself. All other unobserved determinants of Y are continuously

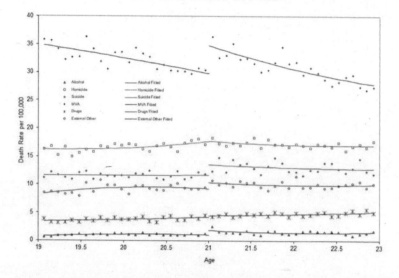

Figure 24. Mortality rates along age running variable [Carpenter and Dobkin, 2009].

related to the running variable X. Does there exist some omitted variable wherein the outcome, would jump at c_0 *even if we disregarded the treatment altogether?* If so, then the continuity assumption is violated and our methods do not require the LATE.

I apologize if I'm beating a dead horse, but continuity is a subtle assumption and merits a little more discussion. The continuity assumption means that $E[Y^1 \mid X]$ wouldn't have jumped at c_0. If it had jumped, then it means something other than the treatment caused it to jump because Y^1 is already under treatment. So an example might be a study finding a large increase in motor vehicle accidents at age 21. I've reproduced a figure from and interesting study on mortality rates for different types of causes [Carpenter and Dobkin, 2009]. I have reproduced one of the key figures in Figure 24. Notice the large discontinuous jump in motor vehicle death rates at age 21. The most likely explanation is that age 21 causes people to drink more, and sometimes even while they are driving.

But this is only a causal effect if motor vehicle accidents don't jump at age 21 for other reasons. Formally, this is *exactly* what is implied by continuity—the absence of simultaneous treatments at the cutoff. For instance, perhaps there is something biological that happens to 21-year-olds that causes them to suddenly become bad drivers. Or maybe 21-year-olds are all graduating from college at age 21, and during celebrations, they get into wrecks. To test this, we might replicate Carpenter and Dobkin [2009] using data from Uruguay, where the drinking age is 18. If we saw a jump in motor vehicle accidents at age 21 in Uruguay, then we might have reason to believe the continuity assumption does not hold in the United States. Reasonably defined placebos can help make the case that the continuity assumption holds, even if it is not a direct test per se.

Sometimes these abstract ideas become much easier to understand with data. Health economist Marcelo Perraillon uses simulated data to teach about the estimation challenges with nonlinearities. I will use a similar approach using Stata and R in the following examples.

```
                               STATA
                         rdd_simulate1.do
1    clear
2    capture log close
3    set obs 1000
4    set seed 1234567
5
6    * Generate running variable. Stata code attributed to Marcelo Perraillon.
7    gen x = rnormal(50, 25)
8    replace x=0 if x < 0
9    drop if x > 100
10   sum x, det
11
12   * Set the cutoff at X=50. Treated if X > 50
13   gen D = 0
14   replace D = 1 if x > 50
15   gen y1 = 25 + 0*D + 1.5*x + rnormal(0, 20)
16
17   * Potential outcome Y1 not jumping at cutoff (continuity)
18   twoway (scatter y1 x if D==0, msize(vsmall) msymbol(circle_hollow)) (scatter y1
     ↪  x if D==1, sort mcolor(blue) msize(vsmall) msymbol(circle_hollow)) (lfit y1 x
     ↪  if D==0, lcolor(red) msize(small) lwidth(medthin) lpattern(solid)) (lfit y1 x,
     ↪  lcolor(dknavy) msize(small) lwidth(medthin) lpattern(solid)), xtitle(Test
     ↪  score (X)) xline(50) legend(off)
19
```

R
rdd_simulate1.R

```
1   library(tidyverse)
2
3   # simulate the data
4   dat <- tibble(
5     x = rnorm(1000, 50, 25)
6   ) %>%
7     mutate(
8       x = if_else(x < 0, 0, x)
9     ) %>%
10    filter(x < 100)
11
12  # cutoff at x = 50
13  dat <- dat %>%
14    mutate(
15      D  = if_else(x > 50, 1, 0),
16      y1 = 25 + 0 * D + 1.5 * x + rnorm(n(), 0, 20)
17    )
18
19  ggplot(aes(x, y1, colour = factor(D)), data = dat) +
20    geom_point(alpha = 0.5) +
21    geom_vline(xintercept = 50, colour = "grey", linetype = 2)+
22    stat_smooth(method = "lm", se = F) +
23    labs(x = "Test score (X)", y = "Potential Outcome (Y1)")
```

Figure 25 shows the results from this simulation. Notice that the value of $E[Y^1 \mid X]$ is changing continuously over X and through c_0. This is an example of the continuity assumption. It means *absent the treatment itself*, the expected potential outcomes would've remained a smooth function of X even as passing c_0. Therefore, if continuity held, then *only* the treatment, triggered at c_0, could be responsible for discrete jumps in $E[Y \mid X]$.

The nice thing about simulations is that we actually observe the potential outcomes *since we made them ourselves*. But in the real world, we don't have data on potential outcomes. If we did, we could test the continuity assumption directly. But remember—by the switching equation, we only observe actual outcomes, never potential outcomes. Thus, since units switch from Y^0 to Y^1 at c_0, we actually

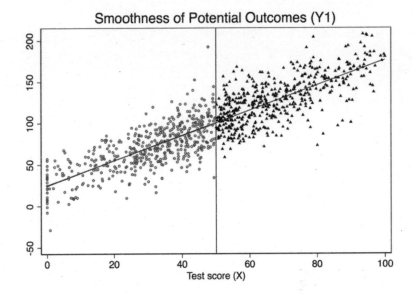

Figure 25. Smoothness of Y^1 across the cutoff illustrated using simulated data. Figure attributed to Marcelo Perraillon.

can't directly evaluate the continuity assumption. This is where institutional knowledge goes a long way, because it can help build the case that nothing else is changing at the cutoff that would otherwise shift potential outcomes.

Let's illustrate this using simulated data. Notice that while Y^1 by construction had not jumped at 50 on the X running variable, Y will. Let's look at the output in Figure 26. Notice the jump at the discontinuity in the outcome, which I've labeled the LATE, or local average treatment effect.

STATA

rdd_simulate2.do

```
1   * Stata code attributed to Marcelo Perraillon.
2   gen y = 25 + 40*D + 1.5*x + rnormal(0, 20)
3   scatter y x if D==0, msize(vsmall) || scatter y x if D==1, msize(vsmall) legend(off)
    ↪  xline(50, lstyle(foreground)) || lfit y x if D ==0, color(red) || lfit y x if D ==1,
    ↪  color(red) ytitle("Outcome (Y)") xtitle("Test Score (X)")
4
```

R
rdd_simulate2.R

```
1   # simulate the discontinuity
2   dat <- dat %>%
3     mutate(
4       y2 = 25 + 40 * D + 1.5 * x + rnorm(n(), 0, 20)
5     )
6
7   # figure 36
8   ggplot(aes(x, y2, colour = factor(D)), data = dat) +
9     geom_point(alpha = 0.5) +
10    geom_vline(xintercept = 50, colour = "grey", linetype = 2) +
11    stat_smooth(method = "lm", se = F) +
12    labs(x = "Test score (X)", y = "Potential Outcome (Y)")
```

Estimation using local and global least squares regressions. I'd like to
now dig into the actual regression model you would use to estimate the
LATE parameter in an RDD. We will first discuss some basic modeling
choices that researchers often make—some trivial, some important.
This section will focus primarily on regression-based estimation.

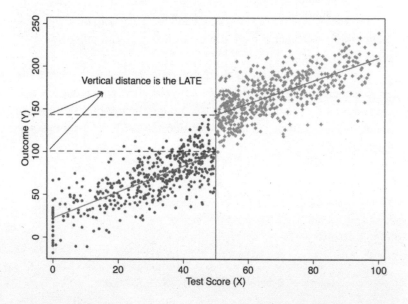

Figure 26. Estimated LATE using simulated data. Figure attributed to Marcelo
Perraillon.

While not necessary, it is nonetheless quite common for authors to transform the running variable X by recentering at c_0:

$$Y_i = \alpha + \beta(X_i - c_0) + \delta D_i + \varepsilon_i$$

This doesn't change the interpretation of the treatment effect—only the interpretation of the intercept. Let's use Card et al. [2008] as an example. Medicare is triggered when a person turns 65. So recenter the running variable (age) by subtracting 65:

$$\begin{aligned}
Y &= \beta_0 + \beta_1(Age - 65) + \beta_2 Edu + \varepsilon \\
&= \beta_0 + \beta_1 Age - \beta_1 65 + \beta_2 Edu + \varepsilon \\
&= (\beta_0 - \beta_1 65) + \beta_1 Age + \beta_2 Edu + \varepsilon \\
&= \alpha + \beta_1 Age + \beta_2 Edu + \varepsilon
\end{aligned}$$

where $\alpha = \beta_0 + \beta_1 65$. All other coefficients, notice, have the same interpretation except for the intercept.

Another practical question relates to nonlinear data-generating processes. A nonlinear data-generating process could easily yield false positives if we do not handle the specification carefully. Because sometimes we are fitting local linear regressions around the cutoff, we could spuriously pick up an effect simply for no other reason than that we imposed linearity on the model. But if the underlying data-generating process is nonlinear, then it may be a spurious result due to misspecification of the model. Consider an example of this nonlinearity in Figure 27.

STATA

rdd_simulate3.do

```
1   * Stata code attributed to Marcelo Perraillon.
2   drop y y1 x* D
3   set obs 1000
4   gen x = rnormal(100, 50)
5   replace x=0 if x < 0
6   drop if x > 280
7   sum x, det
8
9   * Set the cutoff at X=140. Treated if X > 140
10  gen D = 0
```

(continued)

STATA (continued)

```
11    replace D = 1 if x > 140
12    gen x2 = x*x
13    gen x3 = x*x*x
14    gen y = 10000 + 0*D - 100*x +x2 + rnormal(0, 1000)
15    reg y D x
16
17    scatter y x if D==0, msize(vsmall) || scatter y x ///
18      if D==1, msize(vsmall) legend(off) xline(140, ///
19      lstyle(foreground)) ylabel(none) || lfit y x ///
20      if D ==0, color(red) || lfit y x if D ==1, ///
21      color(red) xtitle("Test Score (X)") ///
22      ytitle("Outcome (Y)")
23
24    * Polynomial estimation
25    capture drop y
26    gen y = 10000 + 0*D - 100*x +x2 + rnormal(0, 1000)
27    reg y D x x2 x3
28    predict yhat
29
30    scatter y x if D==0, msize(vsmall) || scatter y x ///
31      if D==1, msize(vsmall) legend(off) xline(140, ///
32      lstyle(foreground)) ylabel(none) || line yhat x ///
33      if D ==0, color(red) sort || line yhat x if D==1, ///
34      sort color(red) xtitle("Test Score (X)") ///
35      ytitle("Outcome (Y)")
```

R

rdd_simulate3.R

```
1    # simultate nonlinearity
2    dat <- tibble(
3      x = rnorm(1000, 100, 50)
4    ) %>%
5      mutate(
6      x = case_when(x < 0 ~ 0, TRUE ~ x),
7      D = case_when(x > 140 ~ 1, TRUE ~ 0),
8      x2 = x*x,
9      x3 = x*x*x,
10     y3 = 10000 + 0 * D - 100 * x + x2 + rnorm(1000, 0, 1000)
11   ) %>%
12     filter(x < 280)
```

(continued)

R *(continued)*
13
14
15 ggplot(aes(x, y3, colour = factor(D)), data = dat) +
16   geom_point(alpha = 0.2) +
17   geom_vline(xintercept = 140, colour = "grey", linetype = 2) +
18   stat_smooth(method = "lm", se = F) +
19   labs(x = "Test score (X)", y = "Potential Outcome (Y)")
20
21 ggplot(aes(x, y3, colour = factor(D)), data = dat) +
22   geom_point(alpha = 0.2) +
23   geom_vline(xintercept = 140, colour = "grey", linetype = 2) +
24   stat_smooth(method = "loess", se = F) +
25   labs(x = "Test score (X)", y = "Potential Outcome (Y)")
26

I show this both visually and with a regression. As you can see in Figure 27, the data-generating process was nonlinear, but when with straight lines to the left and right of the cutoff, the trends in the running variable generate a spurious discontinuity at the cutoff. This shows up in a regression as well. When we fit the model using a least squares regression controlling for the running variable, we estimate a causal effect though there isn't one. In Table 40, the estimated effect of D on Y is large and highly significant, even though the true effect is zero. In this situation, we would need some way to model the nonlinearity below and above the cutoff to check whether, even given the nonlinearity, there had been a jump in the outcome at the discontinuity.

Suppose that the nonlinear relationships is

$$E[Y_i^0 \mid X_i] = f(X_i)$$

for some reasonably smooth function $f(X_i)$. In that case, we'd fit the regression model:

$$Y_i = f(X_i) + \delta D_i + \eta_i$$

Since $f(X_i)$ is counterfactual for values of $X_i > c_0$, how will we model the nonlinearity? There are two ways of approximating $f(X_i)$. The traditional approaches let $f(X_B ei)$ equal a *pth*-order polynomial:

$$Y_i = \alpha + \beta_1 x_i + \beta_2 x_i^2 + \cdots + \beta_p x_i^p + \delta D_i + \eta_i$$

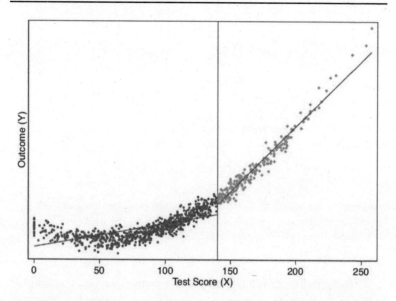

Figure 27. Simulated nonlinear data from Stata. Figure attributed to Marcelo Perraillon.

Table 40. Estimated effect of D on Y using OLS controlling for linear running variable.

Dependent variable	Y
Treatment (D)	6580.16***
	(305.88)

Higher-order polynomials can lead to overfitting and have been found to introduce bias [Gelman and Imbens, 2019]. Those authors recommend using local linear regressions with linear and quadratic forms only. Another way of approximating $f(X_i)$ is to use a nonparametric kernel, which I discuss later.

Though Gelman and Imbens [2019] warn us about higher-order polynomials, I'd like to use an example with pth-order polynomials, mainly because it's not uncommon to see this done today. I'd also like you to know some of the history of this method and understand better what old papers were doing. We can generate this function, $f(X_i)$, by allowing the X_i terms to differ on both sides of the cutoff by including them both individually and interacting them with D_i. In that case, we have:

$$E[Y_i^0 \mid X_i] = \alpha + \beta_{01}\tilde{X}_i + \cdots + \beta_{0p}\tilde{X}_i^p$$
$$E[Y_i^1 \mid X_i] = \alpha + \delta + \beta_{11}\tilde{X}_i + \cdots + \beta_{1p}\tilde{X}_i^p$$

where $\tilde{X}_i$ is the recentered running variable (i.e., $X_i - c_0$). Centering at c_0 ensures that the treatment effect at $X_i = X_0$ is the coefficient on D_i in a regression model with interaction terms. As Lee and Lemieux [2010] note, allowing different functions on both sides of the discontinuity should be the main results in an RDD paper.

To derive a regression model, first note that the observed values must be used in place of the potential outcomes:

$$E[Y \mid X] = E[Y^0 \mid X] + \Big(E[Y^1 \mid X] - E[Y^0 \mid X]\Big)D$$

Your regression model then is

$$Y_i = \alpha + \beta_{01}\tilde{X}_i + \cdots + \beta_{0p}\tilde{X}_i^p + \delta D_i + \beta_1^* D_i\tilde{X}_i + \cdots + \beta_p^* D_i\tilde{X}_i^p + \varepsilon_i$$

where $\beta_1^* = \beta_{11} - \beta_{01}$, and $\beta_p^* = \beta_{1p} - \beta_{0p}$. The equation we looked at earlier was just a special case of the above equation with $\beta_1^* = \beta_p^* = 0$. The treatment effect at c_0 is δ. And the treatment effect at $X_i - c_0 > 0$ is $\delta + \beta_1^* c + \cdots + \beta_p^* c^p$. Let's see this in action with another simulation.

STATA
rdd_simulate4.do

```
1   * Stata code attributed to Marcelo Perraillon.
2   capture drop y
3   gen y = 10000 + 0*D - 100*x +x2 + rnormal(0, 1000)
4   reg y D##c.(x x2 x3)
5   predict yhat
6
7   scatter y x if D==0, msize(vsmall) || scatter y x ///
8       if D==1, msize(vsmall) legend(off) xline(140, ///
9       lstyle(foreground)) ylabel(none) || line yhat x ///
10      if D ==0, color(red) sort || line yhat x if D==1, ///
11      sort color(red) xtitle("Test Score (X)") ///
12      ytitle("Outcome (Y)")
```

Table 41. Estimated effect of *D* on *Y* using OLS controlling for linear and quadratic running variable.

Dependent variable	Y
Treatment (D)	-43.24
	(147.29)

```
                              R
                      rdd_simulate4.R
1    library(stargazer)
2
3    dat <- tibble(
4      x = rnorm(1000, 100, 50)
5    ) %>%
6      mutate(
7        x = case_when(x < 0 ~ 0, TRUE ~ x),
8        D = case_when(x > 140 ~ 1, TRUE ~ 0),
9        x2 = x*x,
10       x3 = x*x*x,
11       y3 = 10000 + 0 * D - 100 * x + x2 + rnorm(1000, 0, 1000)
12     ) %>%
13     filter(x < 280)
14
15   regression <- lm(y3 ~ D*., data = dat)
16
17   stargazer(regression, type = "text")
18
19   ggplot(aes(x, y3, colour = factor(D)), data = dat) +
20     geom_point(alpha = 0.2) +
21     geom_vline(xintercept = 140, colour = "grey", linetype = 2) +
22     stat_smooth(method = "loess", se = F) +
23     labs(x = "Test score (X)", y = "Potential Outcome (Y)")
```

Let's look at the output from this exercise in Figure 28 and Table 41. As you can see, once we model the data using a quadratic (the cubic ultimately was unnecessary), there is no estimated treatment effect at the cutoff. There is also no effect in our least squares regression.

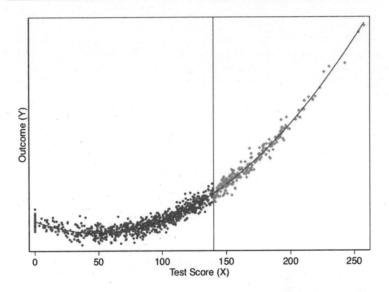

Figure 28. Simulated nonlinear data from Stata. Figure attributed to Marcelo Perraillon.

Nonparametric kernels. But, as we mentioned earlier, Gelman and Imbens [2019] have discouraged the use of higher-order polynomials when estimating local linear regressions. An alternative is to use kernel regression. The nonparametric kernel method has problems because you are trying to estimate regressions at the cutoff point, which can result in a boundary problem (see Figure 29). In this picture, the bias is caused by strong trends in expected potential outcomes throughout the running variable.

While the true effect in this diagram is *AB*, with a certain bandwidth a rectangular kernel would estimate the effect as *A′B′*, which is as you can see a biased estimator. There is systematic bias with the kernel method if the underlying nonlinear function, $f(X)$, is upwards-or downwards-xsloping.

The standard solution to this problem is to run local linear nonparametric regression [Hahn et al., 2001]. In the case described above, this would substantially reduce the bias. So what is that? Think of kernel regression as a weighted regression restricted to a window (hence

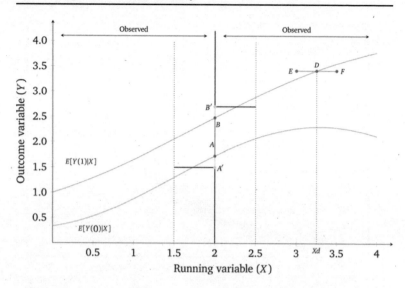

Figure 29. Boundary problem.

"local"). The kernel provides the weights to that regression.[7] A rectangular kernel would give the same result as taking $E[Y]$ at a given bin on X. The triangular kernel gives more importance to the observations closest to the center.

The model is some version of:

$$(\widehat{a},\widehat{b}) =_{a,b} \sum_{i=1}^{n} \left(y_i - a - b(x_i - c_0)\right)^2 K\left(\frac{x_i - c_o}{h}\right) 1(x_i > c_0) \qquad (6.4)$$

While estimating this in a given window of width h around the cutoff is straightforward, what's not straightforward is knowing how large or small to make the bandwidth. This method is sensitive to the choice of bandwidth, but more recent work allows the researcher to estimate *optimal* bandwidths [Calonico et al., 2014; Imbens and Kalyanaraman, 2011]. These may even allow for bandwidths to vary left and right of the cutoff.

---

7 Stata's poly command estimates kernel-weighted local polynomial regression.

Medicare and universal health care. Card et al. [2008] is an example of a sharp RDD, because it focuses on the provision of universal health-care insurance for the elderly—Medicare at age 65. What makes this a policy-relevant question? Universal insurance has become highly relevant because of the debates surrounding the Affordable Care Act, as well as several Democratic senators supporting Medicare for All. But it is also important for its sheer size. In 2014, Medicare was 14% of the federal budget at $505 billion.

Approximately 20% of non-elderly adults in the United States lacked insurance in 2005. Most were from lower-income families, and nearly half were African American or Hispanic. Many analysts have argued that unequal insurance coverage contributes to disparities in health-care utilization and health outcomes across socioeconomic status. But, even among the policies, there is heterogeneity in the form of different copays, deductibles, and other features that affect use. Evidence that better insurance causes better health outcomes is limited because health insurance suffers from deep selection bias. Both supply and demand for insurance depend on health status, confounding observational comparisons between people with different insurance characteristics.

The situation for elderly looks very different, though. Less than 1% of the elderly population is uninsured. Most have fee-for-service Medicare coverage. And that transition to Medicare occurs sharply at age 65—the threshold for Medicare eligibility.

The authors estimate a reduced form model measuring the causal effect of health insurance status on health-care usage:

$$y_{ija} = X_{ija}\alpha + f_k(\alpha;\beta) + \sum_k C_{ija}^k \delta^k + u_{ija}$$

where i indexes individuals, j indexes a socioeconomic group, a indexes age, u_{ija} indexes the unobserved error, y_{ija} health care usage, X_{ija} a set of covariates (e.g., gender and region), $f_j(\alpha;\beta)$ a smooth function representing the age profile of outcome y for group j, and C_{ija}^k ($k = 1,2,\ldots,K$) are characteristics of the insurance coverage held by the individual such as copayment rates. The problem with estimating this model, though, is that insurance coverage is endogenous: $cov(u,C) \neq 0$. So the authors use as identification of the age threshold for Medicare eligibility at 65, which they argue is credibly exogenous variation in insurance status.

Suppose health insurance coverage can be summarized by two dummy variables: C_{ija}^1 (any coverage) and C_{ija}^2 (generous insurance). Card et al. [2008] estimate the following linear probability models:

$$C_{ija}^1 = X_{ija}\beta_j^1 + g_j^1(a) + D_a\pi_j^1 + v_{ija}^1$$
$$C_{ija}^2 = X_{ija}\beta_j^2 + g_j^2(a) + D_a\pi_j^2 + v_{ija}^2$$

where β_j^1 and β_j^2 are group-specific coefficients, $g_j^1(a)$ and $g_j^2(a)$ are smooth age profiles for group j, and D_a is a dummy if the respondent is equal to or over age 65. Recall the reduced form model:

$$y_{ija} = X_{ija}\alpha + f_k(a;\beta) + \sum_k C_{ija}^k \delta^k + u_{ija}$$

Combining the C_{ija} equations, and rewriting the reduced form model, we get:

$$y_{ija} = X_{ija}\left(\alpha_j + \beta_j^1\delta^1 + \beta_j^2\delta^2\right)h_j(a) + D_a\pi_j^y + v_{ija}^y$$

where $h(a) = f_j(a) + \delta^1 g_j^1(a) + \delta^2 g_j^2(a)$ is the reduced form age profile for group j, $\pi_j^y = \pi_j^1\delta^1 + \pi_j^2\delta^2$ and $v_{ija}^y = u_{ija} + v_{ija}^1\delta^1 + v_{ija}^2\delta^2$ is the error term. Assuming that the profiles $f_j(a)$, $g_j(a)$, and $g_j^2(a)$ are continuous at age 65 (i.e., the continuity assumption necessary for identification), then any discontinuity in y is due to insurance. The magnitudes will depend on the size of the insurance changes at age 65 (π_j^1 and π_j^2) and on the associated causal effects (δ^1 and δ^2).

For some basic health-care services, such as routine doctor visits, it may be that the only thing that matters is insurance. But, in those situations, the implied discontinuity in Y at age 65 for group j will be proportional to the change in insurance status experienced by that group. For more expensive or elective services, the generosity of the coverage may matter—for instance, if patients are unwilling to cover the required copay or if the managed care program won't cover the service. This creates a potential identification problem in interpreting the discontinuity in y for any one group. Since π_j^y is a linear combination of the discontinuities in coverage and generosity, δ^1 and δ^2 can be estimated by a regression across groups:

$$\pi_j^y = \delta^0 + \delta^1 \pi_j^1 + \delta_j^2 \pi_j^2 + e_j$$

where e_j is an error term reflecting a combination of the sampling errors in π_j^y, π_j^1 and, π_j^2.

Card et al. [2008] use a couple of different data sets—one a standard survey and the other administrative records from hospitals in three states. First, they use the 1992–2003 National Health Interview Survey (NHIS). The NHIS reports respondents' birth year, birth month, and calendar quarter of the interview. Authors used this to construct an estimate of age in quarters at date of interview. A person who reaches 65 in the interview quarter is coded as age 65 and 0 quarters. Assuming a uniform distribution of interview dates, one-half of these people will be 0–6 weeks younger than 65 and one-half will be 0–6 weeks older. Analysis is limited to people between 55 and 75. The final sample has 160,821 observations.

The second data set is hospital discharge records for California, Florida, and New York. These records represent a complete census of discharges from all hospitals in the three states except for federally regulated institutions. The data files include information on age in months at the time of admission. Their sample selection criteria is to drop records for people admitted as transfers from other institutions and limit people between 60 and 70 years of age at admission. Sample sizes are 4,017,325 (California), 2,793,547 (Florida), and 3,121,721 (New York).

Some institutional details about the Medicare program may be helpful. Medicare is available to people who are at least 65 and have worked forty quarters or more in covered employment or have a spouse who did. Coverage is available to younger people with severe kidney disease and recipients of Social Security Disability Insurance. Eligible individuals can obtain Medicare hospital insurance (Part A) free of charge and medical insurance (Part B) for a modest monthly premium. Individuals receive notice of their impending eligibility for Medicare shortly before they turn 65 and are informed they have to enroll in it and choose whether to accept Part B coverage. Coverage begins on the first day of the month in which they turn 65.

There are five insurance-related variables: probability of Medicare coverage, any health insurance coverage, private coverage, two or more forms of coverage, and individual's primary health insurance is managed care. Data are drawn from the 1999–2003 NHIS, and for each characteristic, authors show the incidence rate at age 63–64 and the change at age 65 based on a version of the C_K equations that include a quadratic in age, fully interacted with a post-65 dummy as well as controls for gender, education, race/ethnicity, region, and sample year. Alternative specifications were also used, such as a parametric model fit to a narrower age window (age 63–67) and a local linear regression specification using a chosen bandwidth. Both show similar estimates of the change at age 65.

The authors present their findings in Table 42. The way that you read this table is each cell shows the *average treatment effect* for the 65-year-old population that complies with the treatment. We can see, not surprisingly, that the effect of receiving Medicare is to cause a very large increase of being on Medicare, as well as reducing coverage on private and managed care.

Formal identification in an RDD relating to some outcome (insurance coverage) to a treatment (Medicare age-eligibility) that itself depends on some running variable, age, relies on the continuity assumptions that we discussed earlier. That is, we must assume that the conditional expectation functions for both potential outcomes is continuous at age= 65. This means that both $E[Y^0 \mid a]$ and $E[Y^1 \mid a]$ are continuous through age of 65. If that assumption is plausible, then the average treatment effect at age 65 is identified as:

$$\lim_{65 \leftarrow a} E[y^1 \mid a] - \lim_{a \to 65} E[y^0 \mid a]$$

The continuity assumption requires that all other factors, observed and unobserved, that affect insurance coverage are trending smoothly at the cutoff, in other words. But what else changes at age 65 other than Medicare eligibility? Employment changes. Typically, 65 is the traditional age when people retire from the labor force. Any abrupt change in employment could lead to differences in health-care utilization if nonworkers have more time to visit doctors.

The authors need to, therefore, investigate this possible confounder. They do this by testing for any potential discontinuities at age

Table 42. Insurance characteristics just before age 65 and estimated discontinuities at age 65.

	On Medicare	Any insurance	Private coverage	2+ forms coverage	Managed care
Overall sample	59.7	9.5	−2.9	44.1	−28.4
	(4.1)	(0.6)	(1.1)	(2.8)	(2.1)
White non-Hispanic					
Less than high school	58.5	13.0	−6.2	44.5	−25.0
	(4.6)	(2.7)	(3.3)	(4.0)	(4.5)
High school graduate	64.7	7.6	−1.9	51.8	−30.3
	(5.0)	(0.7)	(1.6)	(3.8)	(2.6)
Some college	68.4	4.4	−2.3	55.1	−40.1
	(4.7)	(0.5)	(1.8)	(4.0)	(2.6)
Minority					
High school dropout	44.5	21.5	−1.2	19.4	−8.3
	(3.1)	(2.1)	(2.5)	(1.9)	(3.1)
High school graduate	44.6	8.9	−5.8	23.4	−15.4
	(4.7)	(2.8)	(5.1)	(4.8)	(3.5)
Some college	52.1	5.8	−5.4	38.4	−22.3
	(4.9)	(2.0)	(4.3)	(3.8)	(7.2)
Classified by ethnicity only					
White non-Hispanic	65.2	7.3	−2.8	51.9	−33.6
	(4.6)	(0.5)	(1.4)	(3.5)	(2.3)
Black non-Hispanic	48.5	11.9	−4.2	27.8	−13.5
	(3.6)	(2.0)	(2.8)	(3.7)	(3.7)
Hispanic	44.4	17.3	−2.0	21.7	−12.1
	(3.7)	(3.0)	(1.7)	(2.1)	(3.7)

Note: Entries in each cell are estimated regression discontinuities at age 65 from quadratics in age interacted with a dummy for 65 and older. Other controls such as gender, race, education, region, and sample year are also included. Data is from the pooled 1999–2003 NHIS.

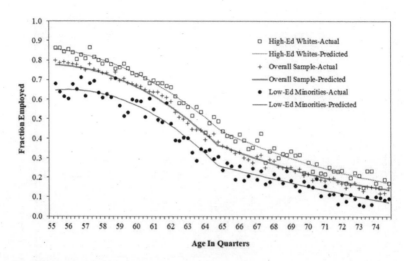

Employment Rates of Men by Age and Demographic Group

Figure 30. Investigating the CPS for discontinuities at age 65 [Card et al., 2008].

65 for confounding variables using a third data set—the March CPS 1996–2004. And they ultimately find no evidence for discontinuities in employment at age 65 (Figure 30).

Next the authors investigate the impact that Medicare had on access to care and utilization using the NHIS data. Since 1997, NHIS has asked four questions. They are:

> "During the past 12 months has medical care been delayed for this person because of worry about the cost?"
>
> "During the past 12 months was there any time when this person needed medical care but did not get it because [this person] could not afford it?"
>
> "Did the individual have at least one doctor visit in the past year?"
>
> "Did the individual have one or more overnight hospital stays in the past year?"

Estimates from this analysis are presented in Table 43. Each cell measures the average treatment effect for the complier population

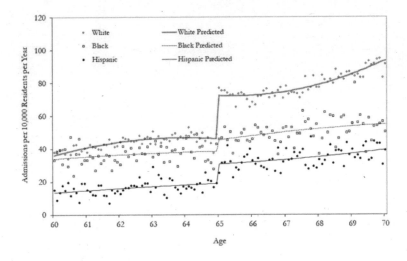

Figure 31. Changes in hospitalizations [Card et al., 2008].

at the discontinuity. Standard errors are in parentheses. There are a few encouraging findings from this table. First, the share of the relevant population who delayed care the previous year fell 1.8 points, and similar for the share who did not get care at all in the previous year. The share who saw a doctor went up slightly, as did the share who stayed at a hospital. These are not very large effects in magnitude, it is important to note, but they are relatively precisely estimated. Note that these effects differed considerably by race and ethnicity as well as education.

Having shown modest effects on care and utilization, the authors turn to examining the kinds of care they received by examining specific changes in hospitalizations. Figure 31 shows the effect of Medicare on hip and knee replacements by race. The effects are largest for whites.

In conclusion, the authors find that universal health-care coverage for the elderly increases care and utilization as well as coverage. In a subsequent study [Card et al., 2009], the authors examined the impact of Medicare on mortality and found slight decreases in mortality rates (see Table 44).

Table 43. Measures of access to care just before 65 and estimated discontinuities at age 65.

	Delayed last year	Did not get care last year	Saw doctor last year	Hospital stay last year
Overall sample	-1.8	-1.3	1.3	1.2
	(0.4)	(0.3)	(0.7)	(0.4)
White non-Hispanic				
Less than high school	-1.5	-0.2	3.1	1.6
	(1.1)	(1.0)	(1.3)	(1.3)
High school graduate	0.3	-1.3	-0.4	0.3
	(2.8)	(2.8)	(1.5)	(0.7)
Some college	-1.5	-1.4	0.0	2.1
	(0.4)	(0.3)	(1.3)	(0.7)
Minority				
High school dropout	-5.3	-4.2	5.0	0.0
	(1.0)	(0.9)	(2.2)	(1.4)
High school graduate	-3.8	1.5	1.9	1.8
	(3.2)	(3.7)	(2.7)	(1.4)
Some college	-0.6	-0.2	3.7	0.7
	(1.1)	(0.8)	(3.9)	(2.0)
Classified by ethnicity only				
White non-Hispanic	-1.6	-1.2	0.6	1.3
	(0.4)	(0.3)	(0.8)	(0.5)
Black non-Hispanic	-1.9	-0.3	3.6	0.5
	(1.1)	(1.1)	(1.9)	(1.1)
Hispanic	-4.9	-3.8	8.2	11.8
	(0.8)	(0.7)	(0.8)	(1.6)

Note: Entries in each cell are estimated regression discontinuities at age 65 from quadratics in age interacted with a dummy for 65 and older. Other controls such as gender, race, education, region and sample year are also included. First two columns are from 1997–2003 NHIS and last two columns are from 1992–2003 NHIS.

Table 44. Regression discontinuity estimates of changes in mortality rates.

	Death rate in					
	7 days	14 days	28 days	90 days	180 days	365 days
Quadratic	−1.1	−1.0	−1.1	−1.2	−1.0	
no controls	(0.2)	(0.2)	(0.3)	(0.3)	(0.4)	(0.4)
Quadratic	−1.0	−0.8	−0.9	−0.9	−0.8	−0.7
plus controls	0.2)	(0.2)	(0.3)	(0.3)	(0.3)	(0.4)
Cubic plus	−0.7	−0.7	−0.6	−0.9	−0.9	−0.4
controls	(0.3)	(0.2)	(0.4)	(0.4)	(0.5)	(0.5)
Local OLS with ad	−0.8	−0.8	−0.8	−0.9	−1.1	−0.8
hoc bandwidths	(0.2)	(0.2)	(0.2)	(0.2)	(0.3)	(0.3)

Note: Dependent variable for death within interval shown in the column heading. Regression estimates at the discontinuity of age 65 for flexible regression models. Standard errors in parentheses.

Inference. As we've mentioned, it's standard practice in the RDD to estimate causal effects using local polynomial regressions. In its simplest form, this amounts to nothing more complicated than fitting a linear specification separately on each side of the cutoff using a least squares regression. But when this is done, you are using only the observations within some pre-specified window (hence "local"). As the true conditional expectation function is probably not linear at this window, the resulting estimator likely suffers from specification bias. But if you can get the window narrow enough, then the bias of the estimator is probably small relative to its standard deviation.

But what if the window cannot be narrowed enough? This can happen if the running variable only takes on a few values, or if the gap between values closest to the cutoff are large. The result could be you simply do not have enough observations close to the cutoff for the local polynomial regression. This also can lead to the heteroskedasticity-robust confidence intervals to undercover the average causal effect because it is not centered. And here's the really bad news—this probably is happening *a lot* in practice.

In a widely cited and very influential study, Lee and Card [2008] suggested that researchers should cluster their standard errors by the running variable. This advice has since become common practice in the empirical literature. Lee and Lemieux [2010], in a survey article on proper RDD methodology, recommend this practice, just to name one example. But in a recent study, Kolesár and Rothe [2018] provide extensive theoretical and simulation-based evidence that clustering on the running variable is perhaps one of the *worst* approaches you could take. In fact, clustering on the running variable can actually be substantially worse than heteroskedastic-robust standard errors.

As an alternative to clustering and robust standard errors, the authors propose two alternative confidence intervals that have guaranteed coverage properties under various restrictions on the conditional expectation function. Both confidence intervals are "honest," which means they achieve correct coverage uniformly over all conditional expectation functions in large samples. These confidence intervals are currently unavailable in Stata as of the time of this writing, but they can be implemented in R with the RDHonest package.[8] R users are encouraged to use these confidence intervals. Stata users are encouraged to switch (grudgingly) to R so as to use these confidence intervals. Barring that, Stata users should use the heteroskedastic robust standard errors. But whatever you do, don't cluster on the running variable, as that is nearly an unambiguously bad idea.

A separate approach may be to use randomization inference. As we noted, Hahn et al. [2001] emphasized that the conditional expected potential outcomes must be continuous across the cutoff for a regression discontinuity design to identify the local average treatment effect. But Cattaneo et al. [2015] suggest an alternative assumption which has implications for inference. They ask us to consider that perhaps around the cutoff, in a short enough window, the treatment was assigned to units randomly. It was effectively a coin flip which side of the cutoff someone would be for a small enough window around the cutoff. Assuming there exists a neighborhood around the cutoff where this randomization-type condition holds, then this assumption may be viewed as an approximation of a randomized experiment around the cutoff. Assuming this is plausible, we can proceed as if only those observations closest to the discontinuity were randomly assigned,

8 RDHonest is available at https://github.com/kolesarm/RDHonest.

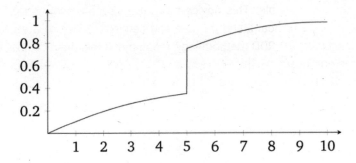

Figure 32. Vertical axis is the probability of treatment for each value of the running variable.

which leads naturally to randomization inference as a methodology for conducting exact or approximate *p*-values.

The Fuzzy RD Design. In the sharp RDD, treatment was *determined* when $X_i \geq c_0$. But that kind of deterministic assignment does not always happen. Sometimes there is a discontinuity, but it's not entirely deterministic, though it nonetheless is associated with a discontinuity in treatment assignment. When there is an increase in the *probability* of treatment assignment, we have a *fuzzy* RDD. The earlier paper by Hoekstra [2009] had this feature, as did Angrist and Lavy [1999]. The formal definition of a probabilistic treatment assignment is

$$\lim_{X_i \to c_0} \Pr\left(D_i = 1 \mid X_i = c_0\right) \neq \lim_{c_0 \leftarrow X_i} \Pr\left(D_i = 1 \mid X_i = c_0\right) \qquad (6.5)$$

In other words, the conditional probability is discontinuous as X approaches c_0 in the limit. A visualization of this is presented from Imbens and Lemieux [2008] in Figure 32.

The identifying assumptions are the same under fuzzy designs as they are under sharp designs: they are the continuity assumptions. For identification, we must assume that the conditional expectation of the potential outcomes (e.g., $E[Y^0 \mid X < c_0]$) is changing smoothly through c_0. What changes at c_0 is the treatment assignment probability. An illustration of this identifying assumption is in Figure 33.

Estimating some average treatment effect under a fuzzy RDD is very similar to how we estimate a local average treatment effect with

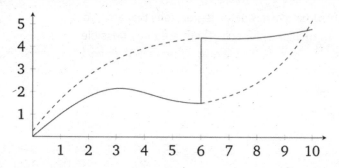

Figure 33. Potential and observed outcomes under a fuzzy design.

instrumental variables. I will cover instrumental variables in more detail later in the book, but for now let me tell you about estimation under fuzzy designs using IV. One can estimate several ways. One simple way is a type of Wald estimator, where you estimate some causal effect as the ratio of a reduced form difference in mean outcomes around the cutoff and a reduced form difference in mean treatment assignment around the cutoff.

$$\delta_{\text{Fuzzy RDD}} = \frac{\lim_{X \to c_0} E[Y \mid X = c_0] - \lim_{X_0 \leftarrow X} E[Y \mid X = c_0]}{\lim_{X \to c_0} E[D \mid X = c_0] - \lim_{X_0 \leftarrow X} E[D \mid X = c_0]} \quad (6.6)$$

The assumptions for identification here are the same as with any instrumental variables design: all the caveats about exclusion restrictions, monotonicity, SUTVA, and the strength of the first stage.[9]

But one can also estimate the effect using a two-stage least squares model or similar appropriate model such as limited-information maximum likelihood. Recall that there are now two events: the first event is when the running variable exceeds the cutoff, and the second event is when a unit is placed in the treatment. Let Z_i be an indicator for when X exceeds c_0. One can use both Z_i and the interaction terms as instruments for the treatment D_i. If one uses only Z_i as an instrumental variable, then it is a "just identified" model, which usually has good finite sample properties.

---

9 I discuss these assumptions and diagnostics in greater detail later in the chapter on instrument variables.

Let's look at a few of the regressions that are involved in this instrumental variables approach. There are three possible regressions: the first stage, the reduced form, and the second stage. Let's look at them in order. In the case just identified (meaning only one instrument for one endogenous variable), the first stage would be:

$$D_i = \gamma_0 + \gamma_1 X_i + \gamma_2 X_i^2 + \cdots + \gamma_p X_i^p + \pi Z_i + \zeta_{1i}$$

where π is the causal effect of Z_i on the conditional probability of treatment. The fitted values from this regression would then be used in a second stage. We can also use both Z_i and the interaction terms as instruments for D_i. If we used Z_i and all its interactions, the estimated first stage would be:

$$D_i = \gamma_{00} + \gamma_{01}\tilde{X}_i + \gamma_{02}\tilde{X}_i^2 + \cdots + \gamma_{0p}\tilde{X}_i^p + \pi Z_i + \gamma_1^*\tilde{X}_i Z_i + \gamma_2^*\tilde{X}_i Z_i + \cdots + \gamma_p^* Z_i + \zeta_{1i}$$

We would also construct analogous first stages for $\tilde{X}_i D_i, \ldots, \tilde{X}_i^p D_i$.

If we wanted to forgo estimating the full IV model, we might estimate the reduced form only. You'd be surprised how many applied people prefer to simply report the reduced form and not the fully specified instrumental variables model. If you read Hoekstra [2009], for instance, he favored presenting the reduced form—that second figure, in fact, was a picture of the reduced form. The reduced form would regress the outcome Y onto the instrument and the running variable. The form of this fuzzy RDD reduced form is:

$$Y_i = \mu + \kappa_1 X_i + \kappa_2 X_i^2 + \cdots + \kappa_p X_i^p + \delta\pi Z_i + \zeta_{2i}$$

As in the sharp RDD case, one can allow the smooth function to be different on both sides of the discontinuity by interacting Z_i with the running variable. The reduced form for this regression is:

$$Y_i = \mu + \kappa_{01} X_i \tilde{X}_i + \kappa_{02} X_i \tilde{X}_i^2 + \cdots + \kappa_{0p} X_i \tilde{X}_i^p$$
$$+ \delta\pi Z_i + \kappa_{01} X_i^* \tilde{X}_i Z_i + \kappa_{02} X_i^* \tilde{X}_i Z_i + \cdots + \kappa_{0p} X_i^* Z_i + \zeta_{1i}$$

But let's say you wanted to present the estimated effect of the treatment on some outcome. That requires estimating a first stage, using fitted values from that regression, and then estimating a second stage on those fitted values. This, and only this, will identify the causal effect of the treatment on the outcome of interest. The reduced form only estimates the causal effect of the instrument on the outcome. The second-stage model with interaction terms would be the same as before:

$$Y_i = \alpha + \beta_{01}\tilde{x}_i + \beta_{02}\tilde{x}_i^2 + \cdots + \beta_{0p}\tilde{x}_i^p$$
$$+ \delta\widehat{D}_i + \beta_1^*\widehat{D}_i\tilde{x}_i + \beta_2^*\widehat{D}_i\tilde{x}_i^2 + \cdots + \beta_p^*\widehat{D}_i\tilde{x}_i^p + \eta_i$$

Where $\tilde{x}$ are now not only normalized with respect to c_0 but are also fitted values obtained from the first-stage regressions.

As Hahn et al. [2001] point out, one needs the same assumptions for identification as one needs with IV. As with other binary instrumental variables, the fuzzy RDD is estimating the local average treatment effect (LATE) [Imbens and Angrist, 1994], which is the average treatment effect for the compliers. In RDD, the compliers are those whose treatment status changed as we moved the value of x_i from just to the left of c_0 to just to the right of c_0.

Challenges to Identification

The requirement for RDD to estimate a causal effect are the continuity assumptions. That is, the expected potential outcomes change smoothly as a function of the running variable through the cutoff. In words, this means that the only thing that causes the outcome to change abruptly at c_0 is the treatment. But, this can be violated in practice if any of the following is true:

1. The assignment rule is known in advance.
2. Agents are interested in adjusting.
3. Agents have time to adjust.
4. The cutoff is endogenous to factors that independently cause potential outcomes to shift.
5. There is nonrandom heaping along the running variable.

Examples include retaking an exam, self-reporting income, and so on. But some other unobservable characteristic change could happen at the threshold, and this has a direct effect on the outcome. In other words, the cutoff is endogenous. An example would be age thresholds used for policy, such as when a person turns 18 years old and faces more severe penalties for crime. This age threshold triggers the treatment (i.e., higher penalties for crime), but is also correlated with variables that affect the outcomes, such as graduating from high school and voting rights. Let's tackle these problems separately.

McCrary's density test. Because of these challenges to identification, a lot of work by econometricians and applied microeconomists has gone toward trying to figure out solutions to these problems. The most influential is a density test by Justin McCrary, now called the McCrary density test [2008]. The McCrary density test is used to check whether units are sorting on the running variable. Imagine that there are two rooms with patients in line for some life-saving treatment. Patients in room A will receive the life-saving treatment, and patients in room B will *knowingly* receive nothing. What would you do if you were in room B? Like me, you'd probably stand up, open the door, and walk across the hall to room A. There are natural incentives for the people in room B to get into room A, and the only thing that would keep people in room B from sorting into room A is if doing so were impossible.

But, let's imagine that the people in room B had successfully sorted themselves into room A. What would that look like to an outsider? If they were successful,. then room A would have more patients than room B. In fact, in the extreme, room A is crowded and room B is empty. This is the heart of the McCrary density test, and when we see such things at the cutoff, we have some suggestive evidence that people are sorting on the running variable. This is sometimes called manipulation.

Remember earlier when I said we should think of continuity as the null because nature doesn't make jumps? If you see a turtle on a fencepost, it probably didn't get there itself. Well, the same goes for the density. If the null is a continuous density through the cutoff, then bunching in the density at the cutoff is a sign that someone is moving over to the cutoff—probably to take advantage of the rewards that await there. Sorting on the sorting variable is a testable prediction under the null of a continuous density. Assuming a continuous distribution of units, sorting on the running variable means that units are moving just on the other side of the cutoff. Formally, if we assume a desirable treatment D and an assignment rule $X \geq c_0$, then we expect individuals will sort into D by choosing X such that $X \geq c_0$—so long as they're able. If they do, then it could imply selection bias insofar as their sorting is a function of potential outcomes.

The kind of test needed to investigate whether manipulation is occurring is a test that checks whether there is bunching of units at the cutoff. In other words, we need a *density test*. McCrary [2008] suggests a formal test where under the null, the density should be continuous at the cutoff point. Under the alternative hypothesis, the density should increase at the kink.[10] I've always liked this test because it's a really simple statistical test based on a theory that human beings are optimizing under constraints. And if they are optimizing, that makes for testable predictions—like a discontinuous jump in the density at the cutoff. Statistics built on behavioral theory can take us further.

To implement the McCrary density test, partition the assignment variable into bins and calculate frequencies (i.e., the number of observations) in each bin. Treat the frequency counts as the dependent variable in a local linear regression. If you can estimate the conditional expectations, then you have the data on the running variable, so in principle you can always do a density test. I recommend the package rddensity,[11] which you can install for R as well.[12] These packages are based on Cattaneo et al. [2019], which is based on local polynomial regressions that have less bias in the border regions.

This is a high-powered test. You need a lot of observations at c_0 to distinguish a discontinuity in the density from noise. Let me illustrate in Figure 34 with a picture from McCrary [2008] that shows a situation with and without manipulation.

Covariate balance and other placebos. It has become common in this literature to provide evidence for the credibility of the underlying identifying assumptions, at least to some degree. While the assumptions cannot be directly tested, indirect evidence may be persuasive. I've

---

10 In those situations, anyway, where the treatment is desirable to the units.

11 https://sites.google.com/site/rdpackages/rddensity.

12 http://cran.r-project.org/web/packages/rdd/rdd.eps.

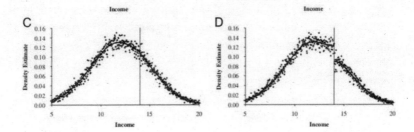

Figure 34. A picture with and without a discontinuity in the density. Reprinted from *Journal of Econometrics*, 142, J. McCrary, "Manipulation of the Running Variable in the Regression Discontinuity Design: A Design Test," 698–714. © 2008, with permission from Elsevier.

already mentioned one such test—the McCrary density test. A second test is a covariate balance test. For RDD to be valid in your study, there must not be an observable discontinuous change in the average values of reasonably chosen covariates around the cutoff. As these are pretreatment characteristics, they should be invariant to change in treatment assignment. An example of this is from Lee et al. [2004], who evaluated the impact of Democratic vote share just at 50%, on various demographic factors (Figure 35).

This test is basically what is sometimes called a *placebo* test. That is, you are looking for there to be no effects where there shouldn't be any. So a third kind of test is an extension of that—just as there shouldn't be effects at the cutoff on pretreatment values, there shouldn't be effects on the outcome of interest at arbitrarily chosen cutoffs. Imbens and Lemieux [2008] suggest looking at one side of the discontinuity, taking the median value of the running variable in that section, and pretending it was a discontinuity, c_0'. Then test whether there is a discontinuity in the outcome at c_0'. You do *not* want to find anything.

Nonrandom heaping on the running variable. Almond et al. [2010] is a fascinating study. The authors are interested in estimating the causal effect of medical expenditures on health outcomes, in part because many medical technologies, while effective, may not justify the costs associated with their use. Determining their effectiveness is challenging given that medical resources are, we hope, optimally assigned to

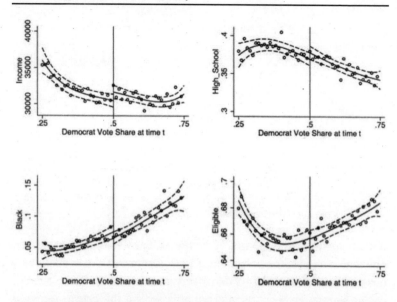

Figure 35. Panels refer to (top left to bottom right) district characteristics: real income, percentage high school degree, percentage black, and percentage eligible to vote. Circles represent the average characteristic within intervals of 0.01 in Democratic vote share. The continuous line represents the predicted values from a fourth-order polynomial in vote share fitted separately for points above and below the 50% threshold. The dotted line represents the 95% confidence interval. Reprinted from Lee, D. S., Moretti, E., and Butler, M. J. (2004). "Do Voters Affect or Elect Policies: Evidence from the U.S. House." *Quarterly Journal of Economics*, 119(3):807–859. Permission from Oxford University Press.

patients based on patient potential outcomes. To put it a different way, if the physician perceives that an intervention will have the best outcome, then that is likely a treatment that will be assigned to the patient. This violates independence, and more than likely, if the endogeneity of the treatment is deep enough, controlling for selection directly will be tough, if not impossible. As we saw with our earlier example of the perfect doctor, such nonrandom assignment of interventions can lead to confusing correlations. Counterintuitive correlations may be nothing more than selection bias.

But Almond et al. [2010] had an ingenious insight—in the United States, it is typically the case that babies with a very low birth weight receive heightened medical attention. This categorization is called the "very low birth weight" range, and such low birth weight is quite dangerous for the child. Using administrative hospital records linked to mortality data, the authors find that the 1-year infant mortality decreases by around 1 percentage point when the child's birth weight is just below the 1,500-gram threshold compared to those born just above. Given the mean 1-year mortality of 5.5%, this estimate is sizable, suggesting that the medical interventions triggered by the very-low-birth-weight classification have benefits that far exceed their costs.

Barreca et al. [2011] and Barreca et al. [2016] highlight some of econometric issues related to what they call "heaping" on the running variable. Heaping is when there is an excess number of units at certain points along the running variable. In this case, it appeared to be at regular 100-gram intervals and was likely caused by a tendency for hospitals to round to the nearest integer. A visualization of this problem can be seen in the original Almond et al. [2010], which I reproduce here in Figure 36. The long black lines appearing regularly across the birthweight distribution are excess mass of children born at those numbers. This sort of event is unlikely to occur naturally in nature, and it is almost certainly caused by either sorting or rounding. It could be due to less sophisticated scales or, more troubling, to staff rounding a child's birth weight to 1,500 grams in order to make the child eligible for increased medical attention.

Almond et al. [2010] attempt to study this more carefully using the conventional McCrary density test and find no clear, statistically significant evidence for sorting on the running variable at the 1,500-gram cutoff. Satisfied, they conduct their main analysis, in which they find a causal effect of around a 1-percentage-point reduction in 1-year mortality.

The focus of Barreca et al. [2011] and Barreca et al. [2016] is very much on the heaping phenomenon shown in Figure 36. Part of the strength of their work, though, is their illustration of some of the shortcomings of a conventional McCrary density test. In this case, the data heap at 1,500 grams appears to be babies whose mortality rates are unusually high. These children are outliers compared to units to *both* the immediate left and the immediate right. It is important to note that such events would not occur naturally; there is no reason to believe that

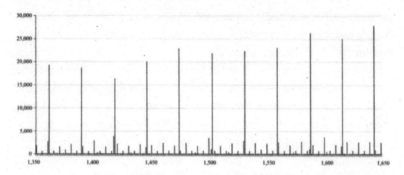

Figure 36. Distribution of births by gram. Reprinted from Almond, D., Doyle, J. J., Kowalski, A., and Williams, H. (2010). "Estimating Returns to Medical Care: Evidence from at-risk Newborns." *The Quarterly Journal of Economics*, 125(2):591–634. Permission from Oxford University Press.

nature would produce heaps of children born with outlier health defects every 100 grams. The authors comment on what might be going on:

> This [heaping at 1,500 grams] may be a signal that poor-quality hospitals have relatively high propensities to round birth weights but is also consistent with manipulation of recorded birth weights by doctors, nurses, or parents to obtain favorable treatment for their children. Barreca et al. [2011] show that this nonrandom heaping leads one to conclude that it is "good" to be strictly less than any 100-g cutoff between 1,000 and 3,000 grams.

Since estimation in an RDD compares means as we approach the threshold from either side, the estimates should not be sensitive to the observations at the thresholds itself. Their solution is a so-called "donut hole" RDD, wherein they remove units in the vicinity of 1,500 grams and reestimate the model. Insofar as units are dropped, the parameter we are estimating at the cutoff has become an even more unusual type of local average treatment effect that may be even less informative about the average treatment effects that policy makers are desperate to know. But the strength of this rule is that it allows for the possibility that units at the heap differ markedly due to selection bias from those in the surrounding area. Dropping these units reduces the sample size by around 2% but has very large effects on 1-year mortality, which is approximately 50% lower than what was found by Almond et al. [2010].

These companion papers help us better understand some of the ways in which selection bias can creep into the RDD. Heaping is not the end of the world, which is good news for researchers facing such a problem. The donut hole RDD can be used to circumvent some of the problems. But ultimately this solution involves dropping observations, and insofar as your sample size is small relative to the number of heaping units, the donut hole approach could be infeasible. It also changes the parameter of interest to be estimated in ways that may be difficult to understand or explain. Caution with nonrandom heaping along the running variable is probably a good thing.

Replicating a Popular Design: The Close Election

Within RDD, there is a particular kind of design that has become quite popular, the close-election design. Essentially, this design exploits a feature of American democracies wherein winners in political races are declared when a candidate gets the minimum needed share of votes. Insofar as very close races represent exogenous assignments of a party's victory, which I'll discuss below, then we can use these close elections to identify the causal effect of the winner on a variety of outcomes. We may also be able to test political economy theories that are otherwise nearly impossible to evaluate.

The following section has two goals. First, to discuss in detail the close election design using the classic Lee et al. [2004]. Second, to show how to implement the close-election design by replicating several parts of Lee et al. [2004].

Do Politicians or Voters Pick Policies? The big question motivating Lee et al. (2004) has to do with whether and in which way voters affect policy. There are two fundamentally different views of the role of elections in a representative democracy: convergence theory and divergence theory.

The convergence theory states that heterogeneous voter ideology forces each candidate to moderate his or her position (e.g., similar to the median voter theorem):

> Competition for votes can force even the most partisan Republicans and Democrats to moderate their policy choices. In the extreme case, competition may be so strong that it leads to "full policy convergence": opposing parties are forced to adopt identical policies. [Lee et al. 2004, 807]

Divergence theory is a slightly more commonsense view of political actors. When partisan politicians cannot credibly commit to certain policies, then convergence is undermined and the result can be full policy "divergence." Divergence is when the winning candidate, after taking office, simply pursues her most-preferred policy. In this extreme case, voters are unable to compel candidates to reach any kind of policy compromise, and this is expressed as two opposing candidates choosing very different policies under different counterfactual victory scenarios.

Lee et al. [2004] present a model, which I've simplified. Let R and D be candidates in a congressional race. The policy space is a single dimension where D's and R's policy preferences in a period are quadratic loss functions, $u(l)$ and $v(l)$, and l is the policy variable. Each player has some bliss point, which is his or her most preferred location along the unidimensional policy range. For Democrats, it's $l^* = c(> 0)$, and for Republicans it's $l_* = 0$. Here's what this means.

Ex ante, voters expect the candidate to choose some policy and they expect the candidate to win with probability $P(x^e, y^e)$, where x^e and y^e are the policies chosen by Democrats and Republicans, respectively. When $x^e > y^e$, then $\frac{\partial P}{\partial x^e} > 0, \frac{\partial P}{\partial y^e} < 0$.

P^* represents the underlying popularity of the Democratic Party, or put differently, the probability that D would win if the policy chosen x equaled the Democrat's bliss point c.

The solution to this game has multiple Nash equilibria, which I discuss now.

1. Partial/complete convergence: Voters affect policies.
   - The key result under this equilibrium is $\frac{\partial x^*}{\partial P^*} > 0$.
   - Interpretation: If we dropped more Democrats into the district from a helicopter, it would exogenously increase P^* and this would result in candidates changing their policy positions, i.e., $\frac{\partial x^*}{\partial P^*} > 0$.
2. Complete divergence: Voters elect politicians with fixed policies who do whatever they want to do.[13]
   - Key result is that more popularity has no effect on policies. That is, $\frac{\partial x^*}{\partial P^*} = 0$.
   - An exogenous shock to P^* (i.e., dropping Democrats into the district) does *nothing* to equilibrium policies. Voters elect politicians who then do whatever they want because of their fixed policy preferences.

The potential roll-call voting record outcomes of the candidate following some election is

$$RC_t = D_t x_t + (1 - D_t) y_t$$

where D_t indicates whether a Democrat won the election. That is, only the winning candidate's policy is observed. This expression can be transformed into regression equations:

$$RC_t = \alpha_0 + \pi_0 P_t^* + \pi_1 D_t + \varepsilon_t$$
$$RC_{t+1} = \beta_0 + \pi_0 P_{t+1}^* + \pi_1 D_{t+1} + \varepsilon_{t+1}$$

where α_0 and β_0 are constants.

This equation can't be directly estimated because we never observe P^*. But suppose we could randomize D_t. Then D_t would be independent of P_t^* and ε_t. Then taking conditional expectations with respect to D_t, we get:

---

13 The honey badger doesn't care. It takes what it wants. See https://www.youtube.com/watch?v=4r7wHMg5Yjg.

$$\underbrace{E[RC_{t+1} \mid D_t = 1] - E[RC_{t+1} \mid D_t = 0]}_{\text{Observable}} = \pi_0[P_{t+1}^{*D} - P_{t+1}^{*R}]$$

$$+ \underbrace{\pi_1[P_{t+1}^{D} - P_{t+1}^{R}]}_{\text{Observable}} \qquad (6.7)$$

$$= \underbrace{\gamma}_{\text{Total effect of initial win on future roll call votes}}$$

$$\underbrace{E[RC_t \mid D_t = 1] - E[RC_t \mid D_t = 0]}_{\text{Observable}} = \pi_1 \qquad (6.8)$$

$$\underbrace{E[D_{t+1} \mid D_t = 1] - E[D_{t+1} \mid D_t = 0]}_{\text{Observable}} = P_{t+1}^{D} - P_{t+1}^{R} \qquad (6.9)$$

The "elect" component is $\pi_1[P_{t+1}^{D} - P_{t+1}^{R}]$ and is estimated as the difference in mean voting records between the parties at time t. The fraction of districts won by Democrats in $t+1$ is an estimate of $[P_{t+1}^{D} - P_{t+1}^{R}]$. Because we can estimate the total effect, γ, of a Democrat victory in t on RC_{t+1}, we can net out the elect component to implicitly get the "effect" component.

But random assignment of D_t is crucial. For without it, this equation would reflect π_1 *and* selection (i.e., Democratic districts have more liberal bliss points). So the authors aim to randomize D_t using a RDD, which I'll now discuss in detail.

Replication exercise. There are two main data sets in this project. The first is a measure of how liberal an official voted. This is collected from the Americans for Democratic Action (ADA) linked with House of Representatives election results for 1946–1995. Authors use the ADA score for all US House representatives from 1946 to 1995 as their voting record index. For each Congress, the ADA chose about twenty-five high-profile roll-call votes and created an index varying from 0 to 100 for each representative. Higher scores correspond to a more "liberal" voting record. The running variable in this study is the vote share. That is the share of all votes that went to a Democrat. ADA scores are then linked to election returns data during that period.

Recall that we need randomization of D_t. The authors have a clever solution. They will use arguably exogenous variation in Democratic

Table 45. Original results based on ADA scores—close elections sample.

Dependent variable	ADA_{t+1}	ADA_t	DEM_{t+1}
Estimated gap	21.2	47.6	0.48
	(1.9)	(1.3)	(0.02)

Note: Standard errors in parentheses. The unit of observation is a district-congressional session. The sample includes only observations where the Democrat vote share at time t is strictly between 48% and 52%. The estimated gap is the difference in the average of the relevant variable for observations for which the Democrat vote share at time t is strictly between 50% and 52% and observations for which the Democrat vote share at time t is strictly between 48% and 50%. Time t and $t+1$ refer to congressional sessions. ADA_t is the adjusted ADA voting score. Higher ADA scores correspond to more liberal roll-call voting records. Sample size is 915.

wins to check whether convergence or divergence is correct. If convergence is true, then Republicans and Democrats who just barely won should vote almost identically, whereas if divergence is true, they should vote differently at the margins of a close race. This "at the margins of a close race" is crucial because the idea is that it is at the margins of a close race that the distribution of voter preferences is the same. And if voter preferences are the same, but policies diverge at the cutoff, then it suggests politicians and not voters are driving policy making.

The exogenous shock comes from the discontinuity in the running variable. At a vote share of just above 0.5, the Democratic candidate wins. They argue that just around that cutoff, random chance determined the Democratic win—hence the random assignment of D_t [Cattaneo et al., 2015]. Table 45 is a reproduction of Cattaneo et al.'s main results. The effect of a Democratic victory increases liberal voting by 21 points in the next period, 48 points in the current period, and the probability of reelection by 48%. The authors find evidence for both divergence and incumbency advantage using this design. Let's dig into the data ourselves now and see if we can find where the authors are getting these results. We will examine the results around Table 45 by playing around with the data and different specifications.

STATA

lmb_1.do

```
1   use https://github.com/scunning1975/mixtape/raw/master/lmb-data.dta, clear
2   * Stata code attributed to Marcelo Perraillon.
3   * Replicating Table 1 of Lee, Moretti and Butler (2004)
4   reg score lagdemocrat   if lagdemvoteshare>.48 & lagdemvoteshare<.52,
    ↪  cluster(id)
5   reg score democrat      if lagdemvoteshare>.48 & lagdemvoteshare<.52,
    ↪  cluster(id)
6   reg democrat lagdemocrat if lagdemvoteshare>.48 & lagdemvoteshare<.52,
    ↪  cluster(id)
```

R

lmb_1.R

```
1   library(tidyverse)
2   library(haven)
3   library(estimatr)
4
5   read_data <- function(df)
6   {
7    full_path <- paste("https://raw.github.com/scunning1975/mixtape/master/",
8             df, sep = "")
9    df <- read_dta(full_path)
10   return(df)
11  }
12
13  lmb_data <- read_data("lmb-data.dta")
14
15  lmb_subset <- lmb_data %>%
16   filter(lagdemvoteshare>.48 & lagdemvoteshare<.52)
17
18  lm_1 <- lm_robust(score ~ lagdemocrat, data = lmb_subset, clusters = id)
19  lm_2 <- lm_robust(score ~ democrat, data = lmb_subset, clusters = id)
20  lm_3 <- lm_robust(democrat ~ lagdemocrat, data = lmb_subset, clusters = id)
21
22  summary(lm_1)
23  summary(lm_2)
24  summary(lm_3)
```

Table 46. Replicated results based on ADA scores—close elections sample.

Dependent variable	ADA_{t+1}	ADA_t	DEM_{t+1}
Estimated gap	21.28***	47.71***	0.48***
	(1.95)	(1.36)	(0.03)
N	915	915	915

Note: Cluster robust standard errors in parentheses. * $p<0.10$, ** $p<0.05$, *** $p<0.01$

We reproduce regression results from Lee, Moretti, and Butler in Table 46. While the results are close to Lee, Moretti, and Butler's original table, they are slightly different. But ignore that for now. The main thing to see is that we used regressions limited to the window right around the cutoff to estimate the effect. These are local regressions in the sense that they use data close to the cutoff. Notice the window we chose—we are only using observations between 0.48 and 0.52 vote share. So this regression is estimating the coefficient on D_t right around the cutoff. What happens if we use all the data?

STATA
lmb_2.do

```
1   * Stata code attributed to Marcelo Perraillon.
2   reg score lagdemocrat, cluster(id)
3   reg score democrat, cluster(id)
4   reg democrat lagdemocrat, cluster(id)
```

R
lmb_2.R

```
1   #using all data (note data used is lmb_data, not lmb_subset)
2
3   lm_1 <- lm_robust(score ~ lagdemocrat, data = lmb_data, clusters = id)
4   lm_2 <- lm_robust(score ~ democrat, data = lmb_data, clusters = id)
5   lm_3 <- lm_robust(democrat ~ lagdemocrat, data = lmb_data, clusters = id)
6
7   summary(lm_1)
8   summary(lm_2)
9   summary(lm_3)
```

Table 47. Results based on ADA scores—full sample.

Dependent variable	ADA_{t+1}	ADA_t	DEM_{t+1}
Estimated gap	31.50***	40.76***	0.82***
	(0.48)	(0.42)	(0.01)
N	13,588	13,588	13,588

Note: Cluster robust standard errors in parentheses. * $p<0.10$, ** $p<0.05$, *** $p<0.01$

Notice that when we use all of the data, we get somewhat different effects (Table 47). The effect on future ADA scores gets larger by 10 points, but the contemporaneous effect gets smaller. The effect on incumbency, though, increases considerably. So here we see that simply running the regression yields different estimates when we include data far from the cutoff itself.

Neither of these regressions included controls for the running variable though. It also doesn't use the recentering of the running variable. So let's do both. We will simply subtract 0.5 from the running variable so that values of 0 are where the vote share equals 0.5, negative values are Democratic vote shares less than 0.5, and positive values are Democratic vote shares above 0.5. To do this, type in the following lines:

STATA
lmb_3.do

```
1  * Re-center the running variable. Stata code attributed to Marcelo Perraillon.
2  gen demvoteshare_c = demvoteshare - 0.5
3  reg score lagdemocrat demvoteshare_c, cluster(id)
4  reg score democrat demvoteshare_c, cluster(id)
5  reg democrat lagdemocrat demvoteshare_c, cluster(id)
```

```
                                    R
                              Imb_3.R
1   Imb_data <- Imb_data %>%
2     mutate(demvoteshare_c = demvoteshare - 0.5)
3
4   lm_1 <- lm_robust(score ~ lagdemocrat + demvoteshare_c, data = Imb_data,
    ↪  clusters = id)
5   lm_2 <- lm_robust(score ~ democrat + demvoteshare_c, data = Imb_data, clusters
    ↪  = id)
6   lm_3 <- lm_robust(democrat ~ lagdemocrat + demvoteshare_c, data = Imb_data,
    ↪  clusters = id)
7
8   summary(lm_1)
9   summary(lm_2)
10  summary(lm_3)
11
```

We report our analysis from the programming in Table 48. While the incumbency effect falls closer to what Lee et al. [2004] find, the effects are still quite different.

It is common, though, to allow the running variable to vary on either side of the discontinuity, but how exactly do we implement that? Think of it—we need for a regression line to be on either side, which means necessarily that we have *two* lines left and right of the discontinuity. To do this, we need an interaction—specifically an interaction of the running variable with the treatment variable. To implement this in Stata, we can use the code shown in lmb_4.do.

Table 48. Results based on ADA scores—full sample.

Dependent variable	ADA_{t+1}	ADA_t	DEM_{t+1}
Estimated gap	33.45***	58.50***	0.55***
	(0.85)	(0.66)	(0.01)
N	13,577	13,577	13,577

Note: Cluster robust standard errors in parentheses. *$p < 0.10$. **$p < 0.05$. ***$p < 0.01$.

STATA
lmb_4.do

```
1   * Stata code attributed to Marcelo Perraillon.
2   xi: reg score i.lagdemocrat*demvoteshare_c, cluster(id)
3   xi: reg score i.democrat*demvoteshare_c, cluster(id)
4   xi: reg democrat i.lagdemocrat*demvoteshare_c, cluster(id)
```

R
lmb_4.R

```
1   lm_1 <- lm_robust(score ~ lagdemocrat*demvoteshare_c,
2           data = lmb_data, clusters = id)
3   lm_2 <- lm_robust(score ~ democrat*demvoteshare_c,
4           data = lmb_data, clusters = id)
5   lm_3 <- lm_robust(democrat ~ lagdemocrat*demvoteshare_c,
6           data = lmb_data, clusters = id)
7
8   summary(lm_1)
9   summary(lm_2)
10  summary(lm_3)
11
```

In Table 49, we report the global regression analysis with the running variable interacted with the treatment variable. This pulled down the coefficients somewhat, but they remain larger than what was found when we used only those observations within 0.02 points of the 0.5. Finally, let's estimate the model with a quadratic.

STATA
lmb_5.do

```
1   * Stata code attributed to Marcelo Perraillon.
2   gen demvoteshare_sq = demvoteshare_c^2
3   xi: reg score lagdemocrat##c.(demvoteshare_c demvoteshare_sq), cluster(id)
4   xi: reg score democrat##c.(demvoteshare_c demvoteshare_sq), cluster(id)
5   xi: reg democrat lagdemocrat##c.(demvoteshare_c demvoteshare_sq),
    ↪   cluster(id)
```

```
                                    R
                              Imb_5.R
1    Imb_data %>%
2      mutate(demvoteshare_sq = demvoteshare_c^2)
3
4    lm_1 <- lm_robust(score ~ lagdemocrat*demvoteshare_c +
     ↪   lagdemocrat*demvoteshare_sq,
5              data = Imb_data, clusters = id)
6    lm_2 <- lm_robust(score ~ democrat*demvoteshare_c +
     ↪   democrat*demvoteshare_sq,
7              data = Imb_data, clusters = id)
8    lm_3 <- lm_robust(democrat ~ lagdemocrat*demvoteshare_c +
     ↪   lagdemocrat*demvoteshare_sq,
9              data = Imb_data, clusters = id)
10
11   summary(lm_1)
12   summary(lm_2)
13   summary(lm_3)
14
```

Including the quadratic causes the estimated effect of a democratic victory on future voting to fall considerably (see Table 50). The effect on contemporaneous voting is smaller than what Lee et al. [2004] find, as is the incumbency effect. But the purpose here is simply to illustrate the standard steps using global regressions.

But notice, we are still estimating *global* regressions. And it is for that reason that the coefficient is larger. This suggests that there exist strong outliers in the data that are causing the distance at c_0 to spread more widely. So a natural solution is to again limit our analysis to a smaller window. What this does is drop the observations far away from c_0 and omit the influence of outliers from our estimation at the cutoff. Since we used $+/- -0.02$ last time, we'll use $+/- -0.05$ this time just to mix things up.

Table 49. Results based on ADA scores—full sample with linear interactions.

Dependent variable	ADA_{t+1}	ADA_t	DEM_{t+1}
Estimated gap	30.51***	55.43 ***	0.53***
	(0.82)	(0.64)	(0.01)
N	13,577	13,577	13,577

Note: Cluster robust standard errors in parentheses. *$p < 0.10$, **$p < 0.05$, ***$p < 0.01$

Table 50. Results based on ADA scores—full sample with linear and quadratic interactions.

Dependent variable	ADA_{t+1}	ADA_t	DEM_{t+1}
Estimated gap	13.03***	44.40 ***	0.32***
	(1.27)	(0.91)	(1.74)
N	13,577	13,577	13,577

Note: Cluster robust standard errors in parentheses. $*p < 0.10$, $**p < 0.05$, $***p < 0.01$

STATA
lmb_6.do

```
1   * Use 5 points from cutoff. Stata code attributed to Marcelo Perraillon.
2   xi: reg score lagdemocrat##c.(demvoteshare_c demvoteshare_sq) if
    ↪ lagdemvoteshare>.45 & lagdemvoteshare<.55, cluster(id)
3   xi: reg score democrat##c.(demvoteshare_c demvoteshare_sq) if
    ↪ lagdemvoteshare>.45 & lagdemvoteshare<.55, cluster(id)
4   xi: reg democrat lagdemocrat##c.(demvoteshare_c demvoteshare_sq) if
    ↪ lagdemvoteshare>.45 & lagdemvoteshare<.55, cluster(id)
```

R
lmb_6.R

```
1    lmb_data %>%
2      filter(demvoteshare > .45 & demvoteshare < .55) %>%
3      mutate(demvoteshare_sq = demvoteshare_c^2)
4
5    lm_1 <- lm_robust(score ~ lagdemocrat*demvoteshare_c +
     ↪ lagdemocrat*demvoteshare_sq,
6            data = lmb_data, clusters = id)
7    lm_2 <- lm_robust(score ~ democrat*demvoteshare_c +
     ↪ democrat*demvoteshare_sq,
8            data = lmb_data, clusters = id)
9    lm_3 <- lm_robust(democrat ~ lagdemocrat*demvoteshare_c +
     ↪ lagdemocrat*demvoteshare_sq,
10           data = lmb_data, clusters = id)
11
12   summary(lm_1)
13   summary(lm_2)
14   summary(lm_3)
15
```

Table 51. Results based on ADA scores—close election sample with linear and quadratic interactions.

Dependent variable	ADA_{t+1}	ADA_t	DEM_{t+1}
Estimated gap	3.97***	46.88***	0.12***
	(1.49)	(1.54)	(0.02)
N	2,441	2,441	2,441

Note: Cluster robust standard errors in parentheses. $*p < 0.10$, $**p < 0.05$, $***p < 0.01$

As can be seen in Table 51, when we limit our analysis to $+/-$ 0.05 around the cutoff, we are using more observations away from the cutoff than we used in our initial analysis. That's why we only have 2,441 observations for analysis as opposed to the 915 we had in our original analysis. But we also see that including the quadratic interaction pulled the estimated size on future voting down considerably, even when using the smaller sample.

But putting that aside, let's talk about all that we just did. First we fit a model without controlling for the running variable. But then we included the running variable, introduced in a variety of ways. For instance, we interacted the variable of Democratic vote share with the democratic dummy, as well as including a quadratic. In all this analysis, we extrapolated trends lines from the running variable beyond the support of the data to estimate local average treatment effects right at the cutoff.

But we also saw that the inclusion of the running variable in any form tended to reduce the effect of a victory for Democrats on future Democratic voting patterns, which was interesting. Lee et al. [2004] original estimate of around 21 is attenuated considerably when we include controls for the running variable, even when we go back to estimating very local flexible regressions. While the effect remains significant, it is considerably smaller, whereas the immediate effect remains quite large.

But there are still other ways to explore the impact of the treatment at the cutoff. For instance, while Hahn et al. [2001] clarified assumptions about RDD—specifically, continuity of the conditional expected potential outcomes—they also framed estimation as a nonparametric problem and emphasized using local polynomial regressions. What exactly does this mean though in practice?

Nonparametric methods mean a lot of different things to different people in statistics, but in RDD contexts, the idea is to estimate a model that doesn't assume a functional form for the relationship between the outcome variable (Y) and the running variable (X). The model would be something like this:

$$Y = f(X) + \varepsilon$$

A very basic method would be to calculate $E[Y]$ for each bin on X, like a histogram. And Stata has an option to do this called cmogram, created by Christopher Robert. The program has a lot of useful options, and we can re-create important figures from Lee et al. [2004]. Figure 37 shows the relationship between the Democratic win (as a function of the running variable, Democratic vote share) and the candidates, second-period ADA score.

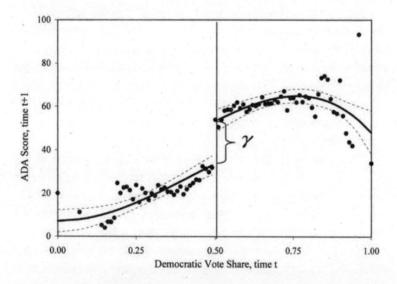

Figure 37. Showing total effect of initial win on future ADA scores. Reprinted from Lee, D. S., Moretti, E., and Butler, M. J. (2004). "Do Voters Affect or Elect Policies: Evidence from the U.S. House." *Quarterly Journal of Economics*, 119(3):807–859. Permission from Oxford University Press.

To reproduce this, there are a few options. You could manually create this figure yourself using either the "twoway" command in Stata or "ggplot" in R. But I'm going to show you using the canned cmogram routine that was created, as it's a quick-and-dirty way to get some information about the data.

STATA

lmb_7.do

```
1   * Stata code attributed to Marcelo Perraillon.
2   ssc install cmogram
3   cmogram score lagdemvoteshare, cut(0.5) scatter line(0.5) qfitci
4   cmogram score lagdemvoteshare, cut(0.5) scatter line(0.5) lfit
5   cmogram score lagdemvoteshare, cut(0.5) scatter line(0.5) lowess
```

R

lmb_7.R

```
1   #aggregating the data
2   categories <- lmb_data$lagdemvoteshare
3
4   demmeans <- split(lmb_data$score, cut(lmb_data$lagdemvoteshare, 100)) %>%
5    lapply(mean) %>%
6    unlist()
7
8   agg_lmb_data <- data.frame(score = demmeans, lagdemvoteshare = seq(0.01,1,
    ↪   by = 0.01))
9
10  #plotting
11  lmb_data <- lmb_data %>%
12   mutate(gg_group = case_when(lagdemvoteshare > 0.5 ~ 1, TRUE ~ 0))
13
14  ggplot(lmb_data, aes(lagdemvoteshare, score)) +
15   geom_point(aes(x = lagdemvoteshare, y = score), data = agg_lmb_data) +
16   stat_smooth(aes(lagdemvoteshare, score, group = gg_group), method = "lm",
17        formula = y ~ x + I(x^2)) +
18   xlim(0,1) + ylim(0,100) +
19   geom_vline(xintercept = 0.5)
20
```

(continued)

R (continued)

```
21   ggplot(lmb_data, aes(lagdemvoteshare, score)) +
22     geom_point(aes(x = lagdemvoteshare, y = score), data = agg_lmb_data) +
23     stat_smooth(aes(lagdemvoteshare, score, group = gg_group), method = "loess")
     ↪  +
24     xlim(0,1) + ylim(0,100) +
25     geom_vline(xintercept = 0.5)
26
27   ggplot(lmb_data, aes(lagdemvoteshare, score)) +
28     geom_point(aes(x = lagdemvoteshare, y = score), data = agg_lmb_data) +
29     stat_smooth(aes(lagdemvoteshare, score, group = gg_group), method = "lm") +
30     xlim(0,1) + ylim(0,100) +
31     geom_vline(xintercept = 0.5)
```

Figure 38 shows the output from this program. Notice the similarities between what we produced here and what Lee et al. [2004] produced in their figure. The only differences are subtle changes in the binning used for the two figures.

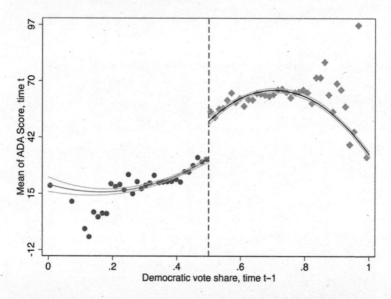

Figure 38. Using cmogram with quadratic fit and confidence intervals. Reprinted from Lee, D. S., Moretti, E., and Butler, M. J. (2004). "Do Voters Affect or Elect Policies: Evidence from the U.S. House." *Quarterly Journal of Economics*, 119(3):807–859.

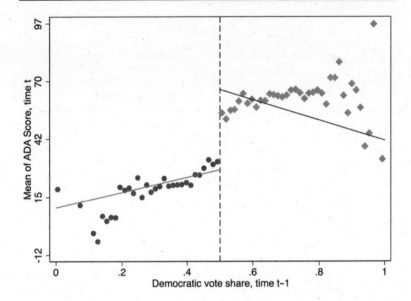

Figure 39. Using cmogram with linear fit. Reprinted from Lee, D. S., Moretti, E., and Butler, M. J. (2004). "Do Voters Affect or Elect Policies: Evidence from the U.S.
House." *Quarterly Journal of Economics*, 119(3):807–859.

We have options other than a quadratic fit, though, and it's useful to
compare this graph with one in which we only fit a linear model. Now,
because there are strong trends in the running variable, we probably
just want to use the quadratic, but let's see what we get when we use
simpler straight lines.

Figure 39 shows what we get when we only use a linear fit of the
data left and right of the cutoff. Notice the influence that outliers far
from the actual cutoff play in the estimate of the causal effect at the
cutoff. Some of this would go away if we restricted the bandwidth to
be shorter distances to and from the cutoff, but I leave it to you to
do that.

Finally, we can use a lowess fit. A lowess fit more or less crawls
through the data and runs small regression on small cuts of data. This
can give the figure a zigzag appearance. We nonetheless show it in
Figure 40.

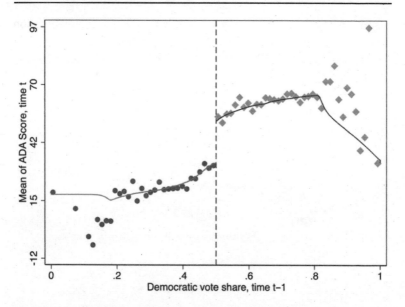

Figure 40. Using cmogram with lowess fit. Reprinted from Lee, D. S., Moretti, E., and Butler, M. J. (2004). "Do Voters Affect or Elect Policies: Evidence from the U.S.
House." *Quarterly Journal of Economics*, 119(3):807–859.

If there don't appear to be any trends in the running variable, then
the polynomials aren't going to buy you much. Some very good papers
only report a linear fit because there weren't very strong trends to begin
with. For instance, consider Carrell et al. [2011]. Those authors are
interested in the causal effect of drinking on academic test outcomes
for students at the Air Force Academy. Their running variable is the
precise age of the student, which they have because they know the
student's date of birth and they know the date of every exam taken at
the Air Force Academy. Because the Air Force Academy restricts students' social life, there is a starker increase in drinking at age 21 on its
campus than might be the case for a more a typical university campus.
They examined the causal effect of drinking age on normalized grades
using RDD, but because there weren't strong trends in the data, they
presented a graph with only a linear fit. Your choice should be in large
part based on what, to your eyeball, is the best fit of the data.

Hahn et al. [2001] have shown that one-sided kernel estimation such as lowess may suffer from poor properties because the point of interest is at the boundary (i.e., the discontinuity). This is called the "boundary problem." They propose using local linear nonparametric regressions instead. In these regressions, more weight is given to the observations at the center.

You can also estimate kernel-weighted local polynomial regressions. Think of it as a weighted regression restricted to a window like we've been doing (hence the word "local") where the chosen kernel provides the weights. A rectangular kernel would give the same results as $E[Y]$ at a given bin on X, but a triangular kernel would give more importance to observations closest to the center. This method will be sensitive to the size of the bandwidth chosen. But in that sense, it's similar to what we've been doing. Figure 41 shows this visually.

STATA
lmb_8.do

```
1   * Stata code attributed to Marcelo Perraillon.
2   capture drop sdem* x1 x0
3   lpoly score demvoteshare if democrat == 0, nograph kernel(triangle) gen(x0
    ↪  sdem0) bwidth(0.1)}
4   lpoly score demvoteshare if democrat == 1, nograph kernel(triangle) gen(x1
    ↪  sdem1) bwidth(0.1)}
5   scatter sdem1 x1, color(red) msize(small) || scatter sdem0 x0, msize(small)
    ↪  color(red) xline(0.5,lstyle(dot)) legend(off) xtitle("Democratic vote share")
    ↪  ytitle("ADA score")
6
```

R
lmb_8.R

```
1   library(tidyverse)
2   library(stats)
3
4   smooth_dem0 <- lmb_data %>%
5     filter(democrat == 0) %>%
6     select(score, demvoteshare)
7   smooth_dem0 <- as_tibble(ksmooth(smooth_dem0$demvoteshare,
    ↪  smooth_dem0$score,
8              kernel = "box", bandwidth = 0.1))
```

(continued)

R *(continued)*

```
9
10
11   smooth_dem1 <- lmb_data %>%
12     filter(democrat == 1) %>%
13     select(score, demvoteshare) %>%
14     na.omit()
15   smooth_dem1 <- as_tibble(ksmooth(smooth_dem1$demvoteshare,
     ↪   smooth_dem1$score,
16                      kernel = "box", bandwidth = 0.1))
17
18   ggplot() +
19     geom_smooth(aes(x, y), data = smooth_dem0) +
20     geom_smooth(aes(x, y), data = smooth_dem1) +
21     geom_vline(xintercept = 0.5)
22
23
24
25
```

A couple of final things. First, recall the continuity assumption. Because the continuity assumption specifically involves continuous conditional expectation functions of the potential outcomes through-out the cutoff, it therefore is *untestable*. That's right—it's an untestable assumption. But, what we can do is check for whether there are changes in the conditional expectation functions for other exogenous covariates that cannot or should not be changing as a result of the cut-off. So it's very common to look at things like race or gender around the cutoff. You can use these same methods to do that, but I do not do them here. Any RDD paper will always involve such placebos; even though they are not direct tests of the continuity assumption, they are indirect tests. Remember, when you are publishing, your readers aren't as familiar with this thing you're studying, so your task is explain to readers what you know. Anticipate their objections and the sources of their skepticism. Think like them. Try to put yourself in a stranger's shoes. And then test those skepticisms to the best of your ability.

Second, we saw the importance of bandwidth selection, or win-dow, for estimating the causal effect using this method, as well as the

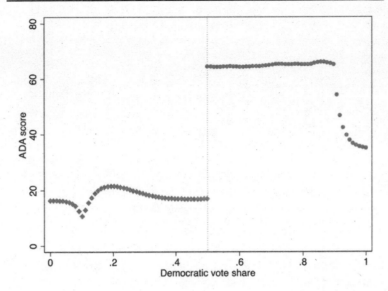

Figure 41. Local linear nonparametric regressions.

importance of selection of polynomial length. There's always a trade-off when choosing the bandwidth between bias and variance—the shorter the window, the lower the bias, but because you have less data, the variance in your estimate increases. Recent work has been focused on optimal bandwidth selection, such as Imbens and Kalyanaraman [2011] and Calonico et al. [2014]. The latter can be implemented with the user-created rdrobust command. These methods ultimately choose optimal bandwidths that may differ left and right of the cutoff based on some bias-variance trade-off. Let's repeat our analysis using this nonparametric method. The coefficient is 46.48 with a standard error of 1.24.

STATA

lmb_9.do

```
1   * Stata code attributed to Marcelo Perraillon.
2   ssc install rdrobust, replace
3   rdrobust score demvoteshare, c(0.5)
```

R
lmb_9.R

```
1   library(tidyverse)
2   library(rdrobust)
3
4   rdr <- rdrobust(y = lmb_data$score,
5            x = lmb_data$demvoteshare, c = 0.5)
6   summary(rdr)
```

This method, as we've repeatedly said, is data-greedy because it gobbles up data at the discontinuity. So ideally these kinds of methods will be used when you have large numbers of observations in the sample so that you have a sizable number of observations at the discontinuity. When that is the case, there should be some harmony in your findings across results. If there isn't, then you may not have sufficient power to pick up this effect.

Finally, we look at the implementation of the McCrary density test. We will implement this test using local polynomial density estimation [Cattaneo et al., 2019]. This requires installing two files in Stata. Visually inspecting the graph in Figure 42, we see no signs that there was manipulation in the running variable at the cutoff.

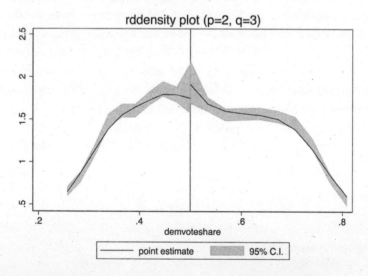

Figure 42. McCrary density test using local linear nonparametric regressions.

STATA

lmb_10.do

```
1   * McCrary density test. Stata code attributed to Marcelo Perraillon.
2   net install rddensity,
      ↳ from(https://sites.google.com/site/rdpackages/rddensity/stata) replace
3   net install lpdensity,
      ↳ from(https://sites.google.com/site/nppackages/lpdensity/stata) replace
4   rddensity demvoteshare, c(0.5) plot
```

R

lmb_10.R

```
1   library(tidyverse)
2   library(rddensity)
3   library(rdd)
4
5   DCdensity(lmb_data$demvoteshare, cutpoint = 0.5)
6
7   density <- rddensity(lmb_data$demvoteshare, c = 0.5)
8   rdplotdensity(density, lmb_data$demvoteshare)
```

Concluding remarks about close-election designs. Let's circle back to the close-election design. The design has since become practically a cottage industry within economics and political science. It has been extended to other types of elections and outcomes. One paper I like a lot used close gubernatorial elections to examine the effect of Democratic governors on the wage gap between workers of different races [Beland, 2015]. There are dozens more.

But a critique from Caughey and Sekhon [2011] called into question the validity of Lee's analysis on the House elections. They found that bare winners and bare losers in US House elections differed considerably on pretreatment covariates, which had not been formally evaluated by Lee et al. [2004]. And that covariate imbalance got even worse in the closest elections. Their conclusion is that the sorting problems got more severe, not less, in the closest of House races, suggesting that these races could not be used for an RDD.

At first glance, it appeared that this criticism by Caughey and Sekhon [2011] threw cold water on the entire close-election design, but we since know that is not the case. It appears that the Caughey and Sekhon [2011] criticism may have been only relevant for a subset of House races but did not characterize other time periods or other types of races. Eggers et al. [2014] evaluated 40,000 close elections, including the House in other time periods, mayoral races, and other types of races for political offices in the US and nine other countries. No other case that they encountered exhibited the type of pattern described by Caughey and Sekhon [2011]. Eggers et al. (2014) conclude that the assumptions behind RDD in the close-election design are likely to be met in a wide variety of electoral settings and is perhaps one of the best RD designs we have going forward.

Regression Kink Design

Many times, the concept of a running variable shifting a unit into treatment and in turn causing a jump in some outcome is sufficient. But there are some instances in which the idea of a "jump" doesn't describe what happens. A couple of papers by David Card and coauthors have extended the regression discontinuity design in order to handle these different types of situations. The most notable is Card et al. [2015], which introduced a new method called regression kink design, or RKD. The intuition is rather simple. Rather than the cutoff causing a discontinuous jump in the treatment variable at the cutoff, it changes the first derivative, which is known as a kink. Kinks are often embedded in policy rules, and thanks to Card et al. [2015], we can use kinks to identify the causal effect of a policy by exploiting the jump in the first derivative.

Card et al.'s [2015] paper applies the design to answer the question of whether the level of unemployment benefits affects the length of time spent unemployed in Austria. Unemployment benefits are based on income in a base period. There is then a minimum benefit level that isn't binding for people with low earnings. Then benefits are 55% of the earnings in the base period. There is a maximum benefit level that is then adjusted every year, which creates a discontinuity in the schedule.

Figure 43 shows the relationship between base earnings and unemployment benefits around the discontinuity. There's a visible kink in the empirical relationship between average benefits and base

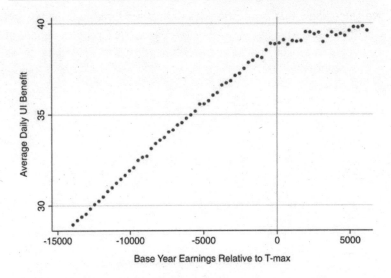

Figure 43. RKD kinks. Reprinted from Card, D., Lee, D. S., Pei, Z., and Weber, A. (2015). "Inference on Causal Effects in a Generalized Regression Kink Design." *Econometrica*, 84(6):2453–2483. Copyright © 2015 Wiley. Used with permission from John Wiley and Sons.

earnings. You can see this in the sharp decline in the slope of the function as base-year earnings pass the threshold. Figure 44 presents a similar picture, but this time of unemployment duration. Again, there is a clear kink as base earnings pass the threshold. The authors conclude that increases in unemployment benefits in the Austrian context exert relatively large effects on unemployment duration.

Conclusion

The regression discontinuity design is often considered a winning design because of its upside in credibly identifying causal effects. As with all designs, its credibility only comes from deep institutional knowledge, particularly surrounding the relationship between the running variable, the cutoff, treatment assignment, and the outcomes themselves. Insofar as one can easily find a situation in which a running variable passing some threshold leads to units being siphoned off

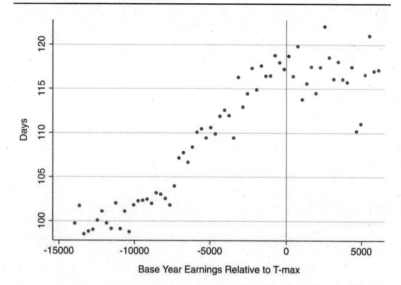

Figure 44. Unemployment duration. Reprinted from Card, D., Lee, D. S., Pei, Z., and Weber, A. (2015). "Inference on Causal Effects in a Generalized Regression Kink Design." *Econometrica*, 84(6):2453–2483. Copyright © 2015 Wiley. Used with permission from John Wiley and Sons.

into some treatment, then if continuity is believable, you're probably sitting on a great opportunity, assuming you can use it to do something theoretically interesting and policy relevant to others.

Regression discontinuity design opportunities abound, particularly within firms and government agencies, for no other reason than that these organizations face scarcity problems and must use some method to ration a treatment. Randomization is a fair way to do it, and that is often the method used. But a running variable is another method. Routinely, organizations will simply use a continuous score to assign treatments by arbitrarily picking a cutoff above which everyone receives the treatment. Finding these can yield a cheap yet powerfully informative natural experiment. This chapter attempted to lay out the basics of the design. But the area continues to grow at a lightning pace. So I encourage you to see this chapter as a starting point, not an ending point.

Instrumental Variables

I made "Sunday Candy,"
I'm never going to hell.
I met Kanye West,
I'm never going to fail.

Chance the Rapper

Just as Archimedes said, "Give me a fulcrum, and I shall move the world," you could just as easily say that with a good-enough instrument, you can identify any causal effect. But, while that is hyperbole, for reasons we will soon see, it is nonetheless the case that the instrumental variables (IV) design is potentially one of most important research designs ever devised. It is also unique because it is one of those instances that the econometric estimator was not simply ripped off from statistics (e.g., Eicker-Huber-White standard errors) or imported from some other field (e.g., like regression discontinuity). IV was invented by an economist, and its history is fascinating.

History of Instrumental Variables: Father and Son

Philip Wright was born in 1861 and died in 1934. He received his bachelor's degree from Tufts in 1884 and a master's degree from Harvard in 1887. His son, Sewall Wright, was born in 1889 when Philip was 28. The family moved from Massachusetts to Galesburg, Illinois, where Philip took a position as professor of mathematics and economics at Lombard College. Philip published numerous articles and books in economics over his career, and he published poetry, too. You can see his vita here at https://scholar.harvard.edu/files/stock/files/

wright_cv.jpg.[1] Sewall attended Lombard College and took his college mathematics courses from his father.

In 1913, Philip took a position at Harvard, and Sewall entered there as a graduate student. Philip would later leave for the Brookings Institute, and Sewall would take his first job in the Department of Zoology at the University of Chicago, where he would eventually be promoted to professor in 1930.

Philip was prolific, which, given his sizable teaching and service requirements, is impressive. He published in top journals such as *Quarterly Journal of Economics*, *Journal of the American Statistical Association*, *Journal of Political Economy*, and *American Economic Review*. A common theme across many of his publications was the identification problem. He was acutely aware of it and intent on solving it.

In 1928, Philip was writing a book about animal and vegetable oils. The reason? He believed that recent tariff increases were harming international relations. And he wrote about the damage from the tariffs, which had affected animal and vegetable oils. The book, it turns out, would become a classic—not for tariffs or oils, but for being the first proof for the existence of an instrumental variables estimator.

While his father was publishing like a fiend in economics, Sewall Wright was revolutionizing the field of genetics. He invented path analysis, a precursor to Pearl's directed acyclical graphic models, and he made important contributions to the theory of evolution and genetics. He was a genius. The decision to not follow in the family business (economics) created a bit of tension between the two men, but all evidence suggests that they found each other intellectually stimulating.

In his book on vegetable and oil tariffs, there is an appendix (entitled Appendix B) in which the calculus of an instrumental variables estimator was worked out. Elsewhere, Philip thanked his son for his valuable contributions to what he had written, referring to the path analysis that Sewall had taught him. This path analysis, it turned out, played a key role in Appendix B.

Appendix B showed a solution to the identification problem. So long as the economist is willing to impose some restrictions on the

---

1 Philip had a passion for poetry, and even published some in his life. He also used his school's printing press to publish the first book of poems by the great American poet Carl Sandburg.

problem, then the system of equations can be identified. Specifically, if there is one instrument for supply, and the supply and demand errors are uncorrelated, then the elasticity of demand can be identified.

But who wrote this Appendix B? Either man could've done so. It is a chapter in an economics book, which points to Philip. But it used the path analysis, which points to Sewall. Historians have debated this, even going so far as to accuse Philip of stealing the idea from his son. If Philip stole the idea, by which I mean that when he published Appendix B, he failed to give proper attribution to his son, then it would at the very least have been a strange oversight. In come Stock and Trebbi [2003] to offer their opinions to this debate over authorship.

Stock and Trebbi [2003] tried to determine the authorship of Appendix B using "stylometric analysis." Stylometric analysis had been used in other applications, such as to identify the author of the 1996 political novel *Primary Colors* (Joseph Klein) and the unsigned *Federalist Papers*. But Stock and Trebbi [2003] is easily the best application of stylometric analysis in economics.[2]

The method is akin to contemporary machine learning methods. The authors collected raw data containing the known original academic writings of each man, plus the first chapter and Appendix B of the book in question. All footnotes, graphs, and figures were excluded. Blocks of 1,000 words were selected from the files. Fifty-four blocks were selected: twenty written by Sewall with certainty, twenty-five by Philip, six from Appendix B, and three from chapter 1. Chapter 1 has always been attributed to Philip, but Stock and Trebbi [2003] treat the three blocks as unknown to check whether their model is correctly predicting authorship when authorship is already known.

The stylometric indicators that they used included the frequency of occurrence in each block of 70 function words. The list was taken from a separate study. These 70 function words produced 70 numerical variables, each of which is a count, per 1,000 words, of an individual function word in the block. Some words were dropped (e.g., "things" because they occurred only once), leaving 69 function words.

---

2 But it's easy to have the best paper on a topic in some field when you're also the only paper on that topic in that field.

The second set of stylometric indicators, taken from another study, concerned grammatical constructions. Stock and Trebbi [2003] used 18 grammatical constructions, which were frequency counts. They included things like noun followed by an adverb, total occurrences of prepositions, coordinating conjunction followed by noun, and so on. There was one dependent variable in their analysis, and that was authorship. The independent variables were 87 covariates (69 function word counts and 18 grammatical statistics).

The results of this analysis are absolutely fascinating. For instance, many covariates have very large t-statistics, which would be unlikely if there really were no stylistic differences between the authors and the indicators were independently distributed.

So what do they find? Most interesting is their regression analysis. They write:

> We regressed authorship against an intercept, the first two principal components of the grammatical statistics and the first two principal components of the function word counts, and we attribute authorship depending on whether the predicted value is greater or less than 0.5. [191]

And what did they find? That all of the Appendix B and chapter 1 blocks were assigned to Philip, not Sewall. They did other robustness checks, and all of them still pointed to Philip as the author.

Writing Appendix B and solving the problem that became Appendix B are technically distinct. But I nonetheless love this story for many reasons. First, I love the idea that an econometric estimator as important as instrumental variables has its roots in economics. I'm so accustomed to stories in which the actual econometric estimator was lifted from statistics (Huber-White standard errors) or educational psychology (regression discontinuity) that it's nice to know economists have added their own designs to the canon. But the other part of the story that I love is the father-son component. It's encouraging to know that a father and son can overcome differences through intellectual collaborations such as this. Such relationships are important, and tensions, when they arise, should be vigorously pursued until those tensions dissolve if possible. Relationships, and love more generally, matter after all. And Philip and Sewall give a story of that.

Intuition of Instrumental Variables

Canonical IV DAG. To understand the instrumental variables estimator, it is helpful to start with a DAG that shows a chain of causal effects that contains all the information needed to understand the instrumental variables strategy. First, notice the backdoor path between D and Y: $D \leftarrow U \rightarrow Y$. Furthermore, note that U is unobserved by the econometrician, which causes the backdoor path to remain open. If we have this kind of *selection on unobservables*, then there does not exist a conditioning strategy that will satisfy the backdoor criterion (in our data). But, before we throw up our arms, let's look at how Z operates through these pathways.

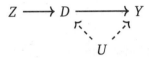

First, there is a mediated pathway from Z to Y via D. When Z varies, D varies, which causes Y to change. But, even though Y is varying when Z varies, notice that Y is only varying *because* D has varied. You sometimes hear people describe this as the "only through" assumption. That is, Z affects Y "only through" D.

Imagine this for a moment though. Imagine D consists of people making choices. Sometimes these choices affect Y, and sometimes these choices are merely correlated with changes in Y due to unobserved changes in U. But along comes some shock, Z, which induces *some* but not all of the people in D to make different decisions. What will happen?

Well, for one, when those people's decisions change, Y will change too, because of the causal effect. But all of the correlation between D and Y in that situation will reflect the causal effect. The reason is that D is a collider along the backdoor path between Z and Y.

But I'm not done with this metaphor. Let's assume that in this D variable, with all these people, only some of the people change their behavior because of D. What then? Well, in that situation, Z is causing a change in Y for just a subset of the population. If the instrument only changes the behavior of women, for instance, then the causal effect of D on Y will only reflect the causal effect of women's choices, not men's choices.

There are two ideas inherent in the previous paragraph that I want to emphasize. First, if there are heterogeneous treatment effects (e.g., men affect Y differently than women do), then our Z shock only identified some of the causal effect of D on Y. And that piece of the causal effect may only be valid for the population of women whose behavior changed in response to Z; it may not be reflective of how men's behavior would affect Y. And second, if Z is inducing some of the change in Y via only a fraction of the change in D, then it's almost as though we have less data to identify that causal effect than we really have.

Here we see two of the difficulties in interpreting instrumental variables and identifying a parameter using instrumental variables. Instrumental variables only identify a causal effect for any group of units whose behaviors are changed as a result of the instrument. We call this the causal effect of the *complier* population; in our example, only women "complied" with the instrument, so we only know its effect for them. And second, instrumental variables are typically going to have larger standard errors, and as such, they will fail to reject in many instances if for no other reason than being underpowered.

Moving along, let's return to the DAG. Notice that we drew the DAG such that Z is independent of U. You can see this because D is a collider along the $Z \rightarrow D \leftarrow U$ path, which implies that Z and U are independent. This is called the "exclusion restriction," which we will discuss in more detail later. But briefly, the IV estimator assumes that Z is independent of the variables that determine Y except for D.

Second, Z is correlated with D, and because of its correlation with D (and D's effect on Y), Z is correlated with Y but only through its effect on D. This relationship between Z and D is called the "first stage" because of the two-stage least squares estimator, which is a kind of IV estimator. The reason it is only correlated with Y via D is because D is a collider along the path $Z \rightarrow D \leftarrow U \rightarrow Y$.

Good instruments should feel weird. How do you know when you have a good instrument? One, it will require prior knowledge. I'd encourage you to write down that prior knowledge into a DAG and use it to reflect on the feasibility of your design. As a starting point, you can contemplate identifying a causal effect using IV only if you can theoretically and logically defend the exclusion restriction, since the exclusion restriction is an untestable assumption. That defense requires theory, and since some people aren't comfortable with theoretical arguments like that, they tend to eschew the use of IV. More and more applied microeconomists are skeptical of IV because they are able to tell limitless stories in which exclusion restrictions do not hold.

But, let's say you think you do have a good instrument. How might you defend it as such to someone else? A necessary but not sufficient condition for having an instrument that can satisfy the exclusion restriction is if people are confused when you tell them about the instrument's relationship to the outcome. Let me explain. No one is likely to be confused when you tell them that you think family size will reduce the labor supply of women. They don't need a Becker model to convince them that women who have more children probably are employed outside the home less often than those with fewer children.

But what would they think if you told them that mothers whose first two children were the same gender were employed outside the home less than those whose two children had a balanced sex ratio? They would probably be confused because, after all, what does the gender composition of one's first two children have to do with whether a woman works outside the home? That's a head scratcher. They're confused because, logically, whether the first two kids are the same gender versus not the same gender doesn't seem on its face to change the incentives a women has to work outside the home, which is based on reservation wages and market wages. *And yet*, empirically it is true that if your first two children are a boy, many families will have a third compared to those who had a boy and a girl first. So what gives?

The gender composition of the first two children matters for a family if they have preferences over diversity of gender. Families where the first two children were boys are more likely to try again in the hopes they'll have a girl. And the same for two girls. Insofar as parents would like to have at least one boy and one girl, then having two boys might cause them to roll the dice for a girl.

And there you see the characteristics of a good instrument. It's weird to a lay person because a good instrument (two boys) only changes the outcome by first changing some endogenous treatment variable (family size) thus allowing us to identify the causal effect of family size on some outcome (labor supply). And so without knowledge of the endogenous variable, relationships between the instrument and the outcome don't make much sense. Why? Because the instrument is irrelevant to the determinants of the outcome except for its effect on the endogenous treatment variable. You also see another quality of the instrument that we like, which is that it's quasi-random.

Before moving along, I'd like to illustrate this "weird instrument" in one more way, using two of my favorite artists: Chance the Rapper and Kanye West. At the start of this chapter, I posted a line from Kanye West's wonderful song "Ultralight Beam" on the underrated *Life of Pablo*. On that song, Chance the Rapper sings:

I made "Sunday Candy," I'm never going to hell.
I met Kanye West, I'm never going to fail.

Several years before "Ultralight Beam," Chance made a song called "Sunday Candy." It's a great song and I encourage you to listen to it. But Chance makes a strange argument here on "Ultralight Beam." He claims that *because* he made "Sunday Candy," *therefore* he won't go to hell. Now even a religious person will find that perplexing, as there is nothing in Christian theology of eternal damnation that would link making a song to the afterlife. This, I would argue, is a "weird instrument" because without knowing the endogenous variable on the mediated path $SC \rightarrow ? \rightarrow H$, the two phenomena don't seem to go together.

But let's say that I told you that after Chance made "Sunday Candy," he got a phone call from his old preacher. The preacher loved the song and invited Chance to come sing it at church. And while revisiting his childhood church, Chance had a religious experience that caused him to convert back to Christianity. Now, and only now, does his statement make sense. It isn't that "Sunday Candy" itself shaped the path of his afterlife, so much as "Sunday Candy" caused a particular event that itself caused his beliefs about the future to change. That the line makes a weird argument is what makes "Sunday Candy" a good instrument.

But let's take the second line—"I met Kanye West, I'm never going to fail." Unlike the first line, this is likely not a good instrument. Why? Because I don't even need to know what variable is along the mediated path $KW \rightarrow ? \rightarrow F$ to doubt the exclusion restriction. If you are a musician, a relationship with Kanye West can possibly make or break your career. Kanye could make your career by collaborating with you on a song or by introducing you to highly talented producers. There is no shortage of ways in which a relationship with Kanye West can cause you to be successful, regardless of whatever unknown endogenous variable we have placed in this mediated path. And since it's easy to tell a story where knowing Kanye West directly causes one's success, knowing Kanye West is likely a *bad instrument*. It simply won't satisfy the exclusion restriction in this context.

Ultimately, good instruments are jarring precisely because of the exclusion restriction—these two things (gender composition and work) don't seem to go together. If they did go together, it would likely mean that the exclusion restriction was violated. But if they don't, then the person is confused, and that is at minimum a possible candidate for a good instrument. This is the commonsense explanation of the "only through" assumption.

Homogeneous Treatment Effects

There are two ways to discuss the instrumental variables design: one in a world where the treatment has the same causal effect for everybody ("homogeneous treatment effects") and one in a world where the treatment effects can differ across the population ("heterogeneous treatment effects"). For homogeneous treatment effects, I will depend on a more traditional approach rather than on potential outcomes notation. When the treatment effect is constant, I don't feel we need potential outcomes notation as much.

Instrumental variables methods are typically used to address omitted variable bias, measurement error, and simultaneity. For instance, quantity and price is determined by the intersection of supply and demand, so any observational correlation between price and quantity is uninformative about the elasticities associated with supply or demand curves. Philip Wright understood this, which was why he investigated the problem so intensely.

I will assume a homogeneous treatment effect of δ which is the same for every person. This means that if college caused my wages to increase by 10%, it also caused your wages to increase by 10%. Let's start by illustrating the problem of omitted variable bias. Assume the classical labor problem where we're interested in the causal effect of schooling on earnings, but schooling is endogenous because of unobserved ability. Let's draw a simple DAG to illustrate this setup.

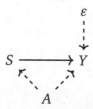

We can represent this DAG with a simple regression. Let the true model of earnings be:

$$Y_i = \alpha + \delta S_i + \gamma A_i + \varepsilon_i$$

where Y is the log of earnings, S is schooling measured in years, A is individual "ability," and ε is an error term uncorrelated with schooling or ability. The reason A is unobserved is simply because the surveyor either forgot to collect it or couldn't collect it and therefore it's missing from the data set.[3] For instance, the CPS tells us nothing about respondents' family background, intelligence, motivation, or non-cognitive ability. Therefore, since ability is unobserved, we have the following equation instead:

$$Y_i = \alpha + \delta S_i + \eta_i$$

where η_i is a composite error term equalling $\gamma A_i + \varepsilon_i$. We assume that schooling is correlated with ability, so therefore it is correlated with η_i, making it endogenous in the second, shorter regression. Only ε_i is uncorrelated with the regressors, and that is by definition.

We know from the derivation of the least squares operator that the estimated value of $\widehat{\delta}$ is:

$$\widehat{\delta} = \frac{C(Y, S)}{V(S)} = \frac{E[YS] - E[Y]E[S]}{V(S)}$$

3 Unobserved ability doesn't mean it's literally unobserved, in other words. It could be just missing from your data set, and therefore is unobserved *to you*.

Plugging in the true value of Y (from the longer model), we get the following:

$$\hat{\delta} = \frac{E[\alpha S + S^2\delta + \gamma\, SA + \varepsilon S] - E(S)E[\alpha + \delta S + \gamma\, A + \varepsilon]}{V(S)}$$

$$= \frac{\delta E(S^2) - \delta E(S)^2 + \gamma\, E(AS) - \gamma\, E(S)E(A) + E(\varepsilon S) - E(S)E(\varepsilon)}{V(S)}$$

$$= \delta + \gamma\, \frac{C(AS)}{V(S)}$$

If $\gamma > 0$ and $C(A,S) > 0$, then $\hat{\delta}$, the coefficient on schooling, is upward biased. And that is probably the case given that it's likely that ability and schooling are positively correlated.

But let's assume that you have found a really great weird instrument Z_i that causes people to get more schooling but that is independent of student ability and the structural error term. It is independent of ability, which means we can get around the endogeneity problem. And it's not associated with the other unobserved determinants of earnings, which basically makes it weird. The DAG associated with this set up would look like this:

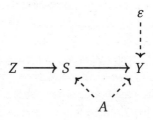

We can use this variable, as I'll now show, to estimate δ. First, calculate the covariance of Y and Z:

$$C(Y,Z) = C(\alpha\delta S + \gamma\, A + \varepsilon, Z)$$

$$= E[(\alpha + \delta S + \gamma\, A + \varepsilon), Z] - E(S)E(Z)$$

$$= \{\alpha E(Z) - \alpha E(Z)\} + \delta\{E(SZ) - E(S)E(Z)\}$$

$$\quad + \gamma\,\{E(AZ) - E(A)E(Z)\} + \{E(\varepsilon Z) - E(\varepsilon)E(Z)\}$$

$$= \delta C(S,Z) + \gamma\, C(A,Z) + C(\varepsilon,Z)$$

Notice that the parameter of interest, δ is on the right side. So how do we isolate it? We can estimate it with the following:

$$\widehat{\delta} = \frac{C(Y,Z)}{C(S,Z)}$$

so long as $C(A,Z) = 0$ and $C(\varepsilon,Z) = 0$.

These zero covariances are the statistical truth contained in the IV DAG from earlier. If ability is independent of Z, then this second covariance is zero. And if Z is independent of the structural error term, ε, then it too is zero. This, you see, is what is meant by the "exclusion restriction": the instrument must be independent of both parts of the composite error term.

But the exclusion restriction is only a necessary condition for IV to work; it is not a sufficient condition. After all, if all we needed was exclusion, then we could use a random number generator for an instrument. Exclusion is not enough. We also need the instrument to be *highly correlated* with the endogenous variable schooling S. And the higher the better. We see that here because we are dividing by $C(S,Z)$, so it necessarily requires that this covariance not be zero.

The numerator in this simple ratio is sometimes called the "reduced form," while the denominator is called the "first stage." These terms are somewhat confusing, particularly the former, as "reduced form" means different things to different people. But in the IV terminology, it is that relationship between the instrument and the outcome itself. The first stage is less confusing, as it gets its name from the two-stage least squares estimator, which we'll discuss next.

When you take the probability limit of this expression, then assuming $C(A,Z) = 0$ and $C(\varepsilon,Z) = 0$ due to the exclusion restriction, you get

$$p\lim \widehat{\delta} = \delta$$

But if Z is not independent of η (either because it's correlated with A or ε), and if the correlation between S and Z is weak, then $\widehat{\delta}$ becomes severely biased in finite samples.

Two-stage least squares. One of the more intuitive instrumental variables estimators is the two-stage least squares (2SLS). Let's review an example to illustrate why it is helpful for explaining some of the IV intuition. Suppose you have a sample of data on Y, S, and Z. For each observation i, we assume the data are generated according to:

$$Y_i = \alpha + \delta S_i + \varepsilon_i$$
$$S_i = \gamma + \beta Z_i + \epsilon_i$$

where $C(Z, \varepsilon) = 0$ and $\beta \neq 0$. The former assumption is the exclusion restriction whereas the second assumption is a non-zero first-stage. Now using our IV expression, and using the result that $\sum_{i=1}^{n}(x_i - \bar{x}) = 0$, we can write out the IV estimator as:

$$\hat{\delta} = \frac{C(Y,Z)}{C(S,Z)}$$

$$= \frac{\frac{1}{n}\sum_{i=1}^{n}(Z_i - \bar{Z})(Y_i - \bar{Y})}{\frac{1}{n}\sum_{i=1}^{n}(Z_i - \bar{Z})(S_i - \bar{S})}$$

$$= \frac{\frac{1}{n}\sum_{i=1}^{n}(Z_i - \bar{Z})Y_i}{\frac{1}{n}\sum_{i=1}^{n}(Z_i - \bar{Z})S_i}$$

When we substitute the true model for Y, we get the following:

$$\hat{\delta} = \frac{\frac{1}{n}\sum_{i=1}^{n}(Z_i - \bar{Z})\{\alpha + \delta S + \varepsilon\}}{\frac{1}{n}\sum_{i=1}^{n}(Z_i - \bar{Z})S_i}$$

$$= \delta + \frac{\frac{1}{n}\sum_{i=1}^{n}(Z_i - \bar{Z})\varepsilon_i}{\frac{1}{n}\sum_{i=1}^{n}(Z_i - \bar{Z})S_i}$$

$$= \delta + \text{"small if } n \text{ is large"}$$

So, let's return to our first description of $\hat{\delta}$ as the ratio of two covariances. With some simple algebraic manipulation, we get the following:

$$\hat{\delta} = \frac{C(Y,Z)}{C(S,Z)}$$

$$= \frac{\frac{C(Z,Y)}{V(Z)}}{\frac{C(Z,S)}{V(Z)}}$$

where the denominator is equal to $\widehat{\beta}$.[4] We can rewrite $\widehat{\beta}$ as:

$$\widehat{\beta} = \frac{C(Z,S)}{V(Z)}$$

$$\widehat{\beta}V(Z) = C(Z,S)$$

Then we rewrite the IV estimator and make a substitution:

$$\widehat{\delta}_{IV} = \frac{C(Z,Y)}{C(Z,S)}$$

$$= \frac{\widehat{\beta}C(Z,Y)}{\widehat{\beta}C(Z,S)}$$

$$= \frac{\widehat{\beta}C(Z,Y)}{\widehat{\beta}^2 V(Z)}$$

$$= \frac{C(\widehat{\beta}Z,Y)}{V(\widehat{\beta}Z)}$$

Notice now what is inside the parentheses: $\widehat{\beta}Z$, which are the fitted values of schooling from the first-stage regression. We are no longer, in other words, using S—we are using its fitted values. Recall that $S = \gamma + \beta Z + \epsilon$; $\widehat{\delta} = \frac{C(\widehat{\beta}ZY)}{V(\widehat{\beta}Z)}$ and let $\widehat{S} = \widehat{\gamma} + \widehat{\beta}Z$. Then the two-stage least squares (2SLS) estimator is:

$$\widehat{\delta}_{IV} = \frac{C(\widehat{\beta}Z,Y)}{V(\widehat{\beta}Z)}$$

$$= \frac{C(\widehat{S},Y)}{V(\widehat{S})}$$

I will now show that $\widehat{\beta}C(Y,Z) = C(\widehat{S},Y)$, and leave it to you to show that $V(\widehat{\beta}Z) = V(\widehat{S})$.

$$C(\widehat{S},Y) = E[\widehat{S}Y] - E[\widehat{S}]E[Y]$$

$$= E\left(Y[\widehat{\gamma} + \widehat{\beta}Z]\right) - E(Y)E(\widehat{\gamma} + \widehat{\beta}Z)$$

$$= \widehat{\gamma}E(Y) + \widehat{\beta}E(YZ) - \widehat{\gamma}E(Y) - \widehat{\beta}E(Y)E(Z)$$

$$= \widehat{\beta}[E(YZ) - E(Y)E(Z)]$$

$$C(\widehat{S},Y) = \widehat{\beta}C(Y,Z)$$

---

4 That is, $S_i = \gamma + \beta Z_i + \epsilon_i$.

Now let's return to something I said earlier—learning 2SLS can help you better understand the intuition of instrumental variables more generally. What does this mean exactly? First, the 2SLS estimator used only the fitted values of the endogenous regressors for estimation. These fitted values were based on all variables used in the model, *including the excludable instrument.* And as all of these instruments are exogenous in the structural model, what this means is that the fitted values themselves have become exogenous too. Put differently, we are using only the variation in schooling that is *exogenous*. So that's kind of interesting, as now we're back in a world where we are identifying causal effects from exogenous changes in schooling caused by our instrument.

But, now the less-exciting news. This exogenous variation in S driven by the instrument is only a subset of the total variation in schooling. Or put differently, IV reduces the variation in the data, so there is less information available for identification, and what little variation we have left comes from only those units who responded to the instrument in the first place. This, it turns out, will be critical later when we relax the homogeneous treatment effects assumption and allow for heterogeneity.

Parental Methamphetamine Abuse and Foster Care

It's helpful to occasionally stop and try to think about real-world applications as much as possible; otherwise the estimators feel very opaque and unhelpful. So to illustrate, I'm going to review one of my own papers with Keith Finlay that sought to estimate the effect that parental methamphetamine abuse had on child abuse and foster care admissions [Cunningham and Finlay, 2012].

It has been claimed that substance abuse, notably illicit drug use, has a negative impact on parenting, causing neglect, but as these all occur in equilibrium, it's possible that the correlation is simply reflective of selection bias. Maybe households with parents who abuse drugs would've had the same negative outcomes had the parents not used drugs. After all, it's not like people are flipping coins when deciding to smoke meth. So let me briefly give you some background to the study so that you better understand the data-generating process.

First, methamphetamine is a toxic poison to the mind and body and highly addictive. Some of the symptoms of meth abuse are increased energy and alertness, decreased appetite, intense euphoria, impaired judgment, and psychosis. Second, the meth epidemic in the United States began on the West Coast, before gradually making its way eastward over the 1990s.

We were interested in the impact that this growth in meth abuse was having on children. Observers and law enforcement had commented, without concrete causal evidence, that the epidemic was causing a growth in foster care admissions. But how could we separate correlation from causality? The solution was contained within how meth itself is produced.

Meth is synthesized from a reduction of ephedrine or pseudoephedrine, which is also the active ingredient in many cold medications, such as Sudafed. Without one of these two precursors, it is impossible to produce the kind of meth people abuse. These precursors had supply chains that could be potentially disrupted because of the concentration of pharmaceutical laboratories. In 2004, nine factories manufactured the bulk of the world's supply of ephedrine and pseudoephedrine. The US Drug Enforcement Agency correctly noted that if it could regulate access to ephedrine and pseudoephedrine, then it could effectively interrupt the production of methamphetamine, and in turn, *hypothetically* reduce meth abuse and its associated social harms.

So, with input from the DEA, Congress passed the Domestic Chemical Diversion Control Act in August 1995, which provided safeguards by regulating the distribution of products that contained ephedrine as the primary medicinal ingredient. But the new legislation's regulations applied to ephedrine, not pseudoephedrine, and since the two precursors were nearly identical, traffickers quickly substituted. By 1996, pseudoephedrine was found to be the primary precursor in almost half of meth lab seizures.

Therefore, the DEA went back to Congress, seeking greater control over pseudoephedrine products. And the Comprehensive Methamphetamine Control Act of 1996 went into effect between October and December 1997. This act required distributors of all forms of pseudoephedrine to be subject to chemical registration. Dobkin and Nicosia

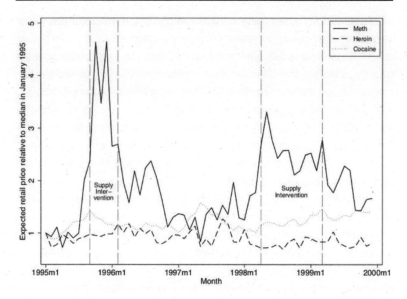

Figure 45. Ratio of median monthly expected retail prices of meth, heroin, and cocaine relative to their respective values in 1995, STRIDE 1995–1999. Reprinted from Cunningham, S. and Finlay, K. (2012). "Parental Substance Abuse and Foster Care: Evidence from Two Methamphetamine Supply Shocks?" *Economic Inquiry*, 51(1):764–782. Copyright © 2012 Wiley. Used with permission from John Wiley and Sons.

[2009] argued that these precursor shocks may very well have been the largest supply shocks in the history of drug enforcement.

We placed a Freedom of Information Act request with the DEA requesting all of their undercover purchases and seizures of illicit drugs going back decades. The data included the price of an undercover purchase, the drug's type, its weight and its purity, as well as the locations in which the purchases occurred. We used these data to construct a price series for meth, heroin, and cocaine. The effect of the two interventions were dramatic. The first supply intervention caused retail (street) prices (adjusted for purity, weight, and inflation) to more than quadruple. The second intervention, while still quite effective at raising relative prices, did not have as large an effect as the first. See Figure 45.

We showed two other drug prices (cocaine and heroin) in addition to meth because we wanted the reader to understand that the

1995 and 1997 shocks were uniquely impacting meth markets. They did not appear to be common shocks affecting all drug markets, in other words. As a result, we felt more confident that our analysis would be able to isolate the effect of methamphetamine, as opposed to substance abuse more generally. The two interventions simply had no effect on cocaine and heroin prices despite causing a massive shortage of meth and raising its retail price. It wouldn't have surprised me if disrupting meth markets had caused a shift in demand for cocaine or heroin and in turn caused its prices to change, yet at first glance in the time series, I'm not finding that. Weird.

We are interested in the causal effect of meth abuse on child abuse, and so our first stage is necessarily a proxy for meth abuse— the number of people entering treatment who listed meth as one of the substances they used in their last episode of substance abuse. As I said before, since a picture is worth a thousand words, I'm going to show you pictures of both the first stage and the reduced form. Why do I do this instead of going directly to the tables of coefficients? Because quite frankly, you are more likely to find those estimates believable if you can see evidence for the first stage and the reduced form in the raw data itself.[5]

In Figure 46, we show the first stage. All of these data come from the Treatment Episode Data Set (TEDS), which includes all people going into treatment for substance abuse at federally funded clinics. Patients list the last three substances used in the most recent "episode." We mark anyone who listed meth, cocaine, or heroin and aggregate by month and state. But first, let's look at the national aggregate in Figure 46. You can see evidence for the effect the two interventions had on meth flows, particularly the ephedrine intervention. Self-admitted meth admissions dropped significantly, as did total meth admissions, but there's no effect on cocaine or heroin. The effect of the pseudoephedrine is not as dramatic, but it appears to cause a break in trend as the growth in meth admissions slows during this period of time. In summary, it appears we have a first stage because, during the interventions, meth admissions declines.

---

5 I wouldn't go so far as to say that presenting pictures of the reduced form and first stage is mandatory in the way that many pictures in an RDD are mandatory, but it's awfully close. Ultimately, seeing is believing.

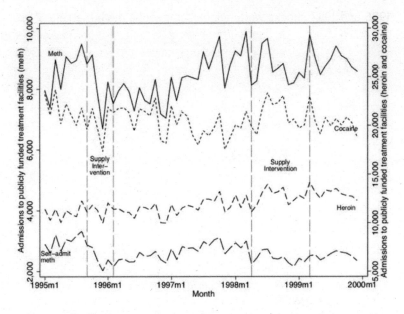

Figure 46. Visual representation of the equivalent of the first stage. Reprinted from Cunningham, S. and Finlay, K. (2012). "Parental Substance Abuse and Foster Care: Evidence from Two Methamphetamine Supply Shocks?" *Economic Inquiry*, 51(1):764–782. Copyright © 2012 Wiley. Used with permission from John Wiley and Sons.

In Figure 47, we show the reduced form—that is, the effect of the price shocks on foster care admissions. Consistent with what we found in our first-stage graphic, the ephedrine intervention in particular had a profoundly negative effect on foster care admissions. They fell from around 8,000 children removed per month to around 6,000, then began rising again. The second intervention also had an effect, though it appears to be milder. The reason we believe that the second intervention had a more modest effect than the first is because (1) the effect on price was about half the size of the first intervention, and (2) domestic meth production was being replaced by Mexican imports of meth over the late 1990s, and the precursor regulations were not applicable in Mexico. Thus, by the end of the 1990s, domestic meth production played a smaller role in total output, and hence the effect on price and admissions was probably smaller.

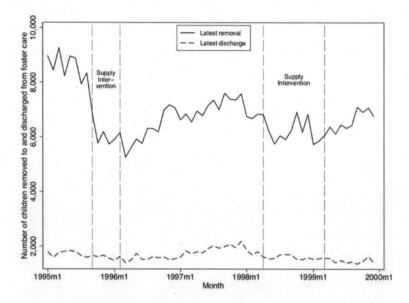

Figure 47. Showing reduced form effect of interventions on children removed from families and placed into foster care. Reprinted from Cunningham, S. and Finlay, K. (2012). "Parental Substance Abuse and Foster Care: Evidence from Two Methamphetamine Supply Shocks?" *Economic Inquiry*, 51(1):764–782. Copyright © 2012 Wiley. Used with permission from John Wiley and Sons.

It's worth reflecting for a moment on the reduced form. Why would rising retail prices of a pure gram of methamphetamine cause a child *not* to be placed in foster care? Prices don't cause child abuse—they're just nominal pieces of information in the world. The only way in which a higher price for meth could reduce foster care admissions is if parents reduced their consumption of methamphetamine, which in turn caused a reduction in harm to one's child. This picture is a key piece of evidence for the reader that this is going on.

In Table 52, I reproduce our main results from my article with Keith. There are a few pieces of key information that all IV tables should have. First, there is the OLS regression. As the OLS regression suffers from endogeneity, we want the reader to see it so that they have something to compare the IV model to. Let's focus on column 1, where the dependent variable is total entry into foster care. We find no effect, interestingly, of meth on foster care when we estimate using OLS.

Table 52. Log latest entry into foster care.

Covariates	Log Latest Entry into Foster Care		Log Latest Entry into Child Neglect		Log Latest Entry into Physical Abuse	
	OLS	2SLS	OLS	2SLS	OLS	2SLS
Log self-referred	0.001	1.54***	0.03	1.03**	0.04	1.49**
Meth treatment rate	(0.02)	(0.59)	(0.02)	(0.41)	(0.03)	(0.62)
Month-of-year fixed effects	Yes	Yes	Yes	Yes	Yes	Yes
State controls	Yes	Yes	Yes	Yes	Yes	Yes
State fixed effects	Yes	Yes	Yes	Yes	Yes	Yes
State linear time trends	Yes	Yes	Yes	Yes	Yes	Yes
First-stage instrument						
Price deviation instrument		−0.0005***		−0.0005***		−0.0005***
		(0.0001)		(0.0001)		(0.0001)
F-statistic for IV in first stage		17.60		17.60		17.60
N	1,343	1,343		1,343		1,343

Note: Log latest entry into foster care is the natural log of the sum of all new foster care admissions by state, race, and month. Models 3 to 10 denote the flow of children into foster care via a given route of admission denoted by the column heading. Models 11 and 12 use the natural log of the sum of all foster care exits by state, race, and month. ***, **, and * denote statistical significance at the 1%, 5%, and 10% levels, respectively.

The second piece of information that one should report in a 2SLS table is the first stage itself. We report the first stage at the bottom of each even-numbered column. As you can see, for each one-unit deviation in price from its long-term trend, meth admissions into treatment (our proxy) fell by -0.0005 log points. This is highly significant at the 1% level, but we check for the strength of the instrument using the F statistic [Staiger and Stock, 1997].[6] We have an F statistic of 17.6, which suggests that our instrument is strong enough for identification.

Finally, let's examine the 2SLS estimate of the treatment effect itself. Notice using only the exogenous variation in log meth admissions, and assuming the exclusion restriction holds in our model, we are able to isolate a causal effect of log meth admissions on log aggregate foster care admissions. As this is a log-log regression, we can interpret the coefficient as an elasticity. We find that a 10% increase in meth admissions for treatment appears to cause around a 15% increase in children being removed from their homes and placed into foster care. This effect is both large and precise. And it was not detectable otherwise (the coefficient was zero).

Why are they being removed? Our data (AFCARS) lists several channels: parental incarceration, child neglect, parental drug use, and physical abuse. Interestingly, we do not find any effect of parental drug use or parental incarceration, which is perhaps counterintuitive. Their signs are negative and their standard errors are large. Rather, we find effects of meth admissions on removals for physical abuse and neglect. Both are elastic (i.e., $\delta > 1$).

What did we learn? First, we learned how a contemporary piece of applied microeconomics goes about using instrumental variables to identify causal effects. We saw the kinds of graphical evidence mustered, the way in which knowledge about the natural experiment and the policies involved helped the authors argue for the exclusion restriction (since it cannot be tested), and the kind of evidence presented from 2SLS, including the first-stage tests for weak instruments. Hopefully seeing a paper at this point was helpful. But the second thing we learned concerned the actual study itself. We learned that for the group

6 I explain the importance of the F statistic later in this chapter, but ordinarily an F test on the excludability of the instrument from the first stage is calculated to check for the instrument's strength.

of meth users whose behavior was changed as a result of rising real prices of a pure gram of methamphetamine (i.e., the complier subpopulation), their meth use was causing child abuse and neglect so severe that it merited removing their children and placing those children into foster care. If you were only familiar with Dobkin and Nicosia [2009], who found no effect of meth on crime using county-level data from California and only the 1997 ephedrine shock, you might incorrectly conclude that there are no social costs associated with meth abuse. But, while meth does not appear to cause crime in California, it does appear to harm the children of meth users and places strains on the foster care system.

The Problem of Weak Instruments

I am not trying to smother you with papers. But before we move back into the technical material itself, I'd like to discuss one more paper. This paper will also help you better understand the weak instrument literature following its publication.

As we've said since the beginning, with example after example, there is a very long tradition in labor economics of building models that can credibly identify the returns to schooling. This goes back to Becker [1994] and the workshop at Columbia that Becker ran for years with Jacob Mincer. This study of the returns to schooling has been an important task given education's growing importance in the distribution of income and wealth since the latter twentieth century with increasing returns to skill in the marketplace [Juhn et al., 1993].

One of the more seminal papers in instrumental variables for the modern period is Angrist and Krueger [1991]. Their idea is simple and clever; a quirk in the United States educational system is that a child enters a grade on the basis of his or her birthday. For a long time, that cutoff was late December. If children were born on or before December 31, then they were assigned to the first grade. But if their birthday was on or after January 1, they were assigned to kindergarten. Thus two people—one born on December 31 and one born on January 1—were exogenously assigned different grades.

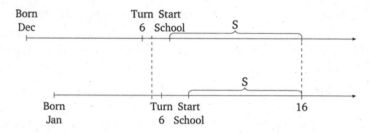

Figure 48. Compulsory schooling start dates by birthdates.

Now there's nothing necessarily relevant here because if those children always stay in school for the duration necessary to get a high school degree, then that arbitrary assignment of start date won't affect high school completion. It'll only affect *when* they get that high school degree. But this is where it gets interesting. For most of the twentieth century, the US had compulsory schooling laws that forced a person to remain in high school until age 16. After age 16, one could legally stop going to school. Figure 48 explains visually this instrumental variable.[7]

Angrist and Krueger had the insight that that small quirk was exogenously assigning more schooling to people born later in the year. The person born in December would reach age 16 with more education than the person born in January, in other words. Thus, the authors uncovered small exogenous variation in schooling. Notice how similar their idea was to regression discontinuity. That's because IV and RDD are conceptually very similar strategies.

Figure 49 shows the first stage, and it is really interesting. Look at all those 3s and 4s at the top of the picture. There's a clear pattern— those with birthdays in the third and fourth quarter have more schooling on average than do those with birthdays in the first and second quarters. That relationship gets weaker as we move into later cohorts, but that is probably because for later cohorts, the price on higher levels of schooling was rising so much that fewer and fewer people were dropping out before finishing their high school degree.

---

7 Angrist and Krueger always made such helpful and effective graphics for their studies, and this paper is a great example.

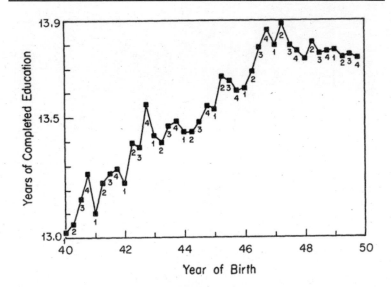

Figure 49. First-stage relationship between quarter of birth and schooling. Reprinted from Cunningham, S. and Finlay, K. (2012). "Parental Substance Abuse and Foster Care: Evidence from Two Methamphetamine Supply Shocks?" *Economic Inquiry*, 51(1):764–782. Copyright © 2012 Wiley. Used with permission from John Wiley and Sons.

Figure 50 shows the reduced-form relationship between quarter of birth and log weekly earnings.[8] You have to squint a little bit, but you can see the pattern—all along the top of the jagged path are 3s and 4s, and all along the bottom of the jagged path are 1s and 2s. Not always, but it's correlated.

Remember what I said about how instruments having a certain ridiculousness to them? That is, you know you have a good instrument if the instrument itself doesn't seem relevant for explaining the outcome of interest because *that's what the exclusion restriction implies*. Why would quarter of birth affect earnings? It doesn't make any obvious, logical sense why it should. But, if I told you that people born later

8 I know, I know. No one has ever accused me of being subtle. But it's an important point—a picture is worth a thousand words. If you can communicate your first stage and reduced form in pictures, you always should, as it will really captivate the reader's attention and be far more compelling than a simple table of coefficients ever could.

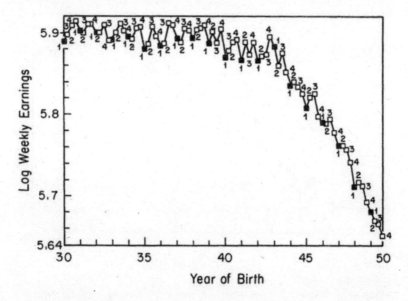

Figure 50. Reduced-form visualization of the relationship between quarter of birth and schooling. Reprinted from Angrist, J. D. and Krueger, A. B. (1991). "Does Compulsory School Attendance Affect Schooling and Earnings?" *Quarterly Journal of Economics*, 106(4):979–1014. Permission from Oxford University Press.

in the year got more schooling than those with less *because of compulsory schooling*, then the relationship between the instrument and the outcome snaps into place. The only reason we can think of as to why the instrument would affect earnings is if the instrument were operating through schooling. Instruments only explain the outcome, in other words, when you understand their effect on the endogenous variable.[9]

Angrist and Krueger use three dummies as their instruments: a dummy for first quarter, a dummy for second quarter, and a dummy for third quarter. Thus, the omitted category is the fourth quarter, which is the group that gets the most schooling. Now ask yourself this: if we regressed years of schooling onto those three dummies, what should

---

9 Having said that, Buckles and Hungerman [2013] found that in fact there are systematic differences in individual attributes that are predictors of ability by birth quarter!

Table 53. Quarter of birth and schooling.

Outcome variable	Birth cohort	Quarter-of-birth effect		
		I	II	III
Total schooling	1930–1939	−0.124	−0.86	−0.015
		(0.017)	(0.017)	(0.016)
	1940–1949	−0.085	−0.035	−0.017
		(0.012)	(0.012)	(0.011)
High school grad	1930–1939	−0.019	−0.020	−0.004
		(0.002)	(0.002)	(0.002)
	1940–1949	−0.015	−0.012	−0.002
		(0.001)	(0.001)	(0.001)
College grad	1930–1939	−0.005	0.003	0.002
		(0.002)	(0.002)	(0.002)
	1940–1949	−0.003	0.004	0.000
		(0.002)	(0.002)	(0.002)

Note: Standard errors in parentheses.

the signs and magnitudes be? That is, what would we expect the relationship between the first quarter (compared to the fourth quarter) and schooling? Let's look at their first-stage results (Table 53).

Table 53 shows the first stage from a regression of the following form:

$$S_i = X\pi_{10} + Z_1\pi_{11} + Z_2\pi_{12} + Z_3\pi_{13} + \eta_1$$

where Z_i is the dummy for the first three quarters, and π_i is the coefficient on each dummy. Now we look at what they produced in Table 53. The coefficients are all *negative* and significant for the total years of education and the high school graduate dependent variables. Notice, too, that the relationship gets much weaker once we move beyond the groups bound by compulsory schooling: the number of years of schooling for high school students (no effect) and probability of being a college graduate (no effect).

Regarding those college non-results. Ask yourself this question: why should we expect quarter of birth to affect the probability of being a high school graduate but not a college grad? What if we had found quarter of birth predicted high school completion, college completion, post-graduate completion, and total years of schooling beyond high

Table 54. Effect of schooling on wages using OLS and 2SLS.

Independent variable	OLS	2SLS
Years of schooling	0.0711	0.0891
	(0.0003)	(0.0161)
9 Year-of-birth dummies	Yes	Yes
8 Region-of-residence dummies	No	No

Note: Standard errors in parentheses. First stage is quarter of birth dummies.

school? Wouldn't it start to seem like this compulsory schooling instrument was not what we thought it was? After all, this quarter of birth instrument really should only impact *high school* completion; since it doesn't bind anyone beyond high school, it shouldn't affect the number of years beyond high school or college completion probabilities. If it did, we might be skeptical of the whole design. But here it didn't, which to me makes it even more convincing that they're identifying a compulsory high school schooling effect.[10]

Now we look at the second stage for both OLS and 2SLS (which the authors label TSLS, but means the same thing). Table 54 shows these results. The authors didn't report the first stage in this table because they reported it in the earlier table we just reviewed. For small values, the log approximates a percentage change, so they are finding a 7.1% return for every additional year of schooling, but with 2SLS it's higher (8.9%). That's interesting, because if it was merely ability bias, then we'd expect the OLS estimate to be *too large*, not too small. So something other than mere ability bias must be going on here.

For whatever it's worth, I am personally convinced at this point that quarter of birth is a valid instrument and that they've identified a causal effect of schooling on earnings, but Angrist and Krueger [1991] want to go further, probably because they want more precision in their estimate. And to get more precision, they load up the first stage with even

---

10 These kinds of falsifications are extremely common in contemporary applied work. This is because many of the identifying assumptions in any research design are simply untestable. And so the burden of proof is on researchers to convince the reader, oftentimes with intuitive and transparent falsification tests.

more instruments. Specifically, they use specifications with 30 dummies (quarter of birth × year) and 150 dummies (quarter of birth × state) as instruments. The idea is that the quarter of birth effect may differ by state and cohort.

But at what cost? Many of these instruments are only now weakly correlated with schooling—in some locations, they have almost no correlation, and for some cohorts as well. We got a flavor of that, in fact, in Table 54, where the later cohorts show less variation in schooling by quarter of birth than the earlier cohorts. What is the effect, then, of reducing the variance in the estimator by loading up the first stage with a bunch of noise?

Bound et al. [1995] is a classic work in what is sometimes called the "weak instrument" literature. It's in this paper that we learn some some very basic problems created by weak instruments, such as the form of 2SLS bias in finite samples. Since Bound et al. [1995] focused on the compulsory schooling application that Angrist and Krueger [1991] had done, I will stick with that example throughout. Let's consider their model with a single endogenous regressor and a simple constant treatment effect. The causal model of interest here is as before:

$$y = \beta s + \varepsilon$$

where y is some outcome and s is some endogenous regressor, such as schooling. Our instrument is Z and the first-stage equation is:

$$s = Z'\pi + \eta$$

Let's start off by assuming that ε and η are correlated. Then estimating the first equation by OLS would lead to biased results, wherein the OLS bias is:

$$E[\widehat{\beta}_{OLS} - \beta] = \frac{C(\varepsilon, s)}{V(s)}$$

We will rename this ratio as $\frac{\sigma_{\varepsilon\eta}}{\sigma_s^2}$. Bound et al. [1995] show that the bias of 2SLS centers on the previously defined OLS bias as the weakness of the instrument grows. Following Angrist and Pischke [2009],

Table 55. Effect of completed schooling on men's log weekly wages.

Independent variable	OLS	2SLS	OLS	2SLS	OLS	2SLS
Years of schooling	0.063	0.142	0.063	0.081	0.063	0.060
	(0.000)	(0.033)	(0.000)	(0.016)	(0.000)	(0.029)
First stage F		13.5		4.8		1.6
Excluded instruments						
Quarter of birth		Yes		Yes		Yes
Quarter of birth × year of birth		No		Yes		Yes
Number of excluded instruments		3		30		28

Note: Standard errors in parentheses. First stage is quarter of birth dummies.

I'll express that bias as a function of the first-stage F statistic:

$$E[\widehat{\beta}_{2SLS} - \beta] \approx \frac{\sigma_{\varepsilon\eta}}{\sigma_\eta^2} \frac{1}{F+1}$$

where F is the population analogy of the F-statistic for the joint significance of the instruments in the first-stage regression. If the first stage is weak, and $F \to 0$, then the bias of 2SLS approaches $\frac{\sigma_{\varepsilon\eta}}{\sigma_\eta^2}$. But if the first stage is very strong, $F \to \infty$, then the 2SLS bias goes to 0.

Returning to our rhetorical question from earlier, what was the cost of adding instruments without predictive power? Adding more weak instruments causes the first-stage F statistic to approach zero and increase the bias of 2SLS.

Bound et al. [1995] studied this empirically, replicating Angrist and Krueger [1991], and using simulations. Table 55 shows what happens once they start adding in controls. Notice that as they do, the F statistic on the excludability of the instruments falls from 13.5 to 4.7 to 1.6. So by the F statistic, they are already running into a weak instrument once they include the 30 quarter of birth × year dummies, and I think that's because as we saw, the relationship between quarter of birth and schooling got smaller for the later cohorts.

Next, they added in the weak instruments—all 180 of them—which is shown in Table 56. And here we see that the problem persists. The

Table 56. Effect of completed schooling on men's log weekly wages controlling for state of birth.

Independent variable	OLS	2SLS	OLS	2SLS
Years of schooling	0.063	0.083	0.063	0.081
	(0.000)	(0.009)	(0.000)	(0.011)
First stage F		2.4		1.9
Excluded instruments				
Quarter of birth		Yes		Yes
Quarter of birth $\times$ year of birth		Yes		Yes
Quarter of birth $\times$ state of birth		Yes		Yes
Number of excluded instruments		180		178

Note: Standard errors in parentheses.

instruments are weak, and therefore the bias of the 2SLS coefficient is close to that of the OLS bias.

But the really damning part of the Bound et al. [1995] was their simulation. The authors write:

> To illustrate that second-stage results do not give us any indication of the existence of quantitatively important finite-sample biases, we reestimated Table 1, columns (4) and (6) and Table 2, columns (2) and (4), using randomly generated information in place of the actual quarter of birth, following a suggestion by Alan Krueger. The means of the estimated standard errors reporting in the last row are quite close to the actual standard deviations of the 500 estimates for each model. . . . It is striking that the second-stage results reported in Table 3 look quite reasonable even with no information about educational attainment in the simulated instruments. They give no indication that the instruments were randomly generated. . . On the other hand, the F statistics on the excluded instruments in the first-stage regressions are always near their expected value of essentially 1 and do give a clear indication that the estimates of the second-stage coefficients suffer from finite-sample biases. (Bound et al., 448)

So, what can you do if you have weak instruments? Unfortunately, not a lot. You can use a just-identified model with your strongest IV.

Second, you can use a limited-information maximum likelihood estimator (LIML). This is approximately median unbiased for over identified constant effects models. It provides the same asymptotic distribution as 2SLS under homogeneous treatment effects but provides a finite-sample bias reduction.

But, let's be real for a second. If you have a weak instrument problem, then you only get so far by using LIML or estimating a just-identified model. The real solution for a weak instrument problem is to *get better instruments*. Under homogeneous treatment effects, you're always identifying the same effect, so there's no worry about a complier only parameter. So you should just continue searching for stronger instruments that simultaneously satisfy the exclusion restriction.[11]

In conclusion, I think we've learned a lot about instrumental variables and why they are so powerful. The estimators based on this design are capable of identifying causal effects when your data suffer from selection on unobservables. Since selection on unobservables is believed to be very common, this is a very useful methodology for addressing that. But, that said, we also have learned some of the design's weaknesses, and hence why some people eschew it. Let's now move to heterogeneous treatment effects so that we can better understand some limitations a bit better.

Heterogeneous Treatment Effects

Now we turn to a scenario where we relax the assumption that treatment effects are the same for every unit. This is where the potential outcomes notation comes in handy. Instead, we will allow for each unit to have a unique response to the treatment, or

$$Y_i^1 - Y_i^0 = \delta_i$$

Note that the treatment effect parameter now differs by individual i. We call this heterogeneous treatment effects.

---

11 Good luck with that. Seriously, good luck because—and I'm going out on a limb here—if you had a better instrument, you'd be using it!

The main questions we have now are: (1) what is IV estimating when we have heterogeneous treatment effects, and (2) under what assumptions will IV identify a causal effect with heterogeneous treatment effects? The reason this matters is that once we introduce heterogeneous treatment effects, we introduce a distinction between the internal validity of a study and its external validity. Internal validity means our strategy identified a causal effect *for the population we studied*. But external validity means the study's finding applied to *different* populations (not in the study). The deal is that under homogeneous treatment effects, there is no tension between external and internal validity because everyone has the same treatment effect. But under heterogeneous treatment effects, there is huge tension; the tension is so great, in fact, that it may even undermine the meaningfulness of the relevance of the estimated causal effect despite an otherwise valid IV design![12]

Heterogeneous treatment effects are built on top of the potential outcomes notation, with a few modifications. Since now we have two arguments—D and Z—we have to modify the notation slightly. We say that Y is a function of D and Z as $Y_i(D_i = 0, Z_i = 1)$, which is represented as $Y_i(0,1)$.

Potential *outcomes* as we have been using the term refers to the Y variable, but now we have a new potential variable—potential *treatment* status (as opposed to observed treatment status). Here are the characteristics:

$$D_i^1 = i\text{'s treatment status when } Z_i = 1$$
$$D_i^0 = i\text{'s treatment status when } Z_i = 0$$

And observed treatment status is based on a treatment status switching equations:

$$D_i = D_i^0 + (D_i^1 - D_i^0)Z_i$$
$$= \pi_0 + \pi_1 Z_i + \phi_i$$

12 My hunch is economists' priors assume heterogeneous treatment effects are the rule and constant treatment effects are the exception, but that many others have the opposite priors. It's actually not obvious to me there is a reason to have any particular priors, but economists' training tends to start with heterogeneity.

where $\pi_{0i} = E[D_i^0]$, $\pi_{1i} = (D_i^1 - D_i^0)$ is the heterogeneous causal effect of the IV on D_i, and $E[\pi_{1i}] =$ the average causal effect of Z_i on D_i.

There are considerably more assumptions necessary for identification once we introduce heterogeneous treatment effects—specifically five assumptions. We now review each of them. And to be concrete, I use repeatedly as an example the effect of military service on earnings using a draft lottery as the instrumental variable [Angrist, 1990]. In that paper, Angrist estimated the returns to military service using as an instrument the person's draft lottery number. The draft lottery number was generated by a random number generator and if a person's number was in a particular range, they were drafted, otherwise they weren't.

First, as before, there is a stable unit treatment value assumption (SUTVA) that states that the potential outcomes for each person i are unrelated to the treatment status of other individuals. The assumption states that if $Z_i = Z_i'$, then $D_i(Z) = D_i(Z')$. And if $Z_i = Z_i'$ and $D_i = D_i'$, then $Y_i(D,Z) = Y_i(D',Z')$. A violation of SUTVA would be if the status of a person at risk of being drafted was affected by the draft status of others at risk of being drafted. Such spillovers violate SUTVA.[13] Not knowing a lot about how that works, I can't say whether Angrist's draft study would've violated SUTVA. But it seems like he's safe to me.

Second, there is the independence assumption. The independence assumption is also sometimes called the "as good as random assignment" assumption. It states that the IV is independent of the potential outcomes and potential treatment assignments. Notationally, it is

$$\left\{ Y_i(D_i^1,1), Y_i(D_i^0,0), D_i^1, D_i^0 \right\} \perp\!\!\!\perp Z_i$$

The independence assumption is sufficient for a causal interpretation of the reduced form:

$$E[Y_i \mid Z_i = 1] - E[Y_i \mid Z_i = 0] = E[Y_i(D_i^1,1) \mid Z_i = 1] - E[Y_i(D_i^0,0) \mid Z_i = 0]$$
$$= E[Y_i(D_i^1,1)] - E[Y_i(D_i^0,0)]$$

---

13 Probably no other identifying assumption is given shorter shrift than SUTVA. Rarely is it mentioned in applied studies, let alone taken seriously.

And many people may actually prefer to work just with the instrument and its reduced form because they find independence satisfying and acceptable. The problem, though, is technically the instrument is not the program you're interested in studying. And there may be many mechanisms leading from the instrument to the outcome that you need to think about (as we will see below). Ultimately, independence is nothing more and nothing less than assuming that the instrument itself is random.

Independence means that the first stage measures the causal effect of Z_i on D_i:

$$E[D_i \mid Z_i = 1] - E[D_i \mid Z_i = 0] = E[D_i^1 \mid Z_i = 1] - E[D_i^0 \mid Z_i = 0]$$
$$= E[D_i^1 - D_i^0]$$

An example of this is if Vietnam conscription for military service was based on randomly generated draft lottery numbers. The assignment of draft lottery number was independent of potential earnings or potential military service because it was "as good as random."

Third, there is the exclusion restriction. The exclusion restriction states that any effect of Z on Y must be via the effect of Z on D. In other words, $Y_i(D_i, Z_i)$ is a function of D_i only. Or formally:

$$Y_i(D_i, 0) = Y_i(D_i, 1) \quad \text{for } D = 0, 1$$

Again, our Vietnam example. In the Vietnam draft lottery, an individual's earnings potential as a veteran or a non-veteran are assumed to be the same regardless of draft eligibility status. The exclusion restriction would be violated if low lottery numbers affected schooling by people avoiding the draft. If this was the case, then the lottery number would be correlated with earnings for at least two cases. One, through the instrument's effect on military service. And two, through the instrument's effect on schooling. The implication of the exclusion restriction is that a random lottery number (independence) does not therefore imply that the exclusion restriction is satisfied. These are different assumptions.

Fourth is the first stage. IV designs require that Z be correlated with the endogenous variable such that

$$E[D_i^1 - D_i^0] \neq 0$$

Z has to have some statistically significant effect on the average probability of treatment. An example would be having a low lottery number. Does it increase the average probability of military service? If so, then it satisfies the first stage requirement. Note, unlike independence and exclusion, the first stage is testable as it is based solely on D and Z, both of which you have data on.

And finally, there is the monotonicity assumption. This is only strange at first glance but is actually quite intuitive. Monotonicity requires that the instrumental variable (weakly) operate in the same direction on all individual units. In other words, while the instrument may have no effect on some people, all those who are affected are affected in the same direction (i.e., positively or negatively, but not both). We write it out like this:

$$\text{Either } \pi_{1i} \geq 0 \text{ for all } i \text{ or } \pi_{1i} \leq 0 \text{ for all } i = 1,\ldots,N$$

What this means, using our military draft example, is that draft eligibility may have no effect on the probability of military service for some people, like patriots, people who love and want to serve their country in the military, but when it does have an effect, it shifts them all into service, or out of service, but not both. The reason we have to make this assumption is that without monotonicity, IV estimators are not guaranteed to estimate a weighted average of the underlying causal effects of the affected group.

If all five assumptions are satisfied, then we have a valid IV strategy. But that being said, while valid, it is not doing what it was doing when we had homogeneous treatment effects. What, then, is the IV strategy estimating under heterogeneous treatment effects? Answer: the local average treatment effect (LATE) of D on Y:

$$\delta_{IV,LATE} = \frac{\text{Effect of } Z \text{ on } Y}{\text{Effect of } Z \text{ on } D}$$

$$= \frac{E[Y_i(D_i^1,1) - Y_i(D_i^0,0)]}{E[D_i^1 - D_i^0]}$$

$$= E[(Y_i^1 - Y_i^0) \mid D_i^1 - D_i^0 = 1]$$

The LATE parameter is the average causal effect of D on Y for those whose treatment status was changed by the instrument, Z. We know that because notice the difference in the last line: $D_i^1 - D_i^0$. So, for those people for whom that is equal to 1, we calculate the difference in potential outcomes. Which means we are only averaging over treatment effects for whom $D_i^1 - D_i^0$. Hence why the parameter we are estimating is "local."

How do we interpret Angrist's estimated causal effect in his Vietnam draft project? Well, IV estimates the average effect of military service on earnings for the subpopulations who enrolled in military service *because of the draft*. These are specifically only those people, though, who *would* not have served otherwise. It doesn't identify the causal effect on patriots who always serve, for instance, because $D_i^1 - D_i^0 = 0$ for patriots. They always serve! $D_i^1 = 1$ *and* $D_i^0 = 1$ for patriots because *they're patriots*! It also won't tell us the effect of military service on those who were exempted from military service for medical reasons because for these people $D_i^1 = 0$ and $D_i^0 = 0$.[14]

The LATE framework has even more jargon, so let's review it now. The LATE framework partitions the population of units with an instrument into potentially four mutually exclusive groups. Those groups are:

1. *Compliers*: This is the subpopulation whose treatment status is affected by the instrument in the correct direction. That is, $D_i^1 = 1$ and $D_i^0 = 0$.
2. *Defiers*: This is the subpopulation whose treatment status is affected by the instrument in the wrong direction. That is, $D_i^1 = 0$ and $D_i^0 = 1$.[15]
3. *Never takers*: This is the subpopulation of units that never take the treatment regardless of the value of the instrument. So, $D_i^1 = D_i^0 = 0$. They simply never take the treatment.[16]

---

14 We have reviewed the properties of IV with heterogeneous treatment effects using a very simple dummy endogenous variable, dummy IV, and no additional controls example. The intuition of LATE generalizes to most cases where we have continuous endogenous variables and instruments, and additional control variables, as well.

15 These are funny people. If they're drafted, they dodge the draft. But if they're not drafted, then they voluntarily enroll. In this context, defiance seems kind of irrational, but that's not always the case.

16 These are draft dodgers. For instance, maybe it's someone whose doctor gave him a bone-spur diagnosis so he could avoid service.

4. *Always takers*: This is the subpopulation of units that always take the treatment regardless of the value of the instrument. So, $D_i^1 = D_i^0 = 1$. They simply always take the instrument.[17]

As outlined above, with all five assumptions satisfied, IV estimates the average treatment effect for compliers, which is the parameter we've called the local average treatment effect. It's local in the sense that it is average treatment effect to the compliers only. Contrast this with the traditional IV pedagogy with homogeneous treatment effects. In that situation, compliers have the same treatment effects as non-compliers, so the distinction is irrelevant. Without further assumptions, LATE is not informative about effects on never-takers or always-takers because the instrument does not affect their treatment status.

Does this matter? Yes, absolutely. It matters because in most applications, we would be mostly interested in estimating the average treatment effect on the whole population, but that's not usually possible with IV.[18]

Now that we have reviewed the basic idea and mechanics of instrumental variables, including some of the more important tests associated with it, let's get our hands dirty with some data. We'll work with a couple of data sets now to help you better understand how to implement 2SLS in real data.

Applications

College in the county. We will once again look at the returns to schooling since it is such a historically popular topic for causal questions in labor. In this application, we will simply estimate a 2SLS model, calculate the first-stage F statistic, and compare the 2SLS results with the OLS results. I will be keeping it simple, because my goal is just to help the reader become familiarized with the procedure.

The data comes from the NLS Young Men Cohort of the National Longitudinal Survey. This data began in 1966 with 5,525 men aged

---

17 These are our patriots.
18 This identification of the LATE under heterogeneous treatment effects material was worked out in Angrist et al. [1996].

14–24 and continued to follow up with them through 1981. These data come from 1966, the baseline survey, and there are a number of questions related to local labor-markets. One of them is whether the respondent lives in the same county as a 4-year (and a 2-year) college.

Card [1995] is interested in estimating the following regression equation:

$$Y_i = \alpha + \delta S_i + \gamma X_i + \varepsilon_i$$

where Y is log earnings, S is years of schooling, X is a matrix of exogenous covariates, and ε is an error term that contains, among other things, unobserved ability. Under the assumption that ε contains ability, and ability is correlated with schooling, then $C(S, \varepsilon) \neq 0$ and therefore schooling is biased. Card [1995] proposes therefore an instrumental variables strategy whereby he will instrument for schooling with the college-in-the-county dummy variable.

It is worth asking ourselves why the presence of a 4-year college in one's county would increase schooling. The main reason I can think of is that the presence of the 4-year college increases the likelihood of going to college by lowering the costs, since the student can live at home. This therefore means that we are selecting on a group of compliers whose behavior is affected by the variable. Some kids, in other words, will always go to college regardless of whether a college is in their county, and some will never go despite the presence of the nearby college. But there may exist a group of compliers who go to college only because their county has a college, and if I'm right that this is primarily picking up people going because they can attend while living at home, then it's necessarily people at some margin who attend only because college became slightly cheaper. This is, in other words, a group of people who are liquidity constrained. And if we believe the returns to schooling for this group are different from those of the always-takers, then our estimates may not represent the ATE. Rather, they would represent the LATE. But in this case, that might actually be an interesting parameter since it gets at the issue of lowering costs of attendance for poorer families.

Here we will do some simple analysis based on Card [1995].

STATA

card.do

```
1   use https://github.com/scunning1975/mixtape/raw/master/card.dta, clear
2   reg lwage  educ  exper black south married smsa
3   ivregress 2sls lwage (educ=nearc4) exper black south married smsa, first
4   reg educ nearc4 exper black south married smsa
5   test nearc4
```

R

card.R

```
1    library(AER)
2    library(haven)
3    library(tidyverse)
4
5    read_data <- function(df)
6    {
7     full_path <- paste("https://raw.github.com/scunning1975/mixtape/master/",
8               df, sep = "")
9     df <- read_dta(full_path)
10    return(df)
11   }
12
13   card <- read_data("card.dta")
14
15   #Define variable
16   #(Y1 = Dependent Variable, Y2 = endogenous variable, X1 = exogenous variable,
     ↪  X2 = Instrument)
17
18   attach(card)
19
20   Y1 <- lwage
21   Y2 <- educ
22   X1 <- cbind(exper, black, south, married, smsa)
23   X2 <- nearc4
24
25   #OLS
26   ols_reg <- lm(Y1 ~ Y2 + X1)
27   summary(ols_reg)
28
29   #2SLS
30   iv_reg = ivreg(Y1 ~ Y2 + X1 | X1 + X2)
31   summary(iv_reg)
32
```

Table 57. OLS and 2SLS regressions of Log Earnings on schooling.

Dependent variable	Log earnings	
	OLS	2SLS
educ	0.071***	0.124**
	(0.003)	(0.050)
exper	0.034***	0.056***
	(0.002)	(0.020)
black	−0.166***	−0.116**
	(0.018)	(0.051)
south	−0.132***	−0.113***
	(0.015)	(0.023)
married	−0.036***	−0.032***
	(0.003)	(0.005)
smsa	0.176***	0.148***
	(0.015)	(0.031)
First-stage instrument		
College in the county		0.327***
Robust standard error		(0.082)
F statistic for IV in first stage		15.767
N	3,003	3,003
Mean dependent variable	6.262	6.262
SD dependent variable	0.444	0.444

Note: Standard errors in parentheses. *$p < 0.10$. **$p < 0.05$. ***$p < 0.01$.

Our results from this analysis have been arranged into Table 57. First, we report our OLS results. For every one year additional of schooling, respondents' earnings increase by approximately 7.1%. Next we estimated 2SLS using the ivregress 2sls command in Stata. Here we find a much larger return to schooling than we had found using OLS—around 75% larger in fact. But let's look at the first stage. We find that the college in the county is associated with 0.327 more years of schooling. This is highly significant ($p < 0.001$). The F statistic exceeds 15, suggesting we don't have a weak instrument problem. The return to schooling associated with this 2SLS estimate is 0.124—that is, for every additional year of schooling, earnings increase by 12.4%. Other covariates are listed if you're interested in studying them as well.

Why would the return to schooling be so much larger for the compliers than for the general population? After all, we showed earlier that if this was simply ability bias, then we'd expect the 2SLS coefficient to be *smaller* than the OLS coefficient, because ability bias implies that the coefficient on schooling is *too large*. Yet we're finding the opposite. So a couple of things it could be. First, it could be that schooling has measurement error. Measurement error would bias the coefficient toward zero, and 2SLS would recover its true value. But I find this explanation to be unlikely, because I don't foresee people really not knowing with accuracy how many years of schooling they currently have. Which leads us to the other explanation, and that is that compliers have larger returns to schooling. But why would this be the case? Assuming that the exclusion restriction holds, then why would compliers, returns be so much larger? We've already established that these people are likely being shifted into more schooling because they live with their parents, which suggests that the college is lowering the marginal cost of going to college. All we are left saying is that for some reason, the higher marginal cost of attending college is causing these people to underinvest in schooling; that in fact their returns are much higher.

Fulton Fish Markets. The second exercise that we'll be doing is based on Graddy [2006]. My understanding is that Graddy collected these data herself by recording prices of fish at the actual Fulton Fish Market. I'm not sure if that is true, but I like to believe it's true. Anyhow, the Fulton Fish Market operated in New York on Fulton Street for 150 years. In November 2005, it moved from Lower Manhattan to a large facility building for the market in the South Bronx. At the time when Graddy (2006) was published, the market was called the New Fulton Fish Market. It's one of the world's largest fish markets, second only to the Tsukiji in Tokyo.

Fish are heterogeneous, highly differentiated products. There are anywhere between one hundred and three hundred varieties of fish sold at the market. There are over fifteen varieties of shrimp alone. Within each variety, there's small fish, large fish, medium fish, fish just caught, fish that have been around a while. There's so much heterogeneity that customers often want to examine fish personally. You get the picture. This fish market functions just like a two-sided platform matching buyers to sellers, which is made more efficient by the thickness the market produces. It's not surprising, therefore, that Graddy found the market such an interesting thing to study.

Let's move to the data. I want us to estimate the price elasticity of demand for fish, which makes this problem much like the problem that Philip Wright faced in that price and quantity are determined simultaneously. The elasticity of demand is a sequence of quantity and price pairs, but with only one pair observed at a given point in time. In that sense, the demand curve is itself a sequence of potential outcomes (quantity) associated with different potential treatments (price). This means the demand curve is itself a real object, but mostly unobserved. Therefore, to trace out the elasticity, we need an instrument that is correlated with supply only. Graddy proposes a few of them, all of which have to do with the weather at sea in the days before the fish arrived to market.

The first instrument is the average maximum wave height in the previous two days. The model we are interested in estimating is:

$$Q = \alpha + \delta P + \gamma X + \varepsilon$$

where Q is log quantity of whiting sold in pounds, P is log average daily price per pound, X are day of the week dummies and a time trend, and ε is the structural error term. Table 58 presents the results from estimating this equation with OLS (first column) and 2SLS (second column). The OLS estimate of the elasticity of demand is −0.549. It could've been anything given price is determined by how many sellers and how many buyers there are at the market on any given day. But when we use the average wave height as the instrument for price, we get a −0.96 price elasticity of demand. A 10% increase in the price causes quantity to decrease by 9.6%. The instrument is strong ($F > 22$). For every one-unit increase in the wave-height, price rose 10%.

I suppose the question we have to ask ourselves, though, is what exactly is this instrument doing to supply. What are higher waves doing exactly? They are making it more difficult to fish, but are they also changing the composition of the fish caught? If so, then it would seem that the exclusion restriction is violated because that would mean the wave height is directly causing fish composition to change, which will directly determine quantities bought and sold.

Table 58. OLS and 2SLS regressions of Log Quantity on Log Price with wave-height instrument.

Dependent variable	Log quantity	
	OLS	2SLS
Log Price	−0.549***	−0.960**
	(0.184)	(0.406)
Monday	−0.318	−0.322
	(0.227)	(0.225)
Tuesday	−0.684***	−0.687***
	(0.224)	(0.221)
Wednesday	−0.535**	−0.520**
	(0.221)	(0.219)
Thursday	0.068	0.106
	(0.221)	(0.222)
Time trend	−0.001	−0.003
	(0.003)	(0.003)
First-stage Instrument		
Average wave height		0.103***
Robust standard error		(0.022)
F statistic for IV in first stage		22.638
N	97	97
Mean dependent variable	8.086	8.086
SD dependent variable	0.765	0.765

Note: Standard errors in parentheses. *$p < 0.10$. **$p < 0.05$. ***$p < 0.01$.

Now let's look at a different instrument: wind speed (Table 59). Specifically, it's the three-day lagged maximum wind speed. We present these results in Table 58. Here we see something we did not see before, which is that this is a weak instrument. The F statistic is less than 10 (approximately 6.5). And correspondingly, the estimated elasticity is twice as large as what we found with wave height. Thus, we know from our earlier discussion of weak instruments that this estimate is likely severely biased, and therefore less reliable than the previous one—even though the previous one itself (1) may not convincingly satisfy the exclusion restriction and (2) is at best a LATE relevant to compliers only. But as we've said, if we think that the compliers'

Table 59. OLS and 2SLS regressions of Log Quantity on Log Price with wind-speed instrument.

Dependent variable	Log quantity	
	OLS	2SLS
Log Price	−0.549***	−1.960**
	(0.184)	(0.873)
Monday	−0.318	−0.332
	(0.227)	(0.281)
Tuesday	−0.684***	−0.696**
	(0.224)	(0.277)
Wednesday	−0.535**	−0.482*
	(0.221)	(0.275)
Thursday	0.068	0.196
	(0.221)	(0.285)
Time trend	−0.001	−0.007
	(0.003)	(0.005)
First Stage Instrument		
Wind Speed		0.017**
Robust standard error		(0.007)
F statistic for IV in first stage		6.581
N	97	97
Mean Dependent Variable	8.086	8.086
Std. Dev. Dependent Variable	0.765	0.765

Note: Standard errors in parentheses. *$p < 0.10$, **$p < 0.05$, ***$p < 0.01$

causal effects are similar to that of the broader population, then the LATE may itself be informative and useful.

Popular IV Designs

Instrumental variables is a strategy to adopt when you have a good instrument, and so in that sense, it is a very general design that can be used in just about any context. But over the years, certain types of IV strategies have been used so many times that they constitute their own designs. And from repetition and reflection, we have a better understanding of how these specific IV designs do and do not work.

Let's discuss three such popular designs: the lottery design, the judge fixed effects design, and Bartik instruments.

Lotteries. Previously, we reviewed the use of IV in identifying causal effects when some regressor is endogenous in observational data. But one particular kind of IV application is randomized trials. In many randomized trials, participation is voluntary among those randomly chosen to be in the treatment group. On the other hand, persons in the control group usually don't have access to the treatment. Only those who are particularly likely to benefit from treatment therefore will probably take up treatment, which almost always leads to positive selection bias. If you compare means between treated and untreated individuals using OLS, you will obtain biased treatment effects even for the randomized trial because of noncompliance. A solution to the problems of least squares in this application is to instrument for Medicaid with whether you were offered treatment and estimate the LATE. Thus even when treatment itself is randomly assigned, it is common for people to use a randomized lottery as an instrument for participation. For a modern example of this, see Baicker et al. [2013], who used the randomized lottery to be enrolled in Oregon's Medicaid program as an instrument for being on Medicaid. Let's discuss the Oregon Medicaid studies now, as they are excellent illustrations of the lottery IV design.

What are the effects of expanding access to public health insurance for low-income adults? Are they positive or negative? Are they large or small? Surprisingly, we have not historically had reliable estimates for these very basic questions because we lacked the kind of experiment needed to make claims one way or another. The limited existing evidence was suggestive, with a lot of uncertainty. Observational studies are confounded by selection into health insurance, and the quasi-experimental evidence tended to only focus on the elderly and small children. There has been only one randomized experiment in a developed country, and it was the RAND health insurance experiment in the 1970s. This was an important, expensive, ambitious experiment, but it only randomized cost-sharing—not coverage itself.

But in the 2000s, Oregon chose to expand its Medicaid program for poor adults by making it more generous. Adults aged 19–64 with income less than 100% of the federal poverty line were eligible so long as they weren't eligible for other similar programs. They also had to be uninsured for fewer than six months and be a legal US resident. The program was called the Oregon Health Plan Standard and it provided comprehensive coverage (but no dental or vision) and minimum cost-sharing. It was similar to other states in payments and management, and the program was closed to new enrollment in 2004.

The expansion is popularly known as the Oregon Medicaid Experiment because the state used a lottery to enroll volunteers. For five weeks, people were allowed to sign up for Medicaid. The state used heavy advertising to make the program salient. There were low barriers to signing up and no eligibility requirements for prescreening. The state in March to October 2008 randomly drew 30,000 people out of a list of 85,000. Those selected were given a chance to apply. If they did apply, then their entire household was enrolled, so long as they returned the application within 45 days. Out of this original 30,000, only 10,000 people were enrolled.

A team of economists became involved with this project early on, out of which several influential papers were written. I'll now discuss some of the main results from Finkelstein et al. [2012] and Baicker et al. [2013]. The authors of these studies sought to study a broad range of outcomes that might be plausibly affected by health insurance—from financial outcomes, to health-care utilization, to health outcomes. The data needed for these outcomes were meticulously collected from third parties. For instance, the pre-randomization demographic information was available from the lottery sign-up. The state administrative records on Medicaid enrollment were also collected and became the primary measure of a first stage (i.e., insurance coverage). And outcomes were collected from administrative sources (e.g., hospital discharge, mortality, credit), mail surveys, and in-person survey and measurement (e.g., blood samples, body mass index, detailed questionnaires).

The empirical framework in these studies is a straightforward IV design. Sometimes they estimated the reduced form, and sometimes they estimated the full 2SLS model. The two-stages were:

$$\text{INSURANCE}_{ihj} = \delta_0 + \delta_1 \text{ LOTTERY}_{ih} + X_{ih}\delta_2 + V_{ih}\delta_3 + \mu_{ihj}$$

$$y_{ihj} = \pi_0 + \pi_1 \widehat{\text{INSURANCE}}_{ih} + X_{ih}\pi_2 + V_{ih}\pi_3 + v_{ihj}$$

Table 60. Effect of lottery on enrollment.

Dependent variable	Full sample	Survey respondents
Ever on Medicaid	0.256	0.290
	(0.004)	(0.007)
Ever on OHP Standard	0.264	0.302
	(0.003)	(0.005)
Number of months on Medicaid	3.355	3.943
	(0.045)	(0.09)

Note: Standard errors in parentheses.

where the first equation is the first stage (insurance regressed onto the lottery outcome plus a bunch of covariates), and the second stage regresses individual-level outcomes onto predicted insurance (plus all those controls). We already know that so long as the first stage is strong, then the F statistic will be large, and the finite sample bias lessens.

The effects of winning the lottery had large effects on enrollment. We can see the results of the first stage in Table 60. They used different samples, but the effect sizes were similar. Winning the lottery raised the probability of being enrolled on Medicaid by 26% and raised the number of months of being on Medicaid from 3.3 to 4 months.

Across the two papers, the authors looked at the effect of Medicaid's health insurance coverage on a variety of outcomes including financial health, mortality, and health-care utilization, but I will review only a few here. In Table 61, the authors present two regression models: column 2 is the intent to treat estimates, which is the reduced form model, and column 3 is the local average treatment effect estimate, which is our full instrumental variables specification. Interestingly, Medicaid increased the number of hospital admissions but had no effect on emergency room visits. The effect on emergency rooms, in fact, is not significant, but the effect on non-emergency-room admissions is positive and significant. This is interesting because it appears that Medicaid is increasing hospital admission without putting additional strain on emergency rooms, which already have scarce resources.

Table 61. Effect of Medicaid on hospital admission.

Dependent variable	ITT	LATE
Any hospital admission	0.5%	2.1%
Hospital admissions through ED	0.2%	0.7%
Hospital admissions not through ED	0.4%	1.6%

Note: Hospital discharge data.

Table 62. Effect of Medicaid on health-care usage.

	ITT	LATE
Have a usual place of care	9.9%	33.9%
Have a personal doctor	8.1%	28.0%
Got all needed healthcare	6.9%	23.9%
Got all needed prescriptions	5.6%	19.5%
Satisfied with quality of care	4.3%	14.2%

Note: Standard errors in parentheses.

What other kinds of health-care utilization are we observing in Medicaid enrollees? Let's look at Table 62, which has five health-care utilization outcomes. Again, I will focus on column 3, which is the LATE estimates. Medicaid enrollees were 34% more likely to have a usual place of care, 28% to have a personal doctor, 24% to complete their health-care needs, 20% more likely to get all needed prescriptions and 14% increased satisfaction with the quality of their care.

But Medicaid is not merely a way to increase access to health care; it also functions effectively as health care insurance in the event of catastrophic health events. And one of the most widely circulated results of the experiment was the finding that Medicaid had on financial outcomes. In Table 63 we see that one of the main effects was reduction in personal debt (by $390) and reducing debt going to debt collection. The authors also found reductions in out-of-pocket medical expenses, and medical expenses, borrowing money or skipping bills for medical expenses, and whether they refused medical treatment due to medical debt.

Table 63. Effect of Medicaid on hospital admission.

Dependent variable	ITT	LATE
Had a bankruptcy	0.2%	0.9%
Had a collection	−1.2%	−4.8%
Had a medical collection	−1.6%	−6.4%
Had non-medical collection	−0.5%	−1.8%
Money owed medical collection	−$99	−$390

Note: Credit records.

Table 64. Effect of Medicaid on hospital admission.

Dependent variable	ITT	LATE
Health good, very good, or excellent	3.9%	13.3%
Health stable or improving	3.3%	11.3%
Depression screen NEGATIVE	2.3%	7.8%
CDC healthy days (physical)	0.381	1.31
CDC healthy days (mental)	0.603	2.08

But the effect on health outcomes was a little unclear from this study. The authors find self-reported health outcomes to be improving, as well as a reduction in depression. They also find more healthy physical and mental health days. But the effects are overall small. Furthermore, they ultimately do not find that Medicaid had any effect on mortality—a result we will return to again in the difference-in-differences chapter.

In conclusion, we see a powerful use of IV in the assignment of lotteries to recipients. The lotteries function as instruments for treatment assignment, which can then be used to estimate some local average treatment effect. This is incredibly useful in experimental designs if only because humans often refuse to comply with their treatment assignment or even participate in the experiment altogether!

Judge fixed effects. A second IV design that has become extremely popular in recent years is the "judge fixed effects" design. You may also hear it called the "leniency design," but because the applications so often involve judges, it seems the former name has stuck. A search

Figure 51. Randomized judge assignment. Although justice is supposedly blind, judges are complex bundles of characteristics that affect their judgments. Artwork by Seth Hahne ©2020.

on Google Scholar for the term yields over 70 hits, with over 50 since 2018 alone.

The concept of the judge fixed effects design is that there exists a narrow pipeline through which all individuals must pass, numerous randomly assigned decision-makers blocking the individuals' passage who assign a treatment to the individuals and discretion among the decision-makers. When all three are there, you probably have the makings of a judge fixed effects design. The reason the method is called the judge fixed effects design is because it has traditionally exploited a feature in American jurisprudence where jurisdictions will randomly assign judges to defendants. In Harris County, Texas, for instance, they used to use a bingo machine to assign defendants to one of dozens of courts [Mueller-Smith, 2015].

The first paper to recognize that there were systematic differences in judge sentencing behavior was an article by Gaudet et al. [1933]. The authors were interested in better understanding what, other than guilt, determined the sentencing outcomes of defendants. They decided to focus on the judge in part because judges were being randomly

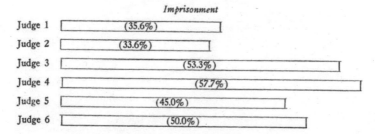

Figure 52. Variation in judge sentencing outcomes. This figure originally appeared in Frederick J. Gaudet, George S. Harris, Charles W. St. John, *Individual Differences in the Sentencing Tendencies of Judges*, 23, J. Crim. L. & Criminology, 811 (1933). Reprinted by special permission of Northwestern University Pritzker School of Law, *Journal of Criminal Law and Criminology*.

rotated to defendants. And since they were being "by chance" rotated to defendants, in a large sample, the characteristics of the defendants should've remained approximately the same across all judges. Any differences in sentencing outcomes, therefore, wouldn't be because of the underlying charge or even the defendant's guilt, but rather, would be connected to the judge. See Figure 51 for a beautiful drawing of this identification strategy based on over 7,000 hand collected cases showing systematic differences in judge sentencing behavior.

Figure 52 is a pretty typical graphic for any paper on judge fixed effects showing the variation in the judge's propensities; it's just kind of interesting that it appears from the very beginning back in 1933. And as you can see in Figure 52, there is weirdly enough a lot of variation in the sentencing propensities across the six judges. Judge 2 imposed imprisonment in only 33.6% of his cases, whereas Judge 4 imposed imprisonment in a whopping 57.7%. And since they are all seeing on average the same defendants, it can't be that Judge 4 is simply seeing worse cases. Rather, there appears to be something systematic, like a tendency for certain judges to always judge defendants more harshly. But why are they like this? Gaudet et al. [1933] offer the following conjecture:

Perhaps the most interesting thing to be noticed in these graphs is the fact that the sentencing tendency of the judge seems to be fairly well determined before he sits on the bench. In other words what determines whether a judge will be severe or lenient is to be found in the environment to which the judge has been subjected previous to his becoming an administrator of sentences.

But the main takeaway is that the authors were the first to discover that the leniency or severity of the judge, and not merely the defendant's own guilt, plays a significant role apparently in the final determination of a case against the defendant. The authors write:

The authors wish to point out that these results tend to show that some of our previous studies in the fields of criminology and penology are based upon very unreliable evidence if our results are typical of sentencing tendencies. In other words, what type of sentence received by a prisoner may be either an indication of the seriousness of his crime or of the severity of the judge. [815]

The next mention of the explicit judge fixed effects design is in the Imbens and Angrist [1994] article decomposing IV into the LATE parameter using potential outcomes notation. At the conclusion of their article, they provide three examples of IV designs that may or may not fit the five identifying assumptions of IV that I discussed earlier. They write:

Example 2 (Administrative Screening): Suppose applicants for a social program are screened by two officials. The two officials are likely to have different admission rates, even if the stated admission criteria are identical. Since the identity of the official is probably immaterial to the response, it seems plausible that Condition 1 [independence] is satisfied. The instrument is binary so Condition 3 is trivially satisfied. However, Condition 2 [monotonicity] requires that if official A accepts applicants with probability $P(0)$, and official B accepts people with probability $(P1) > P(0)$, official B must accept *any* applicant who would have been accepted by official A. This is unlikely to hold if admission is based on a number of criteria. Therefore, in this example we *cannot* use Theorem 1 to identify a local average treatment effect nonparametrically despite the presence of an instrument satisfying Condition 1 [independence]. [472]

While the first time we see the method used for any type of empirical identification is Waldfogel [1995], the first explicit IV strategy is a paper ten years later by Kling [2006], who used randomized judge assignment with judge propensities to instrument for incarceration length. He then linked defendants to employment and earnings records, which he then used to estimate the causal effect of incarceration on labor-market outcomes. He ultimately finds no adverse effects on labor-market consequences from longer sentences in the two states he considers.

But this question was revisited by Mueller-Smith [2015], who used Harris County, Texas, for his location. Harris County has dozens of courts and defendants are randomly assigned to one of them. Mueller-Smith linked defendant outcomes to a variety of labor-market and criminal outcomes, and came to the opposite conclusion as Kling [2006]. Mueller-Smith [2015] finds that incarceration generates net increases in the frequency and severity of recidivism, worsens labor-market outcomes, and increases defendant's dependence on public assistance.

Judicial severity causing adverse consequences on defendants is practically a hallmark of the judge fixed effects literature. Just to name a few such examples, there is the finding that less allowance of Chapter 13 bankruptcy worsens future financial events [Dobbie et al., 2017], racial bias among bail judges [Arnold et al., 2018], pretrial detention having higher rates of guilty pleas, conviction, recidivism, and worsened labor-market outcomes [Dobbie et al., 2018; Leslie and Pope, 2018; Stevenson, 2018], juvenile incarceration worsening high school outcomes and adult recidivism rates [Aizer and Doyle, 2015], foster care raising juvenile delinquency, teen pregnancy, and worsening future employment [Doyle, 2007], foster care increasing adult crime [Doyle, 2008], and countless others. But there are a few exceptions. For instance, Norris et al. [2020] find beneficial effects to children when the marginal siblings and parents are incarcerated.

The three main identifying assumptions that should be on the researcher's mind when attempting to implement a judge fixed effects design are the independence assumption, exclusion restriction, and the monotonicity assumption. Let's discuss them each at a time because, in some scenarios, one of these may be more credible than the other.

The independence assumption seems to be satisfied in many cases because the administrators in question are literally being randomly assigned to individual cases. As such, our instrument—which is sometimes modeled as the average propensity of the judge excluding the case in question or simply as a series of judge fixed effects (which, as I'll mention in a moment, turns out to be equivalent)—easily passes the independence test. But it's possible that strategic behavior on the part of the defendant in response to the strictness of the judge they were assigned can undermine the otherwise random assignment. Consider something that Gaudet et al. [1933] observed in their original study regarding the dynamics of the courtroom when randomly assigned a severe judge:

> The individual tendencies in the sentencing tendencies of judges are evidently recognized by many who are accustomed to observe this sentencing. The authors have been told by several lawyers that some recidivists know the sentencing tendencies of judges so well that the accused will frequently attempt to choose which judge is to sentence them, and further, some lawyers say that they are frequently able to do this. It is said to be done in this way. If the prisoner sees that he is going to be sentenced by Judge X, whom he believes to be severe in his sentencing tendency, he will change his plea from "Guilty" to "Non Vult" or from "Non Vult" to "Not Guilty," etc. His hope is that in this way the sentencing will be postponed and hence he will probably be sentenced by another judge. [812]

There are several approaches one can take to assessing independence. First, checking for balance on pre-treatment covariates is an absolute must. Insofar as this is a randomized experiment, then all observable and unobservable characteristics will be distributed equally across the judges. While we cannot check for balance on unobservables, we can check for balance on observables. Most papers of which I am aware check for covariate balance, usually before doing any actual analysis.

Insofar as you suspect endogenous sorting, you might simply use the original assignment, not the final assignment, for identification. This is because in most cases, we will know the initial judge assignment was random. But this approach may not be feasible in many settings if initial judge or court assignment is not available. Nevertheless, endogenous sorting in response to the severity of the judge could undermine the design by introducing a separate mechanism by which the instrument impacts the final decision (via sorting into the lenient judge's courtroom if possible), and the researcher should attempt to ascertain through conversations with administrators the degree to which this practically occurs in the data.

The violation of exclusion is more often the worry, though, and really should be evaluated case by case. For instance, in Dobbie et al. [2018], the authors are focused on pretrial detention. But pretrial detention is determined by bail set by judges who do not themselves have any subsequent interaction with the next level's randomized judge, and definitely don't have any interaction with the defendant upon the judicial ruling and punishment rendered. So in this case, it does seem like Dobbie et al. [2018] might have a more credible argument that exclusion holds.

But consider a situation where a defendant is randomly assigned a severe judge. In expectation, if the case goes to trial, the defendant faces a higher expected penalty even given a fixed probability of conviction across any judge for no other reason than that the stricter judge will likely choose a harsher penalty and thus drive up the expected penalty. Facing this higher expected penalty, the defense attorney and defendant might decide to accept a lesser plea in response to the judge's anticipated severity, which would violate exclusion since exclusion requires the instrument effect the outcome only through the judge's decision (sentence).

But even if exclusion can be defended, in many situations monotonicity becomes the more difficult case to make for this design. It was explicitly monotonicity that made Imbens and Angrist [1994] skeptical that judge fixed effects could be used to identify the local average treatment effect. This is because the instrument is required to weakly operate the same across all defendants. Either a judge is strict or she isn't, but she can't be both in different circumstances. Yet humans are complex bundles of thoughts and experiences, and those biases may operate in non-transitive ways. For instance, a judge may be lenient, except when the defendant is black or if the offense is a drug charge, in which case they switch and become strict. Mueller-Smith [2015] attempted to overcome potential violations of exclusion and monotonicity through a parametric strategy of simultaneously instrumenting for all observed sentencing dimensions and thus allowing the instruments' effect on sentencing outcomes to be heterogeneous in defendant traits and crime characteristics.

Formal solutions to querying the plausibility of these assumptions have appeared in recent years, though. Frandsen et al. [2019] propose a test for exclusion and monotonicity based on relaxing the monotonicity assumption. This test requires that the average treatment effect among individuals who violate monotonicity be identical to the average treatment effect among some subset of individuals who satisfy it. Their test simultaneously tests for exclusion and monotonicity, so one cannot be sure which violation is driving the test's result unless theoretically one rules out one of the two using a priori information. Their proposed test is based on two observations: that the average outcomes, conditional on judge assignment, should fit a continuous function of judge propensities, and secondly, the slope of that continuous function should be bounded in magnitude by the width of the outcome variable's support. The test itself is relatively straightforward and simply requires examining whether observed outcomes averaged by judges are consistent with such a function. We can see a picture of what it looks like to pass this test in the top panel of Figure 53 versus the bottom panel, which fails the test.

While the authors have made available code and documentation that can be used to implement this test,[19] it is not currently available in R and therefore will not be reviewed here.

In this section, I'd like to accomplish two things. First, I'd like to review an interesting new paper by Megan Stevenson that examined how cash bail affected case outcomes [Stevenson, 2018]. As this is an important policy question, I felt it would be good to review this

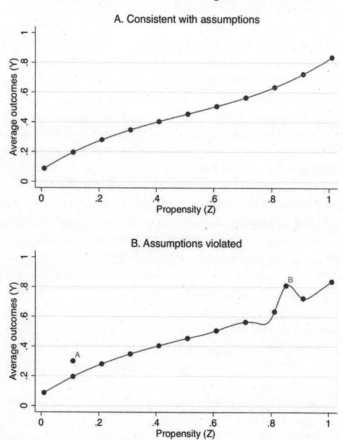

Figure 53. Average outcomes as a function of judge propensities. Frandsen, B. R., Lefgren, L. J., and Leslie, E. C. (2019). Judging judge fixed effects. Working Paper No. 25528, National Bureau of Economic Research, Cambridge, MA. Reprinted with permission from authors.

excellent study. But the second purpose of this section is to replicate her main results so that the reader can see exactly how to implement this instrumental variables strategy.

As with most judge fixed effects papers, Stevenson is working with administrative data for a large city. Large cities are probably the best context due to the large samples which can help ameliorate the finite sample bias of IV. Fortunately, these data are often publicly available and need only be scraped from court records that are in many locations posted online. Stevenson [2018] focuses on Philadelphia, where the natural experiment is the random assignment of bail judges ("magistrates") who unsurprisingly differ widely in their propensity to set bail at affordable levels. In other words, bail judges differ systematically in the price they set for bail, and given a downward-sloping demand curve, more severe judges setting expensive bails will see more defendants unable to pay their bail. As a result, they are forced to remain in detention prior to the trial.

Using a variety of IV estimators, Stevenson [2018] finds that an increase in randomized pretrial detention leads to a 13% increase in the likelihood of receiving a conviction. She argues that this is caused by an increase in guilty pleas among defendants who otherwise would have been acquitted or had their charges dropped—a particularly problematic mechanism, if true. Pretrial detention also led to a 42% increase in the length of the incarceration sentence and a 41% increase in the amount of non-bail fees owed. This provides support for idea that cash bail contributes to a cycle of poverty in which defendants unable to pay their court fees end up trapped in the penal system through higher rates of guilt, higher court fees, and likely higher rates of reoffending [Dobbie et al., 2018].

One might think that the judge fixed effects design is a "just identified" model. Can't we just use as our instrument the average strictness for each judge (excluding the defendant's own case)? Then we have just one instrument for our one endogenous variable, and 2SLS seems like a likely candidate, right? After all, that one instrument would be unique to each individual because each individual would have a unique judge and a unique average strictness if average strictness was calculated as the mean of all judge sentencing excluding the individual under consideration.

The problem is that this is still just a high-dimension instrument. The correct specification is to use the actual judge fixed effects, and depending on your application you may have anywhere from eight (as in Stevenson's case) to hundreds of judges. Insofar as some of these are weak, which they probably will be, you run into a typical kind of overidentification problem where in finite samples you begin moving the point estimates back to centering on the OLS bias as I discussed earlier. This issue is still being resolved by econometricians and is likely to be an active area of research going forward. Some solutions may be to use high-dimension reduction techniques such as LASSO [Gilchrist and Sands, 2016], instrument selection [Donald and Newey, 2001], or perhaps combining individual judges with similar strictness into only one instrument.

Stevenson's data contains 331,971 observations and eight randomly assigned bail judges. Like many papers in the judge fixed effects literature, she uses the jackknife instrumental variables estimator (JIVE) [Angrist et al., 1999]. While 2SLS is the most commonly used IV estimator in applied microeconomics applications, it suffers from finite sample problems when there are weak instruments and the use of many instruments as we showed with the discussion of Bound et al. [1995]. Angrist et al. [1999] proposed an estimator that attempts to eliminate the finite-sample bias of 2SLS called JIVE.[20] These aren't perfect, as their distributions are larger than that of the 2SLS estimator, but they may have an advantage when there are several instruments and some of which are weak (as is likely to occur with judge fixed effects).

JIVE is popularly known as a "leave one out" estimator. Angrist et al. [1999] suggest using all observations in this estimator except for the *i* unit. This is the nice feature for judge fixed effects because ideally the instrument is the mean strictness of the judge *in all other cases*, excluding the particular defendant's case. So JIVE is nice both for its handling of the finite sample bias, and for its construction of the theoretical instrument more generally.

Given the econometrics of judge fixed effects with its many instruments is potentially the frontier of econometrics, my goal here will be somewhat backwards looking. We will simply run through some simple exercises using JIVE so that you can see how historically researchers are estimating their models.

---

20 Best name for an estimator ever.

STATA
bail.do

```
1   use https://github.com/scunning1975/mixtape/raw/master/judge_fe.dta, clear
2
3   global judge_pre judge_pre_1 judge_pre_2 judge_pre_3 judge_pre_4 judge_pre_5
    ↪  judge_pre_6 judge_pre_7 judge_pre_8
4   global demo black age male white
5   global off      fel mis sum F1 F2 F3 F M1 M2 M3 M
6   global prior priorCases priorWI5 prior_felChar  prior_guilt onePrior threePriors
7   global control2      day day2 day3  bailDate t1 t2 t3 t4 t5 t6
8
9
10  * Naive OLS
11  * minimum controls
12  reg guilt jail3 $control2, robust
13  * maximum controls
14  reg guilt jail3 possess robbery DUI1st drugSell aggAss $demo $prior $off
    ↪  $control2 , robust
15
16
17  ** Instrumental variables estimation
18  * 2sls main results
19  * minimum controls
20  ivregress 2sls guilt (jail3= $judge_pre) $control2, robust
21  * maximum controls
22  ivregress 2sls guilt (jail3= $judge_pre) possess robbery DUI1st drugSell aggAss
    ↪  $demo $prior $off $control2 , robust
23
24  * JIVE main results
25  * minimum controls
26  jive guilt (jail3= $judge_pre) $control2, robust
27  * maximum controls
28  jive guilt (jail3= $judge_pre) possess robbery DUI1st drugSell aggAss $demo
    ↪  $prior $off $control2 , robust
```

R

bail.R

```
1   library(tidyverse)
2   library(haven)
3   library(estimatr)
4   library(lfe)
5   library(SteinIV)
6
7   read_data <- function(df)
8   {
9     full_path <- paste("https://raw.github.com/scunning1975/mixtape/master/",
10               df, sep = "")
11    df <- read_dta(full_path)
12    return(df)
13  }
14
15  judge <- read_data("judge_fe.dta")
16
17  #grouped variable names from the data set
18  judge_pre <- judge %>%
19    select(starts_with("judge_")) %>%
20    colnames() %>%
21    subset(., . != "judge_pre_8") %>% # remove one for colinearity
22    paste(., collapse = " + ")
23
24  demo <- judge %>%
25    select(black, age, male, white) %>%
26    colnames() %>%
27    paste(., collapse = " + ")
28
29  off <- judge %>%
30    select(fel, mis, sum, F1, F2, F3, M1, M2, M3, M) %>%
31    colnames() %>%
32    paste(., collapse = " + ")
33
34  prior <- judge %>%
35    select(priorCases, priorWI5, prior_felChar,
36        prior_guilt, onePrior, threePriors) %>%
37    colnames() %>%
38    paste(., collapse = " + ")
39
```

(continued)

R *(continued)*

```
40    control2 <- judge %>%
41      mutate(bailDate = as.numeric(bailDate)) %>%
42      select(day, day2, bailDate,
43          t1, t2, t3, t4, t5) %>% # all but one time period for colinearity
44      colnames() %>%
45      paste(., collapse = " + ")
46
47    #formulas used in the OLS
48    min_formula <- as.formula(paste("guilt ~ jail3 + ", control2))
49    max_formula <- as.formula(paste("guilt ~ jail3 + possess + robbery + DUI1st +
      ↪   drugSell + aggAss",
50                    demo, prior, off, control2, sep = " + "))
51
52    #max variables and min variables
53    min_ols <- lm_robust(min_formula, data = judge)
54    max_ols <- lm_robust(max_formula, data = judge)
55
56    #--- Instrumental Variables Estimations
57    #- 2sls main results
58    #- Min and Max Control formulas
59    min_formula <- as.formula(paste("guilt ~ ", control2, " | 0 | (jail3 ~ 0 +", judge_pre,
      ↪   ")"))
60    max_formula <- as.formula(paste("guilt ~", demo, "+ possess +", prior, "+ robbery
      ↪   +",
61                    off, "+ DUI1st +", control2, "+ drugSell + aggAss | 0 | (jail3 ~ 0
                    ↪   +", judge_pre, ")"))
62    #2sls for min and max
63    min_iv <- felm(min_formula, data = judge)
64    summary(min_iv)
65    max_iv <- felm(max_formula, data = judge)
66    summary(max_iv)
67
68
69
70    #--- JIVE main results
71    #- minimum controls
72    y <- judge %>%
73      pull(guilt)
74
```

(continued)

R *(continued)*

```
75   X_min <- judge %>%
76     mutate(bailDate = as.numeric(bailDate)) %>%
77     select(jail3, day, day2, t1, t2, t3, t4, t5, bailDate) %>%
78     model.matrix(data = .,~.)
79
80   Z_min <- judge %>%
81     mutate(bailDate = as.numeric(bailDate)) %>%
82     select(-judge_pre_8) %>%
83     select(starts_with("judge_pre"), day, day2, t1, t2, t3, t4, t5, bailDate) %>%
84     model.matrix(data = .,~.)
85
86   jive.est(y = y, X = X_min, Z = Z_min)
87
88   #- maximum controls
89   X_max <- judge %>%
90     mutate(bailDate = as.numeric(bailDate)) %>%
91     select(jail3, white, age, male, black,
92          possess, robbery, prior_guilt,
93          prior_guilt, onePrior, priorWI5, prior_felChar, priorCases,
94          DUI1st, drugSell, aggAss, fel, mis, sum,
95          threePriors,
96          F1, F2, F3,
97          M, M1, M2, M3,
98          day, day2, bailDate,
99          t1, t2, t3, t4, t5) %>%
100    model.matrix(data = .,~.)
101
102  Z_max <- judge %>%
103    mutate(bailDate = as.numeric(bailDate)) %>%
104    select(-judge_pre_8) %>%
105    select(starts_with("judge_pre"), white, age, male, black,
106          possess, robbery, prior_guilt,
107          prior_guilt, onePrior, priorWI5, prior_felChar, priorCases,
108          DUI1st, drugSell, aggAss, fel, mis, sum,
109          threePriors,
110          F1, F2, F3,
111          M, M1, M2, M3,
112          day, day2, bailDate,
113          t1, t2, t3, t4, t5) %>%
114    model.matrix(data = .,~.)
115
116  jive.est(y = y, X = X_max, Z = Z_max)
```

Table 65. OLS and IV estimates of detention on guilty plea.

Model	OLS		2SLS		JIVE	
Detention	−0.001	0.029***	0.151**	0.186***	0.162**	0.212***
	(0.002)	(0.002)	(0.065)	(0.064)	(0.070)	(0.076)
N	331,971	331,971	331,971	331,971	331,971	331,971
Mean guilt	0.49	0.49	0.49	0.49	0.49	0.49

Note: First model includes controls for time; second model controls for characteristics of the defendant. Outcome is guilty plea. Heteroskedastic robust standard errors in parentheses. * $p<0.10$, ** $p<0.05$, *** $p<0.01$

These results are pretty interesting. Notice that if we just were to examine this using OLS, you'd conclude there was actually no connection between pre-trial detention and a guilty plea. It was either zero using only time controls, or it raised the probability 3% with our fuller set of controls (mainly demographic controls, prior offenses, and the characteristics of the offense itself). But, when we use IV with the binary judge fixed effects as instruments, the effects change a lot. We end up with estimates ranging from 15 to 21%, and of these probably we should be more focused on JIVE because of its advantages, as previously mentioned. You can examine the strength of the instruments yourself by regressing detention onto the binary instruments to see just how strong the instruments are, but they are very strong. All but two are statistically significant at the 1% level. Of the other two, one has a p-value of 0.076 and the other is weak ($p < 0.25$).

The judge fixed effects design is a very popular form of instrumental variables. It is used whenever there exists a wheel of randomly assigned decision makers assigning a treatment of some kind to other people. Important questions and answers in the area of criminal justice have been examined using this design. When linked with external administrative data sources, researchers have been able to more carefully evaluate the causal effect of criminal justice interventions on long-term outcomes. But the procedure has uniquely sensitive identifying assumptions related to independence, exclusion, and monotonicity that must be carefully contemplated before going forward with the design. Nevertheless, when those assumptions can be credibly defended, it is a powerful estimator of local average treatment effects.

Bartik instruments. Bartik instruments, also known as shift-share instruments, were named after Timothy Bartik, who used them in a careful study of regional labor-markets [Bartik, 1991]. Both Bartik's book and the instrument received wider attention the following year with Blanchard and Katz [1992]. It has been particularly influential in the areas of migration and trade, as well as labor, public, and several other fields. A simple search for the phrase "Bartik instrument" on Google Scholar reveals almost five hundred cites at the time of this writing.

But just as Stigler's law of eponymy promises [Stigler, 1980], Bartik instruments do not originate with Bartik [1991]. Goldsmith-Pinkham et al. [2020] notes that traces of it can be found as early as Perloff [1957] who showed that industry shares could be used to predict income levels. Freeman [1980] also used the change in industry composition as an instrument for labor demand. But due to Bartik's careful empirical analysis using the instrument combined with his detailed exposition of the logic of how the national growth shares created variation in labor-market demand in Appendix 4 of his book, the design has been named after him.

OLS estimates of the effect of employment growth rates on labor-market outcomes are likely hopelessly biased since labor-market outcomes are simultaneously determined by labor supply and labor demand. Bartik therefore suggested using IV to resolve the issue and in Appendix 4 describes the ideal instrument.

Obvious candidates for instruments are variables shifting MSA labor demand. In this book, only one type of demand shifter is used to form instrumental variables: the share effect from a shift-share analysis of each metropolitan area and year-to-year employment change. A shift-share analysis decomposes MSA growth into three components: a national growth component, which calculates what growth would have occurred if all industries in the MSA had grown at the all-industry national average; a share component, which calculates what extra growth would have occurred if each industry in the MSA had grown at that industry's national average; and a shift component, which calculates the extra growth that occurs because industries grow at different rates locally than they do nationally. [Bartik, 1991, 202])

Summarizing all of this, the idea behind a Bartik instrument is to measure the change in a region's labor demand due to changes in the national demand for different industries' products.[21] To make this concrete, let's assume that we are interested in estimating the following wage equation:

$$Y_{l,t} = \alpha + \delta I_{l,t} + \rho X_{l,t} + \varepsilon_{l,t}$$

where $Y_{l,t}$ is log wages in location l (e.g., Detroit) in time period t (e.g., 2000) among native workers, $I_{l,t}$ are immigration flows in region l at time period t and $X_{l,t}$ are controls that include region and time fixed effects, among other things. The parameter δ as elsewhere is some average treatment effect of the immigration flows' effect on native wages. The problem is that it is almost certainly the case that immigration flows are highly correlated with the disturbance term such as the time-varying characteristics of location l (e.g., changing amenities) [Sharpe, 2019].

The Bartik instrument is created by interacting initial "shares" of geographic regions, prior to the contemporaneous immigration flow, with national growth rates. The deviations of a region's growth from the US national average are explained by deviations in the growth prediction variable from the US national average. And deviations of the growth prediction variables from the US national average are due to the shares because the national growth effect for any particular time period is the same for all regions. We can define the Bartik instrument as follows:

$$B_{l,t} = \sum_{k=1}^{K} z_{l,k,t^0} m_{k,t}$$

---

21 Goldsmith-Pinkham et al. [2020] note that many instruments have Bartik features. They describe an instrument as "Bartik-like" if it uses the inner product structure of the endogenous variable to construct an instrument.

where z_{l,k,t^0} are the "initial" t^0 share of immigrants from source country k (e.g., Mexico) in location l (e.g., Detroit) and $m_{k,t}$ is the change in immigration from country k (e.g., Mexico) into the US as a whole. The first term is the share variable and the second term is the shift variable. The predicted flow of immigrants, B, into destination l (e.g., Detroit) is then just a weighted average of the national inflow rates from each country in which weights depend on the initial distribution of immigrants.

Once we have constructed our instrument, we have a two-stage least squares estimator that first regresses the endogenous $I_{l,t}$ onto the controls and our Bartik instrument. Using the fitted values from that regression, we then regress $Y_{l,t}$ onto $\widehat{I}_{l,t}$ to recover the impact of immigration flows onto log wages.

Shifts vs Shares. I'd like to now turn to the identifying assumptions that are unique to this design. There are two perspectives as to what is needed to leverage a Bartik design to identify a causal effect and they separately address the roles of the exogeneity of the shares versus the shifts. Which perspective you take will depend on the ex ante plausibility of certain assumptions. They will also depend on different tools.

Goldsmith-Pinkham et al. [2020] explain the shares perspective. They show that while the shifts affect the strength of the first stage, it is actually the initial shares that provide the exogenous variation. They write that "the Bartik instrument is 'equivalent' to using local industry shares as instruments, and so the exogeneity condition should be interpreted in terms of the shares." Insofar as a researcher's application is exploiting differential exogenous exposure to common shocks, industry specific shocks, or a two-industry scenario, then it is likely that the source of exogeneity comes from the initial shares and not the shifts. This is a type of strict exogeneity assumption where the initial shares are exogenous conditional on observables, such as location fixed effects. What this means in practice is that the burden is on the researcher to argue why they believe the initial shares are indeed exogenous.

But while exogenous shares are sufficient, it turns out they are not necessary for identification of causal effects. Temporal shocks may provide exogeous sources of variation. Borusyak et al. [2019] explain the shifts perspective. They show that exogenous independent shocks to many industries allow a Bartik design to identify causal effects regardless of whether the shares are exogenous so long as the shocks are uncorrelated with the bias of the shares. Otherwise, it may be the shock itself that is creating exogenous variation, in which case the focus on excludability moves away from the initial shares and more towards the national shocks themselves [Borusyak et al., 2019]. The authors write:

> Ultimately, the plausibility of our exogenous shocks framework, as with the alternative framework of Goldsmith-Pinkham et al. [2020] based on exogenous shares, depends on the shift-share IV application. We encourage practitioners to use shift-share instruments based on an *a priori* argument supporting the plausibility of either one of these approaches; various diagnostics and tests of the framework that is most suitable for the setting may then be applied. While [Borusyak et al. [2019]] develops such procedures for the "shocks" view, Goldsmith-Pinkham et al. [2020] provide different tools for the "shares" view. [29]

Insofar as we think about the initial shares as the instruments, and not the shocks, then we are in a world in which those initial shares are measuring differential exogenous exposures to some common shock. As the shares are equilibrium values, based on past labor supply and demand, it may be tough to justify why we should consider them exogenous to the structural unobserved determinants of some future labor-market outcome. But it turns out that that is not the critical piece. A valid Bartik design can be valid even if the shares are correlated indirectly with the levels of the outcomes; they just can't be correlated with the differential changes associated with the national shock itself, which is a subtle but distinct point.

One challenge with Bartik instruments is the sheer number of shifting values. For instance, there are almost four hundred different industries in the United States. Multiplied over many time periods and the exclusion restriction becomes a bit challenging to defend. Goldsmith-Pinkham et al. [2020] provide several suggestions for evaluating the central identifying assumption in this design. For instance, if there is a pre-period, then ironically this design begins to resemble the difference-in-differences design that we will discuss in a subsequent chapter. In that case, we might test for placebos, pre-trends, and so forth.

Another possibility is based on the observation that the Bartik instrument is simply a specific combination of many instruments. In that sense, it bears some resemblance to the judge fixed effects design from earlier in which the judge's propensity was itself a specific combination of many binary fixed effects. With many instruments, other options become available. If the researcher is willing to assume a null of constant treatment effects, then overidentification tests are an option. But overidentification tests can fail if there is treatment heterogeneity as opposed to exclusion not holding. Similar to Borusyak et al. [2019], insofar as one is willing to assume cross-sectional heterogeneity in which treatment effects are constant within a location only, then Goldsmith-Pinkham et al. [2020] provides some diagnostic aids to help evaluate the plausibility of the design itself.

A second result in Goldsmith-Pinkham et al. [2020] is a decomposition of the Bartik estimator into a weighted combination of estimates where each share is an instrument. These weights, called Rotemberg weights, sum to one, and the authors note that higher valued weights indicate that those instruments are responsible for more of the identifying variation in the design itself. These weights provide insight into which of the shares get more weight in the overall estimate, which helps clarify which industry shares should be scrutinized. If regions with high weights pass some basic specification tests, then confidence in the overall identification strategy is more defensible.

Conclusion

In conclusion, instrumental variables are a powerful design for identifying causal effects when your data suffer from selection on unobservables. But even with that in mind, it has many limitations that have in the contemporary period caused many applied researchers to eschew it. First, it only identifies the LATE under heterogeneous treatment effects, and that may or may not be a policy relevant variable.

Its value ultimately depends on how closely the compliers' average treatment effect resembles that of the other subpopulations. Second, unlike RDD, which has only one main identifying assumption (the continuity assumption), IV has up to five assumptions! Thus, you can immediately see why people find IV estimation less credible—not because it fails to identify a causal effect, but rather because it's harder and harder to imagine a pure instrument that satisfies all five conditions.

But all this is to say, IV is an important strategy and sometimes the opportunity to use it will come along, and you should be prepared for when that happens by understanding it and how to implement it in practice. And where can the best instruments be found? Angrist and Krueger [2001] note that the best instruments come from in-depth knowledge of the institutional details of some program or intervention. The things you spend your life studying will in time reveal good instruments. Rarely will you find them from simply downloading a new data set, though. Intimate familiarity is how you find instrumental variables, and there is, alas, no shortcut to achieving that.

Panel Data

That's just the way it is
Things will never be the same
That's just the way it is
Some things will never change.

Tupac Shakur

One of the most important tools in the causal inference toolkit is the panel data estimator. The estimators are designed explicitly for longitudinal data—the repeated observing of a unit over time. Under certain situations, repeatedly observing the same unit over time can overcome a particular kind of omitted variable bias, though not all kinds. While it is possible that observing the same unit over time will not resolve the bias, there are still many applications where it can, and that's why this method is so important. We review first the DAG describing just such a situation, followed by discussion of a paper, and then present a data set exercise in R and Stata.[1]

DAG Example

Before I dig into the technical assumptions and estimation methodology for panel data techniques, I want to review a simple DAG illustrating those assumptions. This DAG comes from Imai and Kim [2017]. Let's say that we have data on a column of outcomes Y_i, which appear in three time periods. In other words, Y_{i1}, Y_{i2}, and Y_{i3} where i indexes a particular unit and $t = 1, 2, 3$ index the time period where

---

1 There's a second reason to learn this estimator. Some of these estimators, such as linear models with time and unit fixed effects, are the modal estimators used in difference-in-differences.

each i unit is observed. Likewise, we have a matrix of covariates, D_i, which also vary over time—D_{i1}, D_{i2}, and D_{i3}. And finally, there exists a single unit-specific unobserved variable, u_i, which varies across units, but which does not vary over time for that unit. Hence the reason that there is no $t = 1,2,3$ subscript for our u_i variable. Key to this variable is (a) it is unobserved in the data set, (b) it is unit-specific, and (c) it does not change over time for a given unit i. Finally there exists some unit-specific time-invariant variable, X_i. Notice that it doesn't change over time, just u_i, but unlike u_i it is observed.

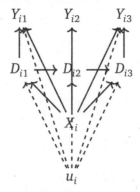

As this is the busiest DAG we've seen so far, it merits some discussion. First, let us note that D_{i1} causes both Y_{i1} as well as the next period's treatment value, D_{i2}. Second, note that an unobserved confounder, u_i, determines all Y and D variables. Consequently, D is endogenous since u_i is unobserved and absorbed into the structural error term of the regression model. Thirdly, there is *no* time-varying unobserved confounder correlated with D_{it}—the only confounder is u_i, which we call the unobserved heterogeneity. Fourth, past outcomes do not directly affect current outcomes (i.e., no direct edge between the Y_{it} variables). Fifth, past outcomes do not directly affect current treatments (i.e., no direct edge from $Y_{i,t-1}$ to D_{it}). And finally, past treatments, $D_{i,t-1}$ do not directly affect current outcomes, Y_{it} (i.e., no direct edge from $D_{i,t-1}$ and Y_{it}). It is under these assumptions that we can use a particular panel method called *fixed effects* to isolate the causal effect of D on Y.[2]

---

2 The fixed-effects estimator, when it includes year fixed effects, is popularly known as the "twoway fixed effects" estimator.

What might an example of this be? Let's return to our story about the returns to education. Let's say that we are interested in the effect of schooling on earnings, and schooling is partly determined by unchanging genetic factors which themselves determine unobserved ability, like intelligence, contentiousness, and motivation [Conley and Fletcher, 2017]. If we observe the same people's time-varying earnings and schoolings over time, then if the situation described by the above DAG describes both the directed edges and *the missing edges*, then we can use panel fixed effects models to identify the causal effect of schooling on earnings.

Estimation

When we use the term "panel data," what do we mean? We mean a data set where we observe the same units (e.g., individuals, firms, countries, schools) over more than one time period. Often our outcome variable depends on several factors, some of which are observed and some of which are unobserved in our data, and insofar as the unobserved variables are correlated with the treatment variable, then the treatment variable is endogenous and correlations are not estimates of a causal effect. This chapter focuses on the conditions under which a correlation between D and Y reflects a causal effect even with unobserved variables that are correlated with the treatment variable. Specifically, if these omitted variables are *constant* over time, then even if they are heterogeneous across units, we can use panel data estimators to consistently estimate the effect of our treatment variable on outcomes.

There are several different kinds of estimators for panel data, but we will in this chapter only cover two: pooled ordinary least squares (POLS) and fixed effects (FE).[3]

---

3 A common third type of panel estimator is the random effects estimator, but in my experience, I have used it less often than fixed effects, so I decided to omit it. Again, this is not because it is unimportant. I just have chosen to do fewer things in more detail based on whether I think they qualify as the most common methods used in the present period by applied empiricists. See Wooldridge [2010] for a more comprehensive treatment, though, of all panel methods including random effects.

First we need to set up our notation. With some exceptions, panel methods are usually based on the traditional notation and not the potential outcomes notation.

Let Y and $D \equiv (D_1, D_2, \ldots, D_k)$ be observable random variables and u be an unobservable random variable. We are interested in the partial effects of variable D_j in the population regression function:

$$E[Y \mid D_1, D_2, \ldots, D_k, u]$$

We observe a sample of $i = 1, 2, \ldots, N$ cross-sectional units for $t = 1, 2, \ldots, T$ time periods (a balanced panel). For each unit i, we denote the observable variables for all time periods as $\{(Y_{it}, D_{it}) : t = 1, 2, \ldots, T\}$.[4] Let $D_{it} \equiv (D_{it1}, D_{it2}, \ldots, D_{itk})$ be a $1 \times K$ vector. We typically assume that the actual cross-sectional units (e.g., individuals in a panel) are identical and independent draws from the population in which case $\{Y_i, D_i, u_i\}_{i=1}^{N} \sim i.i.d.$, or cross-sectional independence. We describe the main observables, then, as $Y_i \equiv (Y_{i1}, Y_{i2}, \ldots, Y_{iT})'$ and $D_i \equiv (D_{i1}, D_{i2}, \ldots, D_{iT})$.

It's helpful now to illustrate the actual stacking of individual units across their time periods. A single unit i will have multiple time periods t

$$Y_i = \begin{pmatrix} Y_{i1} \\ \vdots \\ Y_{it} \\ \vdots \\ Y_{iT} \end{pmatrix}_{T \times 1} \qquad D_i = \begin{pmatrix} D_{i,1,1} & D_{i,1,2} & D_{i,1,j} & \ldots & D_{i,1,K} \\ \vdots & \vdots & \vdots & & \vdots \\ D_{i,t,1} & D_{i,t,2} & D_{i,t,j} & \ldots & D_{i,t,K} \\ \vdots & \vdots & \vdots & & \vdots \\ D_{i,T,1} & D_{i,T,2} & D_{i,T,j} & \ldots & D_{i,T,K} \end{pmatrix}_{T \times K}$$

And the entire panel itself with all units included will look like this:

$$Y = \begin{pmatrix} Y_1 \\ \vdots \\ Y_i \\ \vdots \\ Y_N \end{pmatrix}_{NT \times 1} \qquad D = \begin{pmatrix} D_1 \\ \vdots \\ D_i \\ \vdots \\ D_N \end{pmatrix}_{NT \times K}$$

4 For simplicity, I'm ignoring the time-invariant observations, X_i from our DAG for reasons that will hopefully soon be made clear.

For a randomly drawn cross-sectional unit i, the model is given by

$$Y_{it} = \delta D_{it} + u_i + \varepsilon_{it}, \quad t = 1, 2, \ldots, T$$

As always, we use our schooling-earnings example for motivation. Let Y_{it} be log earnings for a person i in year t. Let D_{it} be schooling for person i in year t. Let δ be the returns to schooling. Let u_i be the sum of all time-invariant person-specific characteristics, such as unobserved ability. This is, as I said earlier, often called the *unobserved heterogeneity*. And let ε_{it} be the time-varying unobserved factors that determine a person's wage in a given period. This is often called the idiosyncratic error. We want to know what happens when we regress Y_{it} on D_{it}.

Pooled OLS. The first estimator we will discuss is the pooled ordinary least squares, or POLS estimator. When we ignore the panel structure and regress Y_{it} on D_{it} we get

$$Y_{it} = \delta D_{it} + \eta_{it}, \quad t = 1, 2, \ldots, T$$

with composite error $\eta_{it} \equiv c_i + \varepsilon_{it}$. The main assumption necessary to obtain consistent estimates for δ is:

$$E[\eta_{it} \mid D_{i1}, D_{i2}, \ldots, D_{iT}] = E[\eta_{it} \mid D_{it}] = 0 \quad \text{for } t = 1, 2, \ldots, T$$

While our DAG did not include ε_{it}, this would be equivalent to assuming that the unobserved heterogeneity, c_i, was uncorrelated with D_{it} for all time periods.

But this is not an appropriate assumption in our case because our DAG explicitly links the unobserved heterogeneity to both the outcome and the treatment in each period. Or using our schooling-earnings example, schooling is likely based on unobserved background factors, u_i, and therefore without controlling for it, we have omitted variable bias and $\widehat{\delta}$ is biased. No correlation between D_{it} and η_{it} *necessarily means* no correlation between the unobserved u_i and D_{it} for all t and that is just probably not a credible assumption. An additional problem is that η_{it} is serially correlated for unit i since u_i is present in each t period. And thus heteroskedastic robust standard errors are also likely too small.

Fixed effects (within Estimator). Let's rewrite our unobserved effects model so that this is still firmly in our minds:

$$Y_{it} = \delta D_{it} + u_i + \varepsilon_{it}; \quad t = 1, 2, \ldots, T$$

If we have data on multiple time periods, we can think of u_i as fixed effects to be estimated. OLS estimation with fixed effects yields

$$\left(\widehat{\delta}, \widehat{u}_1, \ldots, \widehat{u}_N\right) = \arg\min_{b, m_1, \ldots, m_N} \sum_{i=1}^{N} \sum_{t=1}^{T} (Y_{it} - D_{it}b - m_i)^2$$

which amounts to including N individual dummies in regression of Y_{it} on D_{it}.

The first-order conditions (FOC) for this minimization problem are:

$$\sum_{i=1}^{N} \sum_{t=1}^{T} D'_{it}\left(Y_{it} - D_{it}\widehat{\delta} - \widehat{u}_i\right) = 0$$

and

$$\sum_{t=1}^{T} \left(Y_{it} - D_{it}\widehat{\delta} - \widehat{u}_i\right) = 0$$

for $i = 1, \ldots, N$.

Therefore, for $i = 1, \ldots, N$,

$$\widehat{u}_i = \frac{1}{T} \sum_{t=1}^{T} \left(Y_{it} - D_{it}\widehat{\delta}\right) = \overline{Y}_i - \overline{D}_i\widehat{\delta},$$

where

$$\overline{D}_i \equiv \frac{1}{T} \sum_{t=1}^{T} D_{it}; \overline{Y}_i \equiv \frac{1}{T} \sum_{t=1}^{T} Y_{it}$$

Plug this result into the first FOC to obtain:

$$\widehat{\delta} = \left(\sum_{i=1}^{N} \sum_{t=1}^{T} (D_{it} - \overline{D}_i)'(D_{it} - \overline{D}_i)\right)^{-1} \left(\sum_{i=1}^{N} \sum_{t=1}^{T} (D_{it} - \overline{D}_i)'(Y_{it} - \overline{Y})\right)$$

$$\widehat{\delta} = \left(\sum_{i=1}^{N} \sum_{t=1}^{T} \ddot{D}'_{it}\ddot{D}_{it}\right)^{-1} \left(\sum_{i=1}^{N} \sum_{t=1}^{T} \ddot{D}'_{it}\ddot{D}_{it}\right)$$

with time-demeaned variables $\ddot{D}_{it} \equiv D_{it} - \overline{D}, \ddot{Y}_{it} \equiv Y_{it} - \overline{Y}_i$.

In case it isn't clear, though, running a regression with the time-demeaned variables $\ddot{Y}_{it} \equiv Y_{it} - \overline{Y}_i$ and $\ddot{D}_{it} \equiv D_{it} - \overline{D}$ is *numerically equivalent* to a regression of Y_{it} on D_{it} and unit-specific dummy variables. Hence the reason this is sometimes called the "within" estimator, and sometimes called the "fixed effects" estimator. And when year fixed effects are included, the "twoway fixed effects" estimator. They are the same thing.[5]

Even better, the regression with the time-demeaned variables is consistent for δ even when $C[D_{it}, u_i] \neq 0$ because time-demeaning eliminates the unobserved effects. Let's see this now:

$$Y_{it} = \delta D_{it} + u_i + \varepsilon_{it}$$
$$\overline{Y}_i = \delta \overline{D}_i + u_i + \overline{\varepsilon}_i$$
$$(Y_{it} - \overline{Y}_i) = (\delta D_{it} - \delta \overline{D}) + (u_i - u_i) + (\varepsilon_{it} - \overline{\varepsilon}_i)$$
$$\ddot{Y}_{it} = \delta \ddot{D}_{it} + \ddot{\varepsilon}_{it}$$

Where'd the unobserved heterogeneity go?! It was deleted when we time-demeaned the data. And as we said, including individual fixed effects does this time demeaning automatically so that you don't have to go to the actual trouble of doing it yourself manually.[6]

So how do we precisely do this form of estimation? There are three ways to implement the fixed effects (within) estimator. They are:

1. Demean and regress $\ddot{Y}_{it}$ on $\ddot{D}_{it}$ (need to correct degrees of freedom).
2. Regress Y_{it} on D_{it} and unit dummies (dummy variable regression).
3. Regress Y_{it} on D_{it} with canned fixed effects routine in Stata or R.

Later in this chapter, I will review an example from my research. We will estimate a model involving sex workers using pooled OLS, a FE, and a demeaned OLS model. I'm hoping this exercise will help you see the similarities and differences across all three approaches.

---

5 One of the things you'll find over time is that things have different names, depending on the author and tradition, and those names are often completely uninformative.

6 Though feel free to do it if you want to convince yourself that they are numerically equivalent, probably just starting with a bivariate regression for simplicity.

Identifying Assumptions. We kind of reviewed the assumptions necessary to identify δ with our fixed effects (within) estimator when we walked through that original DAG, but let's supplement that DAG intuition with some formality. The main identification assumptions are:

1. $E[\varepsilon_{it} \mid D_{i1}, D_{i2}, \ldots, D_{iT}, u_i] = 0; t = 1, 2, \ldots, T$
   - This means that the regressors are strictly exogenous conditional on the unobserved effect. This allows D_{it} to be arbitrarily related to u_i, though. It only concerns the relationship between D_{it} and ε_{it}, not D_{it}'s relationship to u_i.

2. $rank\left(\sum_{t=1}^{T} E[\ddot{D}'_{it} \ddot{D}_{it}] \right) = K$
   - It shouldn't be a surprise to you by this point that we have a rank condition, because even when we were working with the simpler linear models, the estimated coefficient was always a scaled covariance, where the scaling was by a variance term. Thus regressors must vary over time for at least some i and not be collinear in order that $\widehat{\delta} \approx \delta$.

The properties of the estimator under assumptions 1 and 2 are that $\widehat{\delta}_{FE}$ is consistent ($\underset{N \to \infty}{plim} \, \widehat{\delta}_{FE,N} = \delta$) and $\widehat{\delta}_{FE}$ is unbiased conditional on D.

I only briefly mention inference. But the standard errors in this framework must be "clustered" by panel unit (e.g., individual) to allow for correlation in the ε_{it} for the same person i over time. This yields valid inference so long as the number of clusters is "large."[7]

Caveat #1: Fixed effects cannot solve reverse causality. But, there are still things that fixed effects (within) estimators cannot solve. For instance, let's say we regressed crime rates onto police spending per capita. Becker [1968] argues that increases in the probability of arrest, usually proxied by police per capita or police spending per capita, will reduce crime. But at the same time, police spending per capita is itself

---

7 In my experience, when an econometrician is asked how large is large, they say "the size of your data." But that said, there is a small clusters literature and usually it's thought that fewer than thirty clusters is too small (as a rule of thumb). So it may be that having around thirty to forty clusters is sufficient for the approaching of infinity. This will usually hold in most panel applications such as US states or individuals in the NSLY, etc.

Table 66. Panel estimates of police on crime.

Dependent variable	Between	Within	2SLS (FE)	2SLS (no FE)
Police	0.364	0.413	0.504	0.419
	(0.060)	(0.027)	(0.617)	(0.218)
Controls	Yes	Yes	Yes	Yes

Note: North Carolina county-level data. Standard errors in parentheses.

a function of crime rates. This kind of reverse causality problem shows up in most panel models when regressing crime rates onto police. For instance, see Cornwell and Trumbull [1994]. I've reproduced a portion of this in Table 66. The dependent variable is crime rates by county in North Carolina for a panel, and they find a *positive* correlation between police and crime rates, which is the opposite of what Becker predicts. Does this mean having more police in an area *causes* higher crime rates? Or does it likely reflect the reverse causality problem?

So, one situation in which you wouldn't want to use panel fixed effects is if you have reverse causality or simultaneity bias. And specifically when that reverse causality is very strong in observational data. This would technically violate the DAG, though, that we presented at the start of the chapter. Notice that if we had reverse causality, then $Y \to D$, which is explicitly ruled out by this theoretical model contained in the DAG. But obviously, in the police—crime example, that DAG would be inappropriate, and any amount of reflection on the problem should tell you that that DAG is inappropriate. Thus it requires, as I've said repeatedly, some careful reflection, and writing out exactly what the relationship is between the treatment variables and the outcome variables in a DAG can help you develop a credible identification strategy.

Caveat #2: Fixed effects cannot address time-variant unobserved heterogeneity. The second situation in which panel fixed effects don't buy you anything is if the unobserved heterogeneity is time-varying. In this situation, the demeaning has simply demeaned an unobserved time-variant variable, which is then moved into the composite error term, and which since time-demeaned $\ddot{u}_{it}$ correlated with $\ddot{D}_{it}$, $\ddot{D}_{it}$ remains endogenous. Again, look carefully at the DAG—panel fixed effects are only appropriate if u_i is unchanging. Otherwise it's just another form of omitted variable bias. So, that said, don't just blindly use fixed effects and think that it solves your omitted variable bias problem—in the same way that you shouldn't use matching just because it's convenient to do. You need a DAG, based on an actual economic model, which will allow you to build the appropriate research design. Nothing substitutes for careful reasoning and economic theory, as they are the necessary conditions for good research design.

Returns to marriage and unobserved heterogeneity. When might this be true? Let's use an example from Cornwell and Rupert [1997] in which the authors attempt to estimate the causal effect of marriage on earnings. It's a well-known stylized fact that married men earn more than unmarried men, even controlling for observables. But the question is whether that correlation is causal, or whether it reflects unobserved heterogeneity (i.e., selection bias).

So let's say that we had panel data on individuals. These individuals *i* are observed for four periods *t*. We are interested in the following equation:[8]

$$Y_{it} = \alpha + \delta M_{it} + \beta X_{it} + A_i + \gamma_i + \varepsilon_{it}$$

Let the outcome be their wage Y_{it} observed in each period, and which changes each period. Let wages be a function of marriage. Since people's marital status changes over time, the marriage variable is allowed to change value over time. But race and gender, in most scenarios, do not ordinarily change over time; these are variables which are ordinarily unchanging, or what you may sometimes hear called "time invariant." Finally, the variables A_i and γ_i are variables which are unobserved, vary cross-sectionally across the sample, but do not vary over time. I will call these measures of unobserved ability, which may refer to any fixed endowment in the person, like fixed cognitive ability or noncognitive ability such as "grit." The key here is that it is unit-specific, unobserved, and time-invariant. The ε_{it} is the unobserved determinants of wages which are assumed to be uncorrelated with marriage and other covariates.

Cornwell and Rupert [1997] estimate both a feasible generalized least squares model and three fixed effects models (each of which

8 We use the same notation as used in their paper, as opposed to the $\ddot{Y}$ notation presented earlier.

Table 67. Estimated wage regressions.

Dependent variable	FGLS	Within	Within	Within
Married	0.083	0.056	0.051	0.033
	(0.022)	(0.026)	(0.026)	(0.028)
Education controls	Yes	No	No	No
Tenure	No	No	Yes	Yes
Quadratics in years married	No	No	No	Yes

Note: Standard errors in parentheses.

includes different time-varying controls). The authors call the fixed effects regression a "within" estimator, because it uses the within unit variation for eliminating the confounding. Their estimates are presented in Table 67.

Notice that the FGLS (column 1) finds a strong marriage premium of around 8.3%. But, once we begin estimating fixed effects models, the effect gets smaller and less precise. The inclusion of marriage characteristics, such as years married and job tenure, causes the coefficient on marriage to fall by around 60% from the FGLS estimate, and is no longer statistically significant at the 5% level.

Data Exercise: Survey of Adult Service Providers

Next I'd like to introduce a Stata exercise based on data collection for my own research: a survey of sex workers. You may or may not know this, but the Internet has had a profound effect on sex markets. It has moved sex work indoors while simultaneously breaking the traditional link between sex workers and pimps. It has increased safety and anonymity, too, which has had the effect of causing new entrants. The marginal sex worker has more education and better outside options than traditional US sex workers [Cunningham and Kendall, 2011, 2014, 2016]. The Internet, in sum, caused the marginal sex worker to shift towards women more sensitive to detection, harm, and arrest.

In 2008 and 2009, I surveyed (with Todd Kendall) approximately 700 US Internet-mediated sex workers. The survey was a basic labor-market survey; I asked them about their illicit and legal labor-market experiences, and about demographics. The survey had two parts:

a "static" provider-specific section and a "panel" section. The panel section asked respondents to share information about each of the previous four sessions with clients.[9]

I have created a shortened version of the data set and uploaded it to Github. It includes a few time-invariant provider characteristics, such as race, age, marital status, years of schooling, and body mass index, as well as several time-variant session-specific characteristics including the log of the hourly price, the log of the session length (in hours), characteristics of the client himself, whether a condom was used in any capacity during the session, whether the client was a "regular," etc.

In this exercise, you will estimate three types of models: a pooled OLS model, a fixed effects (FE), and a demeaned OLS model. The model will be of the following form:

$$Y_{is} = \beta_i X_i + \gamma_{is} Z_{is} + u_i + \varepsilon_{is}$$
$$\ddot{Y}_{is} = \gamma_{is} \ddot{Z}_{is} + \ddot{\eta}_{is}$$

where u_i is both unobserved and correlated with Z_{is}.

The first regression model will be estimated with pooled OLS and the second model will be estimated using both fixed effects and OLS. In other words, I'm going to have you estimate the model using canned routines in Stata and R with individual fixed effects, as well as demean the data manually and estimate the demeaned regression using OLS.

Notice that the second regression has a different notation on the dependent and independent variable; it represents the fact that the variables are columns of *demeaned* variables. Thus $\ddot{Y}_{is} = Y_{is} - \overline{Y}_i$. Secondly, notice that the time-invariant X_i variables are missing from the second equation. Do you understand why that is the case? These variables have also been demeaned, but since the demeaning is across time, and since these time-invariant variables do not change over time, the demeaning deletes them from the expression. Notice, also, that the unobserved individual specific heterogeneity, u_i, has disappeared.

---

9 Technically, I asked them to share about the last five sessions, but for this exercise, I have dropped the fifth due to low response rates on the fifth session.

It has disappeared for the same reason that the X_i terms are gone—because the mean of u_i over time is itself, and thus the demeaning deletes it.

Let's examine these models using the following R and Stata programs.

R
sasp.do

```
1   use https://github.com/scunning1975/mixtape/raw/master/sasp_panel.dta,
    ↪   clear
2   tsset id session
3   foreach x of varlist lnw age asq bmi hispanic black other asian schooling cohab
    ↪   married divorced separated age_cl unsafe llength reg asq_cl appearance_cl
    ↪   provider_second asian_cl black_cl hispanic_cl othrace_cl hot massage_cl
4   drop if `x'==.
5   bysort id: gen s=_N
6   keep if s==4
7   foreach x of varlist lnw age asq bmi hispanic black other asian schooling cohab
    ↪   married divorced separated age_cl unsafe llength reg asq_cl appearance_cl
    ↪   provider_second asian_cl black_cl hispanic_cl  othrace_cl hot massage_cl
8
9   egen mean_`x'=mean(`x'), by(id)
10  gen demean_`x'=`x' - mean_`x'
11  drop mean*
12
13  xi: reg lnw  age asq bmi hispanic black other asian schooling cohab married
    ↪   divorced separated age_cl unsafe llength reg asq_cl appearance_cl
    ↪   provider_second asian_cl black_cl hispanic_cl othrace_cl hot massage_cl,
    ↪   robust
14  xi: xtreg lnw  age asq bmi hispanic black other asian schooling cohab married
    ↪   divorced separated age_cl unsafe llength reg asq_cl appearance_cl
    ↪   provider_second asian_cl black_cl hispanic_cl othrace_cl hot massage_cl, fe
    ↪   i(id) robust
15  reg demean_lnw demean_age demean_asq demean_bmi demean_hispanic
    ↪   demean_black demean_other demean_asian demean_schooling
    ↪   demean_cohab demean_married demean_divorced demean_separated
    ↪   demean_age_cl demean_unsafe demean_llength demean_reg
    ↪   demean_asq_cl demean_appearance_cl demean_provider_second
    ↪   demean_asian_cl demean_black_cl demean_hispanic_cl demean_othrace_cl
    ↪   demean_hot demean_massage_cl, robust cluster(id)
```

R

sasp.R

```
1   library(tidyverse)
2   library(haven)
3   library(estimatr)
4   library(plm)
5
6   read_data <- function(df)
7   {
8     full_path <- paste("https://raw.github.com/scunning1975/mixtape/master/",
9               df, sep = "")
10    df <- read_dta(full_path)
11    return(df)
12  }
13
14  sasp <- read_data("sasp_panel.dta")
15
16  #-- Delete all NA
17  sasp <- na.omit(sasp)
18
19  #-- order by id and session
20  sasp <- sasp %>%
21    arrange(id, session)
22
23  #Balance Data
24  balanced_sasp <- make.pbalanced(sasp,
25                    balance.type = "shared.individuals")
26
27  #Demean Data
28  balanced_sasp <- balanced_sasp %>% mutate(
29    demean_lnw = lnw - ave(lnw, id),
30    demean_age = age - ave(age, id),
31    demean_asq = asq - ave(asq, id),
32    demean_bmi = bmi - ave(bmi, id),
33    demean_hispanic = hispanic - ave(hispanic, id),
34    demean_black = black - ave(black, id),
35    demean_other = other - ave(other, id),
36    demean_asian = asian - ave(asian, id),
```

(continued)

R (continued)

```
37    demean_schooling = schooling - ave(schooling, id),
38    demean_cohab = cohab - ave(cohab, id),
39    demean_married = married - ave(married, id),
40    demean_divorced = divorced - ave(divorced, id),
41    demean_separated = separated - ave(separated, id),
42    demean_age_cl = age_cl - ave(age_cl, id),
43    demean_unsafe = unsafe - ave(unsafe, id),
44    demean_llength = llength - ave(llength, id),
45    demean_reg = reg - ave(reg, id),
46    demean_asq_cl = asq_cl - ave(asq_cl, id),
47    demean_appearance_cl = appearance_cl - ave(appearance_cl, id),
48    demean_provider_second = provider_second - ave(provider_second, id),
49    demean_asian_cl = asian_cl - ave(asian_cl, id),
50    demean_black_cl = black_cl - ave(black_cl, id),
51    demean_hispanic_cl = hispanic_cl - ave(hispanic_cl, id),
52    demean_othrace_cl = othrace_cl - ave(lnw, id),
53    demean_hot = hot - ave(hot, id),
54    demean_massage_cl = massage_cl - ave(massage_cl, id)
55    )
56
57    #-- POLS
58    ols <- lm_robust(lnw ~ age + asq + bmi + hispanic + black + other + asian +
  ↪    schooling + cohab + married + divorced + separated +
59        age_cl + unsafe + llength + reg + asq_cl + appearance_cl + provider_second
      ↪    + asian_cl + black_cl + hispanic_cl +
60        othrace_cl + hot + massage_cl, data = balanced_sasp)
61    summary(ols)
62
63
64    #-- FE
65    formula <- as.formula("lnw ~ age + asq + bmi + hispanic + black + other + asian +
  ↪    schooling +
66                cohab + married + divorced + separated +
67                age_cl + unsafe + llength + reg + asq_cl + appearance_cl +
68                provider_second + asian_cl + black_cl + hispanic_cl +
69                othrace_cl + hot + massage_cl")
70
```

(continued)

```
     R (continued)
71   model_fe <- lm_robust(formula = formula,
72          data = balanced_sasp,
73          fixed_effect = ~id,
74          se_type = "stata")
75
76   summary(model_fe)
77
78   #-- Demean OLS
79   dm_formula <- as.formula("demean_lnw ~ demean_age + demean_asq +
 ↪      demean_bmi +
80          demean_hispanic + demean_black + demean_other +
81          demean_asian + demean_schooling + demean_cohab +
82          demean_married + demean_divorced + demean_separated +
83          demean_age_cl + demean_unsafe + demean_llength + demean_reg +
84          demean_asq_cl + demean_appearance_cl +
85          demean_provider_second + demean_asian_cl + demean_black_cl +
86          demean_hispanic_cl + demean_othrace_cl +
87          demean_hot + demean_massage_cl")
88
89   ols_demean <- lm_robust(formula = dm_formula,
90          data = balanced_sasp, clusters = id,
91          se_type = "stata")
92
93   summary(ols_demean)
```

A few comments about this analysis. Some of the respondents left certain questions blank, most likely due to concerns about anonymity and privacy. So I dropped anyone who had missing values for the sake of this exercise. This leaves us with a balanced panel. I have organized the output into Table 68. There's a lot of interesting information in these three columns, some of which may surprise you if only for the novelty of the regressions. So let's talk about the statistically significant ones. The pooled OLS regressions, recall, do not control for unobserved heterogeneity, because by definition those are unobservable. So these are potentially biased by the unobserved heterogeneity, which is a kind of selection bias, but we will discuss them anyhow.

First, a simple scan of the second and third column will show that the fixed effects regression which included (not shown) dummies for the individual herself is equivalent to a regression on the demeaned

Table 68. POLS, FE and Demeaned OLS estimates of the determinants of Log hourly price for a panel of sex workers.

Depvar:	POLS	FE	Demeaned OLS
Unprotected sex with client of any kind	0.013	0.051*	0.051*
	(0.028)	(0.028)	(0.026)
Ln(Length)	−0.308***	−0.435***	−0.435***
	(0.028)	(0.024)	(0.019)
Client was a Regular	−0.047*	−0.037**	−0.037**
	(0.028)	(0.019)	(0.017)
Age of Client	−0.001	0.002	0.002
	(0.009)	(0.007)	(0.006)
Age of Client Squared	0.000	−0.000	−0.000
	(0.000)	(0.000)	(0.000)
Client Attractiveness (Scale of 1 to 10)	0.020***	0.006	0.006
	(0.007)	(0.006)	(0.005)
Second Provider Involved	0.055	0.113*	0.113*
	(0.067)	(0.060)	(0.048)
Asian Client	−0.014	−0.010	−0.010
	(0.049)	(0.034)	(0.030)
Black Client	0.092	0.027	0.027
	(0.073)	(0.042)	(0.037)
Hispanic Client	0.052	−0.062	−0.062
	(0.080)	(0.052)	(0.045)
Other Ethnicity Client	0.156**	0.142***	0.142***
	(0.068)	(0.049)	(0.045)
Met Client in Hotel	0.133***	0.052*	0.052*
	(0.029)	(0.027)	(0.024)
Gave Client a Massage	−0.134***	−0.001	−0.001
	(0.029)	(0.028)	(0.024)
Age of Provider	0.003	0.000	0.000
	(0.012)	(.)	(.)
Age of Provider Squared	−0.000	0.000	0.000
	(0.000)	(.)	(.)
Body Mass Index	−0.022***	0.000	0.000
	(0.002)	(.)	(.)
Hispanic	−0.226***	0.000	0.000
	(0.082)	(.)	(.)
Black	0.028	0.000	0.000
	(0.064)	(.)	(.)
Other	−0.112	0.000	0.000
	(0.077)	(.)	(.)

Table 68. (*Continued*)

Asian	0.086	0.000	0.000
	(0.158)	(.)	(.)
Imputed Years of Schooling	0.020**	0.000	0.000
	(0.010)	(.)	(.)
Cohabitating (Living with a Partner) but Unmarried	−0.054	0.000	0.000
	(0.036)	(.)	(.)
Currently Married and Living with Your Spouse	0.005	0.000	0.000
	(0.043)	(.)	(.)
Divorced and Not Remarried	−0.021	0.000	0.000
	(0.038)	(.)	(.)
Married but Not Currently Living with Your Spouse	−0.056	0.000	0.000
	(0.059)	(.)	(.)
N	1,028	1,028	1,028
Mean of dependent variable	5.57	5.57	0.00

Note: Heteroskedastic robust standard errors in parentheses clustered at the provider level. *$p < 0.10$, **$p < 0.05$, ***$p < 0.01$

data. This should help persuade you that the fixed effects and the demeaned (within) estimators are yielding the same coefficients.

But second, let's dig into the results. One of the first things we observe is that in the pooled OLS model, there is not a compensating wage differential detectable on having unprotected sex with a client.[10] But, notice that in the fixed effects model, unprotected sex has a premium. This is consistent with Rosen [1986] who posited the existence of risk premia, as well as Gertler et al. [2005] who found risk premia for sex workers using panel data. Gertler et al. [2005], though, find a much larger premia of over 20% for unprotected sex, whereas I am finding only a mere 5%. This could be because a large number of the unprotected instances are fellatio, which carries a much lower risk of infection than unprotected receptive intercourse. Nevertheless, it is interesting that unprotected sex, under the assumption of strict exogeneity, appears to cause wages to rise by approximately 5%, which is statistically significant at the 10% level. Given an hourly wage of $262,

10 There were three kinds of sexual encounter—vaginal receptive sex, anal receptive sex, and fellatio. Unprotected sex is coded as any sex act without a condom.

this amounts to a mere 13 additional dollars per hour. The lack of a finding in the pooled OLS model seems to suggest that the unobserved heterogeneity was masking the effect.

Next we look at the session length. Note that I have already adjusted the price the client paid for the length of the session so that the outcome is a log wage, as opposed to a log price. As this is a log-log regression, we can interpret the coefficient on log length as an elasticity. When we use fixed effects, the elasticity increases from -0.308 to -0.435. The significance of this result, in economic terms, though, is that there appear to be "volume discounts" in sex work. That is, longer sessions are more expensive, but at a decreasing rate. Another interesting result is whether the client was a "regular," which meant that she had seen him before in another session. In our pooled OLS model, regulars paid 4.7% less, but this shrinks slightly in our fixed effects model to 3.7% reductions. Economically, this could be lower because new clients pose risks that repeat customers do not pose. Thus, if we expect prices to move closer to marginal cost, the disappearance of some of the risk from the repeated session should lower price, which it appears to do.

Another factor related to price is the attractiveness of the client. Interestingly, this does not go in the direction we may have expected. One might expect that the more attractive the client, the *less* he pays. But in fact it is the opposite. Given other research that finds beautiful people earn more money [Hamermesh and Biddle, 1994], it's possible that sex workers are price-discriminating. That is, when they see a handsome client, they deduce he earns more, and therefore charge him more. This result does not hold up when including fixed effects, though, suggesting that it is due to unobserved heterogeneity, at least in part.

Similar to unprotected sex, a second provider present has a positive effect on price, which is only detectable in the fixed effects model. Controlling for unobserved heterogeneity, the presence of a second provider increases prices by 11.3%. We also see that she discriminates against clients of "other" ethnicities, who pay 14.2% more than White clients. There's a premium associated with meeting in a hotel which is considerably smaller when controlling for provider fixed effects by almost a third. This positive effect, even in the fixed effects model, may simply represent the higher costs associated with meeting in a hotel room. The other coefficients are not statistically significant.

Many of the time-invariant results are also interesting, though. For instance, perhaps not surprisingly, women with higher BMI earn less. Hispanics earn less than White sex workers. And women with more schooling earn more, something which is explored in greater detail in Cunningham and Kendall [2016].

Conclusion

In conclusion, we have been exploring the usefulness of panel data for estimating causal effects. We noted that the fixed effects (within) estimator is a very useful method for addressing a very specific form of endogeneity, with some caveats. First, it will eliminate any and all unobserved and observed time-invariant covariates correlated with the treatment variable. So long as the treatment and the outcome varies over time, and there is strict exogeneity, then the fixed effects (within) estimator will identify the causal effect of the treatment on some outcome.

But this came with certain qualifications. For one, the method couldn't handle *time-variant* unobserved heterogeneity. It's thus the burden of the researcher to determine which type of unobserved heterogeneity problem they face, but if they face the latter, then the panel methods reviewed here are not unbiased and consistent. Second, when there exists strong reverse causality pathways, then panel methods are biased. Thus, we cannot solve the problem of simultaneity, such as what Wright faced when estimating the price elasticity of demand, using the fixed effects (within) estimator. Most likely, we are going to have to move into a different framework when facing that kind of problem.

Still, many problems in the social sciences may credibly be caused by a time-invariant unobserved heterogeneity problem, in which case the fixed effects (within) panel estimator is useful and appropriate.

Difference-
in-
Differences

*What's the difference between
me and you?
About five bank accounts, three
ounces, and two vehicles.*

Dr. Dre

The difference-in-differences design is an early quasi-experimental identification strategy for estimating causal effects that predates the randomized experiment by roughly eighty-five years. It has become the single most popular research design in the quantitative social sciences, and as such, it merits careful study by researchers everywhere.[1] In this chapter, I will explain this popular and important research design both in its simplest form, where a group of units is treated at the same time, and the more common form, where groups of units are treated at different points in time. My focus will be on the identifying assumptions needed for estimating treatment effects, including several practical tests and robustness exercises commonly performed, and I will point you to some of the work on difference-in-differences design (DD) being done at the frontier of research. I have included several replication exercises as well.

John Snow's Cholera Hypothesis

When thinking about situations in which a difference-in-differences design can be used, one usually tries to find an instance where a

[1] A simple search on Google Scholar for phrase "difference-in-differences" yields over forty thousand hits.

consequential treatment was given to some people or units but denied to others "haphazardly." This is sometimes called a "natural experiment" because it is based on naturally occurring variation in some treatment variable that affects only some units over time. All good difference-in-differences designs are based on some kind of natural experiment. And one of the most interesting natural experiments was also one of the first difference-in-differences designs. This is the story of how John Snow convinced the world that cholera was transmitted by water, not air, using an ingenious natural experiment [Snow, 1855].

Cholera is a vicious disease that attacks victims suddenly, with acute symptoms such as vomiting and diarrhea. In the nineteenth century, it was usually fatal. There were three main epidemics that hit London, and like a tornado, they cut a path of devastation through the city. Snow, a physician, watched as tens of thousands suffered and died from a mysterious plague. Doctors could not help the victims because they were mistaken about the mechanism that caused cholera to spread between people.

The majority medical opinion about cholera transmission at that time was *miasma*, which said diseases were spread by microscopic poisonous particles that infected people by floating through the air. These particles were thought to be inanimate, and because microscopes at that time had incredibly poor resolution, it would be years before microorganisms would be seen. Treatments, therefore, tended to be designed to stop poisonous dirt from spreading through the air. But tried and true methods like quarantining the sick were strangely ineffective at slowing down this plague.

John Snow worked in London during these epidemics. Originally, Snow—like everyone—accepted the *miasma* theory and tried many ingenious approaches based on the theory to block these airborne poisons from reaching other people. He went so far as to cover the sick with burlap bags, for instance, but the disease still spread. People kept getting sick and dying. Faced with the theory's failure to explain cholera, he did what good scientists do—he changed his mind and began look for a new explanation.

Snow developed a novel theory about cholera in which the active agent was not an inanimate particle but was rather a living organism. This microorganism entered the body through food and drink, flowed through the alimentary canal where it multiplied and generated a poison that caused the body to expel water. With each evacuation, the organism passed out of the body and, importantly, flowed into England's water supply. People unknowingly drank contaminated water from the Thames River, which caused them to contract cholera. As they did, they would evacuate with vomit and diarrhea, which would flow into the water supply again and again, leading to new infections across the city. This process repeated through a multiplier effect which was why cholera would hit the city in epidemic waves.

Snow's years of observing the clinical course of the disease led him to question the usefulness of *miasma* to explain cholera. While these were what we would call "anecdote," the numerous observations and imperfect studies nonetheless shaped his thinking. Here's just a few of the observations which puzzled him. He noticed that cholera transmission tended to follow human commerce. A sailor on a ship from a cholera-free country who arrived at a cholera-stricken port would only get sick after landing or taking on supplies; he would not get sick if he remained docked. Cholera hit the poorest communities worst, and those people were the very same people who lived in the most crowded housing with the worst hygiene. He might find two apartment buildings next to one another, one would be heavily hit with cholera, but strangely the other one wouldn't. He then noticed that the first building would be contaminated by runoff from privies but the water supply in the second building was cleaner. While these observations weren't impossible to reconcile with *miasma*, they were definitely unusual and didn't seem obviously consistent with *miasmis*.

Snow tucked away more and more anecdotal evidence like these. But, while this evidence raised some doubts in his mind, he was not convinced. He needed a smoking gun if he were to eliminate all doubt that cholera was spread by water, not air. But where would he find that evidence? More importantly, what would evidence like that even look like?

Let's imagine the following thought experiment. If Snow was a dictator with unlimited wealth and power, how could he test his theory that cholera is waterborne? One thing he could do is flip a coin over each household member—heads you drink from the contaminated Thames, tails you drink from some uncontaminated source. Once the assignments had been made, Snow could simply compare cholera mortality between the two groups. If those who drank the clean water were less likely to contract cholera, then this would suggest that cholera was waterborne.

Knowledge that physical randomization could be used to identify causal effects was still eighty-five years away. But there were other issues besides ignorance that kept Snow from physical randomization. Experiments like the one I just described are also impractical, infeasible, and maybe even unethical—which is why social scientists so often rely on natural experiments that mimic important elements of randomized experiments. But what natural experiment was there? Snow needed to find a situation where uncontaminated water had been distributed to a large number of people as if by random chance, and then calculate the difference between those those who did and did not drink contaminated water. Furthermore, the contaminated water would need to be allocated to people in ways that were unrelated to the ordinary determinants of cholera mortality, such as hygiene and poverty, implying a degree of balance on covariates between the groups. And then he remembered—a potential natural experiment in London a year earlier had reallocated clean water to citizens of London. Could this work?

In the 1800s, several water companies served different areas of the city. Some neighborhoods were even served by more than one company. They took their water from the Thames, which had been polluted by victims' evacuations via runoff. But in 1849, the Lambeth water company had moved its intake pipes upstream higher up the Thames, above the main sewage discharge point, thus giving its customers uncontaminated water. They did this to obtain cleaner water, but it had the added benefit of being too high up the Thames to be infected with cholera from the runoff. Snow seized on this opportunity. He realized that it had given him a natural experiment that would allow him to test his hypothesis that cholera was waterborne by comparing the households. If his theory was right, then the Lambeth houses should have lower cholera death rates than some other set of households whose water was infected with runoff—what we might call today the explicit counterfactual. He found his explicit counterfactual in the Southwark and Vauxhall Waterworks Company.

Unlike Lambeth, the Southwark and Vauxhall Waterworks Company had *not* moved their intake point upstream, and Snow spent an entire book documenting similarities between the two companies' households. For instance, sometimes their service cut an irregular path through neighborhoods and houses such that the households on either side were very similar; the only difference being they drank different water with different levels of contamination from runoff. Insofar as the kinds of people that each company serviced were observationally equivalent, then perhaps they were similar on the relevant unobservables as well.

Snow meticulously collected data on household enrollment in water supply companies, going door to door asking household heads the name of their utility company. Sometimes these individuals didn't know, though, so he used a saline test to determine the source himself [Coleman, 2019]. He matched those data with the city's data on the cholera death rates at the household level. It was in many ways as advanced as any study we might see today for how he carefully collected, prepared, and linked a variety of data sources to show the relationship between water purity and mortality. But he also displayed scientific ingenuity for how he carefully framed the research question and how long he remained skeptical until the research design's results convinced him otherwise. After combining everthing, he was able to generate extremely persuasive evidence that influenced policymakers in the city.[2]

Snow wrote up all of his analysis in a manuscript entitled *On the Mode of Communication of Cholera* [Snow, 1855]. Snow's main evidence was striking, and I will discuss results based on Table XII and Table IX (not shown) in Table 69. The main difference between my version and his version of Table XII is that I will use his data to estimate a treatment effect using difference-in-differences.

Table XII. In 1849, there were 135 cases of cholera per 10,000 households at Southwark and Vauxhall and 85 for Lambeth. But in 1854, there

---

2 John Snow is one of my personal heroes. He had a stubborn commitment to the truth and was unpersuaded by low-quality causal evidence. That simultaneous skepticism and open-mindedness gave him the willingness to question common sense when common sense failed to provide satisfactory explanations.

Table 69. Modified Table XII (Snow 1854).

Company name	1849	1854
Southwark and Vauxhall	135	147
Lambeth	85	19

were 147 per 100,000 in Southwark and Vauxhall, whereas Lambeth's cholera cases per 10,000 households fell to 19.

While Snow did not explicitly calculate the difference-in-differences, the ability to do so was there [Coleman, 2019]. If we difference Lambeth's 1854 value from its 1849 value, followed by the same after and before differencing for Southwark and Vauxhall, we can calculate an estimate of the ATT equaling 78 fewer deaths per 10,000. While Snow would go on to produce evidence showing cholera deaths were concentrated around a pump on Broad Street contaminated with cholera, he allegedly considered the simple difference-in-differences the more convincing test of his hypothesis.

The importance of the work Snow undertook to understand the causes of cholera in London cannot be overstated. It not only lifted our ability to estimate causal effects with observational data, it advanced science and ultimately saved lives. Of Snow's work on the cause of cholera transmission, Freedman [1991] states:

> The force of [Snow's] argument results from the clarity of the prior reasoning, the bringing together of many different lines of evidence, and the amount of shoe leather Snow was willing to use to get the data. Snow did some brilliant detective work on nonexperimental data. What is impressive is not the statistical technique but the handling of the scientific issues. He made steady progress from shrewd observation through case studies to analyze ecological data. In the end, he found and analyzed a natural experiment. [298]

Estimation

A simple table. Let's look at this example using some tables, which hopefully will help give you an idea of the intuition behind DD, as well as some of its identifying assumptions.[3] Assume that the intervention

---

3 You'll sometimes see acronyms for difference-in-differences like DD, DiD, Diff-in-diff, or even, God forbid, DnD.

Table 70. Compared to what? Different companies.

Company	Outcome
Lambeth	$Y = L + D$
Southwark and Vauxhall	$Y = SV$

Table 71. Compared to what? Before and after.

Company	Time	Outcome
Lambeth	Before	$Y = L$
	After	$Y = L + (T + D)$

is clean water, which I'll write as D, and our objective is to estimate D's causal effect on cholera deaths. Let cholera deaths be represented by the variable Y. Can we identify the causal effect of D if we just compare the post-treatment 1854 Lambeth cholera death values to that of the 1854 Southwark and Vauxhall values? This is in many ways an obvious choice, and in fact, it is one of the more common naive approaches to causal inference. After all, we have a control group, don't we? Why can't we just compare a treatment group to a control group? Let's look and see.

One of the things we immediately must remember is that the simple difference in outcomes, which is all we are doing here, only collapsed to the ATE if the treatment had been randomized. But it is never randomized in the real world where most choices if not all choices made by real people is endogenous to potential outcomes. Let's represent now the differences between Lambeth and Southwark and Vauxhall with fixed level differences, or fixed effects, represented by L and SV. Both are unobserved, unique to each company, and fixed over time. What these fixed effects mean is that even if Lambeth hadn't changed its water source there, would still be something determining cholera deaths, which is just the time-invariant unique differences between the two companies as it relates to cholera deaths in 1854.

Table 72. Compared to what? Difference in each company's differences.

Companies	Time	Outcome	D_1	D_2
Lambeth	Before	$Y = L$		
	After	$Y = L + T + D$	$T + D$	
				D
Southwark and Vauxhall	Before	$Y = SV$		
	After	$Y = SV + T$	T	

When we make a simple comparison between Lambeth and South-wark and Vauxhall, we get an estimated causal effect equalling $D + (L - SV)$. Notice the second term, $L - SV$. We've seen this before. It's the selection bias we found from the decomposition of the simple difference in outcomes from earlier in the book.

Okay, so say we realize that we cannot simply make cross-sectional comparisons between two units because of selection bias. Surely, though, we can compare a unit to itself? This is sometimes called an interrupted time series. Let's consider that simple before-and-after difference for Lambeth now.

While this procedure successfully eliminates the Lambeth fixed effect (unlike the cross-sectional difference), it doesn't give me an unbiased estimate of D because differences can't eliminate the natural changes in the cholera deaths over time. Recall, these events were oscillating in waves. I can't compare Lambeth before and after $(T + D)$ because of T, which is an omitted variable.

The intuition of the DD strategy is remarkably simple: combine these two simpler approaches so the selection bias and the effect of time are, in turns, eliminated. Let's look at it in the following table.

The first difference, D_1, does the simple before-and-after difference. This ultimately eliminates the unit-specific fixed effects. Then, once those differences are made, we difference the differences (hence the name) to get the unbiased estimate of D.

But there's a a key assumption with a DD design, and that assumption is discernible even in this table. We are assuming that there is no time-variant company specific unobservables. Nothing unobserved in Lambeth households that is changing between these two periods that *also* determines cholera deaths. This is equivalent to assuming that T is the same for all units. And we call this the *parallel trends* assumption. We will discuss this assumption repeatedly as the chapter proceeds, as it is the most important assumption in the design's engine. If you can buy off on the parallel trends assumption, then DD will identify the causal effect.

DD is a powerful, yet amazingly simple design. Using repeated observations on a treatment and control unit (usually several units), we can eliminate the unobserved heterogeneity to provide a credible estimate of the average treatment effect on the treated (ATT) by transforming the data in very specific ways. But when and why does this process yield the correct answer? Turns out, there is more to it than meets the eye. And it is imperative on the front end that you understand what's under the hood so that you can avoid conceptual errors about this design.

The simple 2×2 *DD.* The cholera case is a particular kind of DD design that Goodman-Bacon [2019] calls the 2×2 DD design. The 2×2 DD design has a treatment group k and untreated group U. There is a pre-period for the treatment group, pre(k); a post-period for the treatment group, post(k); a pre-treatment period for the untreated group, pre(U); and a post-period for the untreated group, post(U) So:

$$\widehat{\delta}_{kU}^{2 \times 2} = \left(\overline{y}_k^{\text{post}(k)} - \overline{y}_k^{\text{pre}(k)} \right) - \left(\overline{y}_U^{\text{post}(k)} - \overline{y}_U^{\text{pre}(k)} \right)$$

where $\widehat{\delta}_{kU}$ is the estimated ATT for group k, and $\overline{y}$ is the sample mean for that particular group in a particular time period. The first paragraph differences the treatment group, k, after minus before, the second paragraph differences the untreated group, U, after minus before. And once those quantities are obtained, we difference the second term from the first.

But this is simply the mechanics of calculations. What exactly is this estimated parameter mapping onto? To understand that, we must convert these sample averages into conditional expectations of potential outcomes. But that is easy to do when working with sample averages, as we will see here. First let's rewrite this as a conditional expectation.

$$\widehat{\delta}_{kU}^{2\times2} = \Big(E[Y_k \mid \text{Post}] - E[Y_k \mid \text{Pre}] \Big) - \Big(E[Y_U \mid \text{Post}] - E[Y_U \mid \text{Pre}] \Big)$$

Now let's use the switching equation, which transforms historical quantities of Y into potential outcomes. As we've done before, we'll do a little trick where we add zero to the right-hand side so that we can use those terms to help illustrate something important.

$$\widehat{\delta}_{kU}^{2\times2} = \underbrace{\Big(E[Y_k^1 \mid \text{Post}] - E[Y_k^0 \mid \text{Pre}] \Big) - \Big(E[Y_U^0 \mid \text{Post}] - E[Y_U^0 \mid \text{Pre}] \Big)}_{\text{Switching equation}}$$
$$+ \underbrace{E[Y_k^0 \mid \text{Post}] - E[Y_k^0 \mid \text{Post}]}_{\text{Adding zero}}$$

Now we simply rearrange these terms to get the decomposition of the 2×2 DD in terms of conditional expected potential outcomes.

$$\widehat{\delta}_{kU}^{2\times2} = \underbrace{E[Y_k^1 \mid \text{Post}] - E[Y_k^0 \mid \text{Post}]}_{\text{ATT}}$$
$$+ \underbrace{\Big[E[Y_k^0 \mid \text{Post}] - E[Y_k^0 \mid \text{Pre}] \Big] - \Big[E[Y_U^0 \mid \text{Post}] - E[Y_U^0 \mid \text{Pre}] \Big]}_{\text{Non-parallel trends bias in } 2 \times 2 \text{ case}}$$

Now, let's study this last term closely. This simple 2×2 difference-in-differences will isolate the ATT (the first term) if and only if the second term zeroes out. But why would this second term be zero? It would equal zero if the first difference involving the treatment group, k, equaled the second difference involving the untreated group, U.

But notice the term in the second line. Notice anything strange about it? The object of interest is Y^0, which is some outcome in a world without the treatment. But it's the *post* period, and in the post period, $Y = Y^1$ not Y^0 by the switching equation. Thus, the first term is *counterfactual*. And as we've said over and over, counterfactuals are not observable. This bottom line is often called the parallel trends assumption and it is by definition untestable since we cannot observe this counterfactual conditional expectation. We will return to this again, but for now I simply present it for your consideration.

DD and the Minimum Wage. Now I'd like to talk about more explicit economic content, and the minimum wage is as good a topic as any. The modern use of DD was brought into the social sciences through esteemed labor economist Orley Ashenfelter [1978]. His study was no doubt influential to his advisee, David Card, arguably the greatest labor economist of his generation. Card would go on to use the method in several pioneering studies, such as Card [1990]. But I will focus on one in particular—his now-classic minimum wage study [Card and Krueger, 1994].

Card and Krueger [1994] is an infamous study both because of its use of an explicit counterfactual for estimation, and because the study challenges many people's common beliefs about the negative effects of the minimum wage. It lionized a massive back-and-forth minimum-wage literature that continues to this day.[4] So controversial was this study that James Buchanan, the Nobel Prize winner, called those influenced by Card and Krueger [1994] "camp following whores" in a letter to the editor of the *Wall Street Journal* [Buchanan, 1996].[5]

Suppose you are interested in the effect of minimum wages on employment. Theoretically, you might expect that in competitive labor markets, an increase in the minimum wage would move us up a downward-sloping demand curve, causing employment to fall. But in labor markets characterized by monopsony, minimum wages can increase employment. Therefore, there are strong theoretical reasons to believe that the effect of the minimum wage on employment is ultimately an empirical question depending on many local contextual factors. This is where Card and Krueger [1994] entered. Could they uncover whether minimum wages were ultimately harmful or helpful in some local economy?

---

4 That literature is too extensive to cite here, but one can find reviews of a great deal of the contemporary literature on minimum wages in Neumark et al. [2014] and Cengiz et al. [2019].

5 James Buchanan won the Nobel Prize for his pioneering work on the theory of public choice. He was not, though, a labor economist, and to my knowledge did not have experience estimating causal effects using explicit counterfactuals with observational data. A Google Scholar search for "James Buchanan minimum wage" returned only one hit, the previously mentioned *Wall Street Journal* letter to the editor. I consider his criticism to be ideologically motivated ad hominem and as such unhelpful in this debate.

It's always useful to start these questions with a simple thought experiment: if you had a billion dollars, complete discretion and could run a randomized experiment, how would you test whether minimum wages increased or decreased employment? You might go across the hundreds of local labor markets in the United States and flip a coin—heads, you raise the minimum wage; tails, you keep it at the status quo. As we've done before, these kinds of thought experiments are useful for clarifying both the research design and the causal question.

Lacking a randomized experiment, Card and Krueger [1994] decided on a next-best solution by comparing two neighboring states before and after a minimum-wage increase. It was essentially the same strategy that Snow used in his cholera study and a strategy that economists continue to use, in one form or another, to this day [Dube et al., 2010].

New Jersey was set to experience an increase in the state minimum wage from $4.25 to $5.05 in November 1992, but neighboring Pennsylvania's minimum wage was staying at $4.25. Realizing they had an opportunity to evaluate the effect of the minimum-wage increase by comparing the two states before and after, they fielded a survey of about four hundred fast-food restaurants in both states—once in February 1992 (before) and again in November (after). The responses from this survey were then used to measure the outcomes they cared about (i.e., employment). As we saw with Snow, we see again here that shoe leather is as important as any statistical technique in causal inference.

Let's look at whether the minimum-wage hike in New Jersey in fact raised the minimum wage by examining the distribution of wages in the fast food stores they surveyed. Figure 54 shows the distribution of wages in November 1992 after the minimum-wage hike. As can be seen, the minimum-wage hike was binding, evidenced by the mass of wages at the minimum wage in New Jersey.

As a caveat, notice how effective this is at convincing the reader that the minimum wage in New Jersey was binding. This piece of data visualization is not a trivial, or even optional, strategy to be taken in studies such as this. Even John Snow presented carefully designed maps of the distribution of cholera deaths throughout London. Beautiful pictures displaying the "first stage" effect of the intervention on the

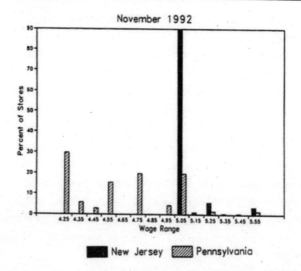

Figure 54. Distribution of wages for NJ and PA in November 1992. Reprinted from
Card, D. and Krueger, A. (1994). "Minimum Wages and Employment: A Case
Study of the Fast-Food Industry in New Jersey and Pennsylvania." *American
Economic Review*, 84:772–793. Reprinted with permission from authors.

treatment are crucial in the rhetoric of causal inference, and few have
done it as well as Card and Krueger.

Let's remind ourselves what we're after—the average causal effect
of the minimum-wage hike on employment, or the ATT. Using our
decomposition of the 2×2 DD from earlier, we can write it out as:

$$\widehat{\delta}^{2\times2}_{NJ,PA} = \underbrace{E[Y^1_{NJ} \mid \text{Post}] - E[Y^0_{NJ} \mid \text{Post}]}_{\text{ATT}}$$

$$+ \underbrace{\Big[E[Y^0_{NJ} \mid \text{Post}] - E[Y^0_{NJ} \mid \text{Pre}] \Big] - \Big[E[Y^0_{PA} \mid \text{Post}] - E[Y^0_{PA} \mid \text{Pre}] \Big]}_{\text{Non-parallel trends bias}}$$

Again, we see the key assumption: the parallel-trends assumption,
which is represented by the first difference in the second line. Insofar as parallel trends holds in this situation, then the second term goes
to zero, and the 2×2 DD collapses to the ATT.

The 2×2 DD requires differencing employment in NJ and PA, then
differencing those first differences. This set of steps estimates the true
ATT so long as the parallel-trends bias is zero. When that is true, $\widehat{\delta}^{2\times2}$

Table 73. Simple DD using sample averages on full-time employment.

Dependent variable	Stores by state		
	PA	NJ	NJ – PA
FTW before	23.3	20.44	−2.89
	(1.35)	(0.51)	(1.44)
FTE after	21.147	21.03	−0.14
	(0.94)	(0.52)	(1.07)
Change in mean FTE	−2.16	0.59	2.76
	(1.25)	(0.54)	(1.36)

Note: Standard errors in parentheses.

is equal to δ^{ATT}. If this bottom line is not zero, though, then simple 2×2 suffers from unknown bias—could bias it upwards, could bias it downwards, could flip the sign entirely. Table 73 shows the results of this exercise from Card and Krueger [1994].

Here you see the result that surprised many people. Card and Krueger [1994] estimate an ATT of +2.76 additional mean full-time-equivalent employment, as opposed to some negative value which would be consistent with competitive input markets. Herein we get Buchanan's frustration with the paper, which is based mainly on a particular model he had in mind, rather than a criticism of the research design the authors used.

While differences in sample averages will identify the ATT under the parallel assumption, we may want to use multivariate regression instead. For instance, if you need to avoid omitted variable bias through controlling for endogenous covariates that vary over time, then you may want to use regression. Such strategies are another way of saying that you will need to close some known critical backdoor. Another reason for the equation is that by controlling for more appropriate covariates, you can reduce residual variance and improve the precision of your DD estimate.

Using the switching equation, and assuming a constant state fixed effect and time fixed effect, we can write out a simple regression model estimating the causal effect of the minimum wage on employment, Y.

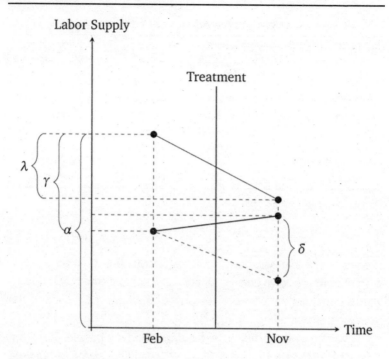

Figure 55. DD regression diagram.

This simple 2×2 is estimated with the following equation:

$$Y_{its} = \alpha + \gamma\, NJ_s + \lambda D_t + \delta(NJ \times D)_{st} + \varepsilon_{its}$$

NJ is a dummy equal to 1 if the observation is from NJ, and D is a dummy equal to 1 if the observation is from November (the post period). This equation takes the following values, which I will list in order according to setting the dummies equal to one and/or zero:

1. PA Pre: α
2. PA Post: $\alpha + \lambda$
3. NJ Pre: $\alpha + \gamma$
4. NJ Post: $\alpha + \gamma + \lambda + \delta$

We can visualize the 2×2 DD parameter in Figure 55.

Now before we hammer the parallel trends assumption for the billionth time, I wanted to point something out here which is a bit subtle. But do you see the δ parameter floating in the air above the November line in the Figure 55? This is the difference between a counterfactual level of employment (the bottom black circle in November on the negatively sloped dashed line) and the actual level of employment (the above black circle in November on the positively sloped solid line) for New Jersey. It is therefore the ATT, because the ATT is equal to

$$\delta = E[Y^1_{NJ,Post}] - E[Y^0_{NJ,Post}]$$

wherein the first is observed (because $Y = Y^1$ in the post period) and the latter is unobserved for the same reason.

Now here's the kicker: OLS will always estimate that δ line *even if the counterfactual slope had been something else*. That's because OLS uses Pennsylvania's change over time to project a point starting at New Jersey's pre-treatment value. When OLS has filled in that missing amount, the parameter estimate is equal to the difference between the observed post-treatment value and that projected value based on the slope of Pennsylvania *regardless of whether that Pennsylvania slope was the correct benchmark for measuring New Jersey's counterfactual slope*. OLS always estimates an effect size using the slope of the untreated group as the counterfactual, regardless of whether that slope is in fact the correct one.

But, see what happens when Pennsylvania's slope is equal to New Jersey's counterfactual slope? Then that Pennsylvania slope used in regression will mechanically estimate the ATT. In other words, only when the Pennsylvania slope is the counterfactual slope for New Jersey will OLS coincidentally identify that true effect. Let's see that here in Figure 56.

Notice the two δ listed: on the left is the true parameter δ^{ATT}. On the right is the one estimated by OLS, $\widehat{\delta}^{OLS}$. The falling solid line is the observed Pennsylvania change, whereas the falling solid line labeled "observed NJ" is the change in observed employment for New Jersey between the two periods.

The true causal effect, δ^{ATT}, is the line from the "observed NJ" point and the "counterfactual NJ" point. But OLS does not estimate this line. Instead, OLS uses the falling Pennsylvania line to draw a parallel line from the February NJ point, which is shown in thin gray. And

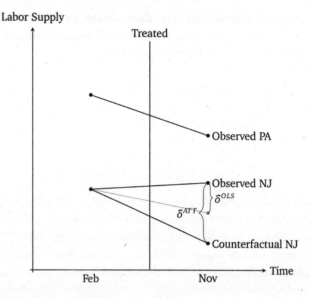

Figure 56. DD regression diagram without parallel trends.

OLS simply estimates the vertical line from the observed NJ point to the post NJ point, which as can be seen underestimates the true causal effect.

Here we see the importance of the parallel trends assumption. The only situation under which the OLS estimate equals the ATT is when the counterfactual NJ just coincidentally lined up with the gray OLS line, which is a line parallel to the slope of the Pennsylvania line. Herein lies the source of understandable skepticism of many who have been paying attention: why should we base estimation on this belief in a coincidence? After all, this is a counterfactual trend, and therefore it is unobserved, given it never occurred. Maybe the counterfactual would've been the gray line, but maybe it would've been some other unknown line. It could've been anything—we just don't know.

This is why I like to tell people that the parallel trends assumption is actually just a restatement of the strict exogeneity assumption we discussed in the panel chapter. What we are saying when we appeal to parallel trends is that we have found a control group who approximates the traveling path of the treatment group *and* that the treatment is not endogenous. If it is endogenous, then parallel trends is always violated because in counterfactual the treatment group would've diverged anyway, regardless of the treatment.

Before we see the number of tests that economists have devised to provide some reasonable confidence in the belief of the parallel trends, I'd like to quickly talk about standard errors in a DD design.

Inference

Many studies employing DD strategies use data from many years—not just one pre-treatment and one post-treatment period like Card and Krueger [1994]. The variables of interest in many of these setups only vary at a group level, such as the state, and outcome variables are often serially correlated. In Card and Krueger [1994], it is very likely for instance that employment in each state is not only correlated within the state but also serially correlated. Bertrand et al. [2004] point out that the conventional standard errors often severely understate the standard deviation of the estimators, and so standard errors are biased downward, "too small," and therefore overreject the null hypothesis. Bertrand et al. [2004] propose the following solutions:

1. Block bootstrapping standard errors.
2. Aggregating the data into one pre and one post period.
3. Clustering standard errors at the group level.

Block bootstrapping. If the block is a state, then you simply sample states with replacement for bootstrapping. Block bootstrap is straightforward and only requires a little programming involving loops and storing the estimates. As the mechanics are similar to that of randomization inference, I leave it to the reader to think about how they might tackle this.

Aggregation. This approach ignores the time-series dimensions altogether, and if there is only one pre and post period and one untreated group, it's as simple as it sounds. You simply average the groups into one pre and post period, and conduct difference-in-differences on those aggregated. But if you have differential timing, it's a bit unusual because you will need to partial out state and year fixed effects before turning the analysis into an analysis involving residualization. Essentially, for those common situations where you have multiple treatment time periods (which we discuss later in greater detail), you would regress the outcome onto panel unit and time fixed effects and any covariates. You'd then obtain the residuals for only the treatment group. You then divide the residuals only into a pre and post period; you are essentially at this point ignoring the never-treated groups. And then you regress the residuals on the after dummy. It's a strange procedure, and does not recover the original point estimate, so I focus instead on the third.

Clustering. Correct treatment of standard errors sometimes makes the number of groups very small: in Card and Krueger [1994], the number of groups is only two. More common than not, researchers will use the third option (clustering the standard errors by group). I have only one time seen someone do all three of these; it's rare though. Most people will present just the clustering solution—most likely because it requires minimal programming.

For clustering, there is no programming required, as most software packages allow for it already. You simply adjust standard errors by clustering at the group level, as we discussed in the earlier chapter, or the level of treatment. For state-level panels, that would mean clustering at the state level, which allows for arbitrary serial correlation in errors within a state over time. This is the most common solution employed.

Inference in a panel setting is independently an interesting area. When the number of clusters is small, then simple solutions like clustering the standard errors no longer suffice because of a growing false positive problem. In the extreme case with only one treatment unit, the over-rejection rate at a significance of 5% can be as high as 80% in simulations even using the wild bootstrap technique which has been suggested for smaller numbers of clusters [Cameron et al., 2008; MacKinnon and Webb, 2017]. In such extreme cases where there is only one treatment group, I have preferred to use randomization inference following Buchmueller et al. [2011].

Providing Evidence for Parallel Trends Through Event Studies and Parallel Leads

A redundant rant about parallel pre-treatment DD coefficients (because I'm worried one was not enough). Given the critical importance of the parallel trends assumption in identifying causal effects with the DD design, and given that one of the observations needed to evaluate the parallel-trends assumption is not available to the researcher, one might throw up their hands in despair. But economists are stubborn, and they have spent decades devising ways to test whether it's reasonable to believe in parallel trends. We now discuss the obligatory test for any DD design—the event study. Let's rewrite the decomposition of the 2×2 DD again.

$$\widetilde{\delta}_{kU}^{2 \times 2} = \underbrace{E[Y_k^1 \mid \text{Post}] - E[Y_k^0 \mid \text{Post}]}_{\text{ATT}}$$

$$+ \underbrace{\Big[E[Y_k^0 \mid \text{Post}] - E[Y_k^0 \mid \text{Pre}] \Big] - \Big[E[Y_U^0 \mid \text{Post}] - E[Y_U^0 \mid \text{Pre}] \Big]}_{\text{Non-parallel trends bias}}$$

We are interested in the first term, ATT, but it is contaminated by selection bias when the second term does not equal zero. Since evaluating the second term requires the counterfactual, $E[Y_k^0 \mid \text{Post}]$, we are unable to do so directly. What economists typically do, instead, is compare placebo pre-treatment leads of the DD coefficient. If DD coefficients in the pre-treatment periods are statistically zero, then the difference-in-differences between treatment and control groups followed a similar trend prior to treatment. And here's the rhetorical art of the design: *if* they had been similar before, *then* why wouldn't they continue to be post-treatment?

But notice that this rhetoric is a kind of proof by assertion. Just because they were similar before does not logically require they be the same after. Assuming that the future is like the past is a form of the gambler's fallacy called the "reverse position." Just because a coin came up heads three times in a row does not mean it will come up heads the fourth time—not without further assumptions. Likewise, we are not obligated to believe that that counterfactual trends would be the same post-treatment because they had been similar pre-treatment without further assumptions about the predictive power of pre-treatment trends. But to make such assumptions is again to make untestable assumptions, and so we are back where we started.

One situation where parallel trends would be obviously violated is if the treatment itself was endogenous. In such a scenario, the assignment of the treatment status would be directly dependent on potential outcomes, and absent the treatment, potential outcomes would've changed regardless. Such traditional endogeneity requires more than merely lazy visualizations of parallel leads. While the test is important, technically pre-treatment similarities are neither necessary nor sufficient to guarantee parallel counterfactual trends [Kahn-Lang and Lang, 2019]. The assumption is not so easily proven. You can never stop being diligent in attempting to determine whether groups of units endogenously selected into treatment, the presence of omitted variable biases, various sources of selection bias, and open backdoor paths. When the structural error term in a dynamic regression model is uncorrelated with the treatment variable, you have strict exogeneity, and that is what gives you parallel trends, and that is what makes you able to make meaningful statements about your estimates.

Checking the pre-treatment balance between treatment and control groups. Now with that pessimism out of the way, let's discuss event study plots because though they are not direct tests of the parallel trends assumption, they have their place because they show that the two groups of units were comparable on dynamics in the pre-treatment period.[6] Such conditional independence concepts have been used profitably throughout this book, and we do so again now.

Authors have tried showing the differences between treatment and control groups a few different ways. One way is to simply show the raw data, which you can do if you have a set of groups who received the treatment at the same point in time. Then you would just visually

---

6 Financial economics also has a procedure called the event study [Binder, 1998], but the way that event study is often used in contemporary causal inference is nothing more than a difference-in-differences design where, instead of a single post-treatment dummy, you saturate a model with leads and lags based on the timing of treatment.

inspect whether the pre-treatment dynamics of the treatment group differed from that of the control group units.

But what if you do not have a single treatment date? What if instead you have differential timing wherein groups of units adopt the treatment at different points? Then the concept of pre-treatment becomes complex. If New Jersey raised its minimum wage in 1992 and New York raised its minimum wage in 1994, but Pennsylvania never raised its minimum wage, the pre-treatment period is defined for New Jersey (1991) and New York (1993), but not Pennsylvania. Thus, how do we go about testing for pre-treatment differences in that case? People have done it in a variety of ways.

One possibility is to plot the raw data, year by year, and simply eyeball. You would compare the treatment group with the never-treated, for instance, which might require a lot of graphs and may also be awkward looking. Cheng and Hoekstra [2013] took this route, and created a separate graph comparing treatment groups with an untreated group for each different year of treatment. The advantage is its transparent display of the raw unadjusted data. No funny business. The disadvantage of this several-fold. First, it may be cumbersome when the number of treatment groups is large, making it practically impossible. Second, it may not be beautiful. But third, this necessarily assumes that the only control group is the never-treated group, which in fact is not true given what Goodman-Bacon [2019] has shown. Any DD is a combination of a comparison between the treatment and the never treated, an early treated compared to a late treated, and a late treated compared to an early treated. Thus only showing the comparison with the never treated is actually a misleading presentation of the underlying mechanization of identification using an twoway fixed-effects model with differential timing.

Anderson et al. [2013] took an alternative, creative approach to show the comparability of states with legalized medical marijuana and states without. As I said, the concept of a pre-treatment period for a control state is undefined when pre-treatment is always in reference to a specific treatment date which varies across groups. So, the authors construct a recentered time path of traffic fatality rates for the control states by assigning random treatment dates to all control counties and then plotting the average traffic fatality rates for each group in years leading up to treatment and beyond. This approach has a few advantages. First, it plots the raw data, rather than coefficients from a regression (as we will see next). Second, it plots that data against controls. But its weakness is that technically, the control series is not in fact *true*. It is chosen so as to give a comparison, but when regressions are eventually run, it will not be based on this series. But the main main shortcoming is that technically it is not displaying any of the control groups that will be used for estimation Goodman-Bacon [2019]. It is not displaying a comparison between the treated and the never treated; it is not a comparison between the early and late treated; it is not a comparison between the late and early treated. While a creative attempt to evaluate the pre-treatment differences in leads, it does not in fact technically show that.

The current way in which authors evaluate the pre-treatment dynamics between a treatment and control group with differential timing is to estimate a regression model that includes treatment leads and lags. I find that it is always useful to teach these concepts in the context of an actual paper, so let's review an interesting working paper by Miller et al. [2019].

Affordable Care Act, expanding Medicaid and population mortality. A provocative new study by Miller et al. [2019] examined the expansion of Medicaid under the Affordable Care Act. They were primarily interested in the effect that this expansion had on population mortality. Earlier work had cast doubt on Medicaid's effect on mortality [Baicker et al., 2013; Finkelstein et al., 2012], so revisiting the question with a larger sample size had value.

Like Snow before them, the authors link data sets on deaths with a large-scale federal survey data, thus showing that shoe leather often goes hand in hand with good design. They use these data to evaluate the causal impact of Medicaid enrollment on mortality using a DD design. Their focus is on the near-elderly adults in states with and without the Affordable Care Act Medicaid expansions and they find a 0.13-percentage-point decline in annual mortality, which is a 9.3% reduction over the sample mean, as a result of the ACA expansion. This effect is a result of a reduction in disease-related deaths and gets larger over time. Medicaid, in this estimation, saved a non-trivial number of lives.

As with many contemporary DD designs, Miller et al. [2019] evaluate the pre-treatment leads instead of plotting the raw data by treatment and control. Post-estimation, they plotted regression coefficients with 95% confidence intervals on their treatment leads and lags. Including leads and lags into the DD model allowed the reader to check both the degree to which the post-treatment treatment effects were dynamic, and whether the two groups were comparable on outcome dynamics pre-treatment. Models like this one usually follow a form like:

$$Y_{its} = \gamma_s + \lambda_t + \sum_{\tau=-q}^{-1} \gamma_\tau D_{s\tau} + \sum_{\tau=0}^{m} \delta_\tau D_{s\tau} + X_{ist} + \varepsilon_{ist}$$

Treatment occurs in year 0. You include q leads or anticipatory effects and m lags or post-treatment effects.

Miller et al. [2019] produce four event studies that when taken together tell the main parts of the story of their paper. This is, quite frankly, the art of the rhetoric of causal inference—visualization of key estimates, such as "first stages" as well as outcomes and placebos. The event study plots are so powerfully persuasive, they will make you a bit jealous, since oftentimes yours won't be nearly so nice. Let's look at the first three. State expansion of Medicaid under the Affordable Care Act increased Medicaid eligibility (Figure 57), which is not altogether surprising. It also caused an increase in Medicaid coverage (Figure 58), and as a consequence reduced the percentage of the uninsured population (Figure 59). All three of these are simply showing that the ACA Medicaid expansion had "bite"—people enrolled and became insured.

There are several features of these event studies that should catch your eye. First, look at Figure 57. The pre-treatment coefficients are nearly on the zero line itself. Not only are they nearly zero in their point estimate, but their standard errors are very small. This means these are very precisely estimated zero differences between individuals in the two groups of states prior to the expansion.

The second thing you see, though, is the elephant in the room. Post-treatment, the probability that someone becomes eligible for Medicaid immediately shoots up to 0.4 and while not as precise as the pre-treatment coefficients, the authors can rule out effects as low as 0.3 to 0.35. These are large increases in eligibility, and the fact that

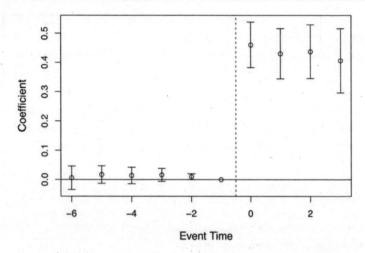

Figure 57. Estimates of Medicaid expansion's effects on eligibility using leads and lags in an event-study model. Miller, S., Altekruse, S., Johnson, N., and Wherry, L. R. (2019). Medicaid and mortality: New evidence from linked survey and administrative data. Working Paper No. 6081, National Bureau of Economic Research, Cambridge, MA. Reprinted with permission from authors.

the coefficients prior to the treatment are basically zero, we find it easy to believe that the risen coefficients post-treatment were caused by the ACA's expansion of Medicaid in states.

Of course, I would not be me if I did not say that *technically* the zeroes pre-treatment do not therefore mean that the post-treatment difference between counterfactual trends and observed trends are zero, but doesn't it seem compelling when you see it? Doesn't it compel you, just a little bit, that the changes in enrollment and insurance status were probably caused by the Medicaid expansion? I daresay a table of coefficients with leads, lags, and standard errors would probably not be as compelling even though it is the identical information. Also, it is only fair that the skeptic refuse these patterns with new evidence of what it is other than the Medicaid expansion. It is not enough to merely hand wave a criticism of omitted variable bias; the critic must be as engaged in this phenomenon as the authors themselves, which is how empiricists earn the right to critique someone else's work.

Similar graphs are shown for coverage—prior to treatment, the two groups of individuals in treatment and control were similar with regards

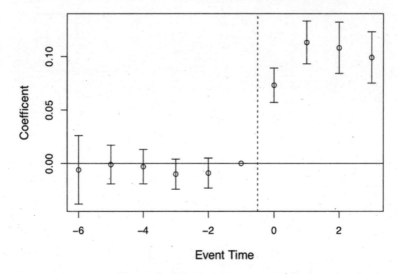

Figure 58. Estimates of Medicaid expansion's effects on coverage using leads and lags in an event-study model. Miller, S., Altekruse, S., Johnson, N., and Wherry, L. R. (2019). Medicaid and mortality: New evidence from linked survey and administrative data. Working Paper No. 6081, National Bureau of Economic Research, Cambridge, MA. Reprinted with permission from authors.

to their coverage and uninsured rate. But post-treatment, they diverge dramatically. Taken together, we have the "first stage," which means we can see that the Medicaid expansion under the ACA had "bite." Had the authors failed to find changes in eligibility, coverage, or uninsured rates, then any evidence from the secondary outcomes would have doubt built in. This is the reason it is so important that you examine the first stage (treatment's effect on usage), as well as the second stage (treatment's effect on the outcomes of interest).

But now let's look at the main result—what effect did this have on population mortality itself? Recall, Miller et al. [2019] linked administrative death records with a large-scale federal survey. So they actually know who is on Medicaid and who is not. John Snow would be proud of this design, the meticulous collection of high-quality data, and all the shoeleather the authors showed.

This event study is presented in Figure 60. A graph like this is the contemporary heart and soul of a DD design, both because it conveys key information regarding the comparability of the treatment and control groups in their dynamics just prior to treatment, and because such

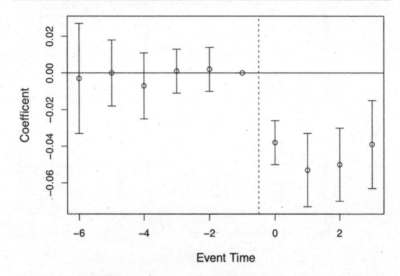

Figure 59. Estimates of Medicaid expansion's effects on the uninsured state using leads and lags in an event-study model. Miller, S., Altekruse, S., Johnson, N., and Wherry, L. R. (2019). Medicaid and mortality: New evidence from linked survey and administrative data. Working Paper No. 6081, National Bureau of Economic Research, Cambridge, MA. Reprinted with permission from authors.

strong data visualization of main effects are powerfully persuasive. It's quite clear looking at it that there was no difference between the trending tendencies of the two sets of state prior to treatment, making the subsequent divergence all the more striking.

But a picture like this is only as important as the thing that it is studying, and it is worth summarizing what Miller et al. [2019] have revealed here. The expansion of ACA Medicaid led to large swaths of people becoming eligible for Medicaid. In turn, they enrolled in Medicaid, which caused the uninsured rate to drop considerably. The authors then find amazingly using linked administrative data on death records that the expansion of ACA Medicaid led to a 0.13 percentage point decline in annual mortality, which is a 9.3 percent reduction over the mean. They go on to try and understand the mechanism (another key feature of this high-quality study) by which such amazing effects may have occurred, and conclude that Medicaid caused near-elderly individuals to receive treatment for life-threatening illnesses. I suspect we will be hearing about this study for many years.

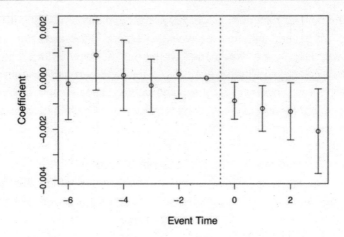

Figure 60. Estimates of Medicaid expansion's effects on on annual mortality using leads and lags in an event study model. Miller, S., Altekruse, S., Johnson, N., and Wherry, L. R. (2019). Medicaid and mortality: New evidence from linked survey and administrative data. Working Paper No. 6081, National Bureau of Economic Research, Cambridge, MA. Reprinted with permission from authors.

The Importance of Placebos in DD

There are several tests of the validity of a DD strategy. I have already discussed one—comparability between treatment and control groups on observable pre-treatment dynamics. Next, I will discuss other credible ways to evaluate whether estimated causal effects are credible by emphasizing the use of placebo falsification.

The idea of placebo falsification is simple. Say that you are finding some negative effect of the minimum wage on low-wage employment. Is the hypothesis true if we find evidence in favor? Maybe, maybe not. Maybe what would really help, though, is if you had in mind an alternative hypothesis and then tried to test that alternative hypothesis. If you cannot reject the null on the alternative hypothesis, then it provides some credibility to your original analysis. For instance, maybe you are picking up something spurious, like cyclical factors or other unobservables not easily captured by a time or state fixed effects. So what can you do?

One candidate placebo falsification might simply be to use data for an alternative type of worker whose wages would not be affected by the binding minimum wage. For instance, minimum wages affect employment and earnings of low-wage workers as these are the workers who literally are hired based on the market wage. Without some serious general equilibrium gymnastics, the minimum wage should not affect the employment of higher wage workers, because the minimum wage is not binding on high wage workers. Since high- and low-wage workers are employed in very different sectors, they are unlikely to be substitutes. This reasoning might lead us to consider the possibility that higher wage workers *might* function as a placebo.

There are two ways you can go about incorporating this idea into our analysis. Many people like to be straightforward and simply fit the same DD design using high wage employment as the outcome. If the coefficient on minimum wages is zero when using high wage worker employment as the outcome, but the coefficient on minimum wages for low wage workers is negative, then we have provided stronger evidence that complements the earlier analysis we did when on the low wage workers. But there is another method that uses the within-state placebo for identification called the difference-in-differences-in-differences ("triple differences"). I will discuss that design now.

Triple differences. In our earlier analysis, we assumed that the only thing that happened to New Jersey after it passed the minimum wage was a common shock, T, but what if there were state-specific time shocks such as NJ_t or PA_t? Then even DD cannot recover the treatment effect. Let's see for ourselves using a modification of the simple minimum-wage table from earlier, which will include the within-state workers who hypothetically were untreated by the minimum wage—the "high-wage workers."

Before the minimum-wage increase, low- and high-wage employment in New Jersey is determined by a group-specific New Jersey fixed effect (e.g., NJ_h). The same is true for Pennsylvania. But after the minimum-wage hike, four things change in New Jersey: national trends cause employment to change by T; New Jersey-specific time shocks change employment by NJ_t; generic trends in low-wage workers change employment by l_t; and the minimum-wage has some unknown effect D. We have the same setup in Pennsylvania except

Table 74. Triple differences design.

States	Group	Period	Outcomes	D_1	D_2	D_3
NJ	Low-wage workers	After	$NJ_l + T + NJ_t + l_t + D$	$T + NJ_t + l_t + D$	$(l_t - h_t) + D$	D
		Before	NJ_l			
	High-wage workers	After	$NJ_h + T + NJ_t + h_t$	$T + NJ_t + h_t$		
		Before	NJ_h			
PA	Low-wage workers	After	$PA_l + T + PA_t + l_t$	$T + PA_t + l_t$	$l_t - h_t$	
		Before	PA_l			
	High-wage workers	After	$PA_h + T + PA_t + h_t$	$T + PA_t + h_t$		
		Before	PA_h			

there is no minimum wage, and Pennsylvania experiences its own time shocks.

Now if we take first differences for each set of states, we only eliminate the state fixed effect. The first difference estimate for New Jersey includes the minimum-wage effect, D, but is also hopelessly contaminated by confounders (i.e., $T + NJ_t + l_t$). So we take a second difference for each state, and doing so, we eliminate two of the confounders: T disappears and NJ_t disappears. But while this DD strategy has eliminated several confounders, it has also introduced new ones (i.e., $(l_t - h_t)$). This is the final source of selection bias that triple differences are designed to resolve. But, by differencing Pennsylvania's second difference from New Jersey, the $(l_t - h_t)$ is deleted and the minimum-wage effect is isolated.

Now, this solution is not without its own set of unique parallel-trends assumptions. But one of the parallel trends here I'd like you to see is the $l_t - h_t$ term. This parallel trends assumption states that the effect can be isolated if the gap between high- and low-wage employment would've evolved similarly in the treatment state counterfactual as it did in the historical control states. And we should probably provide some credible evidence that this is true with leads and lags in an event study as before.

State-mandated maternity benefits. The triple differences design was first introduced by Gruber [1994] in a study of state-level policies providing maternity benefits. I present his main results in Table 75. Notice

Table 75. DDD Estimates of the Impact of State Mandates on Hourly Wages.

Location/year	Pre-law	Post-law	Difference
A. Treatment: Married women, 20–40yo			
Experimental states	1.547	1.513	−0.034
	(0.012)	(0.012)	(0.017)
Control states	1.369	1.397	0.028
	(0.010)	(0.010)	(0.014)
Difference	0.178	0.116	
	(0.016)	(0.015)	
Difference-in-difference		−0.062	
		(0.022)	
B. Control: Over 40 and Single Males 20–40			
Experimental states	1.759	1.748	−0.011
	(0.007)	(0.007)	(0.010)
Control states	1.630	1.627	−0.003
	(0.007)	(0.007)	(0.010)
Difference	1.09	1.21	
	(0.010)	(0.010)	
Difference-in-difference		−0.008	
		(0.014)	
DDD		−0.054	
		(0.026)	

Note: Standard errors in parentheses.

that he uses as his treatment group married women of childbearing age in treatment and control states, but he also uses a set of placebo units (older women and single men 20–40) as within-state controls. He then goes through the differences in means to get the difference-in-differences for each set of groups, after which he calculates the DDD as the difference between these two difference-in-differences.

Ideally when you do a DDD estimate, the causal effect estimate will come from changes in the treatment units, not changes in the control units. That's precisely what we see in Gruber [1994]: the action comes from changes in the married women age 20–40 (−0.062); there's little movement among the placebo units (−0.008). Thus when we calculate the DDD, we know that most of that calculation is coming from the first DD, and not so much from the second. We emphasize this because

DDD is really just another falsification exercise, and just as we would expect no effect had we done the DD on this placebo group, we hope that our DDD estimate is also based on negligible effects among the control group.

What we have done up to now is show how to use sample analogs and simple differences in means to estimate the treatment effect using DDD. But we can also use regression to control for additional covariates that perhaps are necessary to close backdoor paths and so forth. What does that regression equation look like? Both the regression itself, and the data structure upon which the regression is based, are complicated because of the stacking of different groups and the sheer number of interactions involved. Estimating a DDD model requires estimating the following regression:

$$Y_{ijt} = \alpha + \psi X_{ijt} + \beta_1 \tau_t + \beta_2 \delta_j + \beta_3 D_i + \beta_4 (\delta \times \tau)_{jt}$$
$$+ \beta_5 (\tau \times D)_{ti} + \beta_6 (\delta \times D)_{ij} + \beta_7 (\delta \times \tau \times D)_{ijt} + \varepsilon_{ijt}$$

where the parameter of interest is β_7. First, notice the additional subscript, j. This j indexes whether it's the main category of interest (e.g., low-wage employment) or the within-state comparison group (e.g., high-wage employment). This requires a stacking of the data into a panel structure by group, as well as state. Second, the DDD model requires that you include all possible interactions across the group dummy δ_j, the post-treatment dummy τ_t and the treatment state dummy D_i. The regression must include each dummy independently, each individual interaction, and the triple differences interaction. One of these will be dropped due to multicollinearity, but I include them in the equation so that you can visualize all the factors used in the product of these terms.

Abortion legalization and long-term gonorrhea incidence. Now that we know a little about the DD design, it would probably be beneficial to replicate a paper. And since the DDD requires reshaping panel data multiple times, that makes working through a detailed replication even more important. The study we will be replicating is Cunningham and Cornwell [2013], one of my first publications and the third chapter of my dissertation. Buckle up, as this will be a bit of a roller-coaster ride.

Gruber et al. [1999] was the beginning of what would become a controversial literature in reproductive health. They wanted to know the characteristics of the marginal child aborted had that child reached their teen years. The authors found that the marginal counterfactual child aborted was 60% more likely to grow up in a single-parent household, 50% more likely to live in poverty, and 45% more likely to be a welfare recipient. Clearly there were strong selection effects related to early abortion whereby it selected on families with fewer resources.

Their finding about the marginal child led John Donohue and Steven Levitt to wonder if there might be far-reaching effects of abortion legalization given the strong selection associated with its usage in the early 1970s. In Donohue and Levitt [2001], the authors argued that they had found evidence that abortion legalization had also led to massive declines in crime rates. Their interpretation of the results was that abortion legalization had reduced crime by removing high-risk individuals from a birth cohort, and as that cohort aged, those counterfactual crimes disappeared. Levitt [2004] attributed as much as 10% of the decline in crime between 1991 and 2001 to abortion legalization in the 1970s.

This study was, not surprisingly, incredibly controversial—some of it warranted but some unwarranted. For instance, some attacked the paper on ethical grounds and argued the paper was revitalizing the pseudoscience of eugenics. But Levitt was careful to focus only on the scientific issues and causal effects and did not offer policy advice based on his own private views, whatever those may be.

But some of the criticism the authors received was legitimate precisely because it centered on the research design and execution itself. Joyce [2004], Joyce [2009], and Foote and Goetz [2008] disputed the abortion-crime findings—some through replication exercises using different data, some with different research designs, and some through the discovery of key coding errors and erroneous variable construction.

One study in particular challenged the whole enterprise of estimating longrun improvements due to abortion legalization. For instance, Ted Joyce, an expert on reproductive health, cast doubt on the abortion-crime hypothesis using a DDD design [Joyce, 2009]. In addition to challenging Donohue and Levitt [2001], Joyce also threw down a gauntlet. He argued that if abortion legalization had such extreme negative selection as claimed by by Gruber et al. [1999] and Donohue and Levitt [2001], then it shouldn't show up just in crime. It should show up *everywhere*. Joyce writes:

> If abortion lowers homicide rates by 20–30%, then it is likely to have affected an entire spectrum of outcomes associated with well-being: infant health, child development, schooling, earnings and marital status. Similarly, the policy implications are broader than abortion. Other interventions that affect fertility control and that lead to fewer unwanted births—contraception or sexual abstinence—have huge potential payoffs. In short, a causal relationship between legalized abortion and crime has such significant ramifications for social policy and at the same time is so controversial, that further assessment of the identifying assumptions and their robustness to alternative strategies is warranted. [112]

Cunningham and Cornwell [2013] took up Joyce's challenge. Our study estimated the effects of abortion legalization on long-term gonorrhea incidence. Why gonorrhea? For one, single-parent households are a risk factor that lead to earlier sexual activity and unprotected sex, and Levine et al. [1999] found that abortion legalization caused teen childbearing to fall by 12%. Other risky outcomes had been found by numerous authors. Charles and Stephens [2006] reported that children exposed in utero to a legalized abortion regime were less likely to use illegal substances, which is correlated with risky sexual behavior.

My research design differed from Donohue and Levitt [2001] in that they used state-level lagged values of an abortion ratio, whereas I used difference-in-differences. My design exploited the early repeal of abortion in five states in 1970 and compared those states to the states that were legalized under *Roe v. Wade* in 1973. To do this, I needed cohort-specific data on gonorrhea incidence by state and year, but as those data are not collected by the CDC, I had to settle for second best. That second best was the CDC's gonorrhea data broken into five-year age categories (e.g., age 15–19, age 20–24). But this might still be useful because even with aggregate data, it might be possible to test the model I had in mind.

To understand this next part, which I consider the best part of my study, you must first accept a basic view of science that good theories make very specific falsifiable hypotheses. The more specific the hypothesis, the more convincing the theory, because if we find evidence exactly where the theory predicts, a Bayesian is likely to update her beliefs towards accepting the theory's credibility. Let me illustrate what I mean with a brief detour involving Albert Einstein's theory of relativity.

Einstein's theory made several falsifiable hypotheses. One of them involved a precise prediction of the warping of light as it moved past a large object, such as a star. The problem was that testing this theory involved observing distance between stars at night and comparing it to measurements made during the day as the starlight moved past the sun. Problem was, the sun is too bright in the daytime to see the stars, so those critical measurements can't be made. But Andrew Crommelin and Arthur Eddington realized the measurements could be made using an ingenious natural experiment. That natural experiment was an eclipse. They shipped telescopes to different parts of the world under the eclipse's path so that they had multiple chances to make the measurements. They decided to measure the distances of a large cluster of stars passing by the sun when it was dark and then immediately during an eclipse (Figure 61). That test was over a decade after Einstein's work was first published [Coles, 2019]. Think about it for a second—Einstein's theory by deduction is making predictions about phenomena that no one had ever really observed before. If this phenomena turned out to exist, then how couldn't the Bayesian update her beliefs and accept that the theory was credible? Incredibly, Einstein was right—just as he predicted, the apparent position of these stars shifted when moving around the sun. Incredible!

So what does that have to do with my study of abortion legalization and gonorrhea? The theory of abortion legalization having strong selection effects on cohorts makes very specific predictions about the shape of observed treatment effects. And if we found evidence for that shape, we'd be forced to take the theory seriously. So what what were these unusual yet testable predictions exactly?

The testable prediction from the staggered adoption of abortion legalization concerned the age-year-state profile of gonorrhea. The

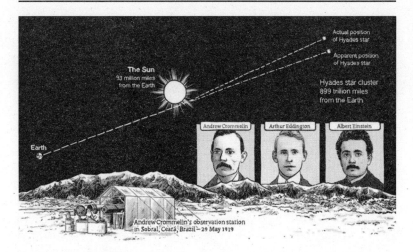

Figure 61. Light bending around the sun, predicted by Einstein, and confirmed in a natural experiment involving an eclipse. Artwork by Seth Hahne ©2020.

early repeal of abortion by five states three years before the rest of the country predicts lower incidence among 15- to 19-year-olds in the repeal states only during the 1986–1992 period relative to their *Roe* counterparts as the treated cohorts aged. That's not really all that special a prediction though. Maybe something happens in those same states fifteen to nineteen years later that isn't controlled for, for instance. What else?

The abortion legalization theory also predicted the *shape* of the observed treatment effects in this particular staggered adoption. Specifically, we should observe nonlinear treatment effects. These treatment effects should be increasingly negative from 1986 to 1989, plateau from 1989 to 1991, then gradually dissipate until 1992. In other words, the abortion legalization hypothesis predicts a parabolic treatment effect as treated cohorts move through the age distribution. All coefficients on the DD coefficients beyond 1992 should be zero and/or statistically insignificant.

I illustrate these predictions in Figure 62. The top horizontal axis shows the year of the panel, the vertical axis shows the age in calendar years, and the cells show the cohort for a given person of a certain age in that given year. For instance, consider a 15-year-old in 1985. She

	CDC Surveillance Data in Calendar Year															
Age in calendar year	1985	1986	1987	1988	1989	1990	1991	1992	1993	1994	1995	1996	1997	1998	1999	2000
15	70	71	72	73	74	75	76	77	78	79	80	81	82	83	84	85
16	69	70	71	72	73	74	75	76	77	78	79	80	81	82	83	84
17	68	69	70	71	72	73	74	75	76	77	78	79	80	81	82	83
18	67	68	69	70	71	72	73	74	75	76	77	78	79	80	81	82
19	66	67	68	69	70	71	72	73	74	75	76	77	78	79	80	81
20	65	66	67	68	69	70	71	72	73	74	75	76	77	78	79	80
21	64	65	66	67	68	69	70	71	72	73	74	75	76	77	78	79
22	63	64	65	66	67	68	69	70	71	72	73	74	75	76	77	78
23	62	63	64	65	66	67	68	69	70	71	72	73	74	75	76	77
24	61	62	63	64	65	66	67	68	69	70	71	72	73	74	75	76
25	60	61	62	63	64	65	66	67	68	69	70	71	72	73	74	75
26	59	60	61	62	63	64	65	66	67	68	69	70	71	72	73	74
27	58	59	60	61	62	63	64	65	66	67	68	69	70	71	72	73
28	57	58	59	60	61	62	63	64	65	66	67	68	69	70	71	72
29	56	57	58	59	60	61	62	63	64	65	66	67	68	69	70	71
Number of cohorts (age 15-19) exposed, reforms in 71, 74: Repeal (1)	0	1	2	3	4	5	5	5	5	5	5	5	5	5	5	5
No Repeal (2)	0	0	0	0	1	2	3	4	5	5	5	5	5	5	5	5
Difference (3)	0	1	2	3	3	3	2	1	0	0	0	0	0	0	0	0

Figure 62. Theoretical predictions of abortion legalization on age profiles of gonorrhea incidence. Reprinted from Cunningham, S. and Cornwell, C. (2013). "The Long-Run Effect of Abortion on Sexually Transmitted Infections." *American Law and Economics Review*, 15(1):381–407. Permission from Oxford University Press.

was born in 1970. A 15-year-old in 1986 was born in 1971. A 15-year-old in 1987 was born in 1972, and so forth. I mark the cohorts who were treated by either repeal or *Roe* in different shades of gray.

The theoretical predictions of the staggered rollout is shown at the bottom of Figure 62. In 1986, only one cohort (the 1971 cohort) was treated and only in the repeal states. Therefore, we should see small declines in gonorrhea incidence among 15-year-olds in 1986 relative to *Roe* states. In 1987, two cohorts in our data are treated in the repeal states relative to *Roe*, so we should see larger effects in absolute value than we saw in 1986. But from 1988 to 1991, we should at most see only three *net* treated cohorts in the repeal states because starting in 1988, the *Roe* state cohorts enter and begin erasing those differences. Starting in 1992, the effects should get smaller in absolute value until

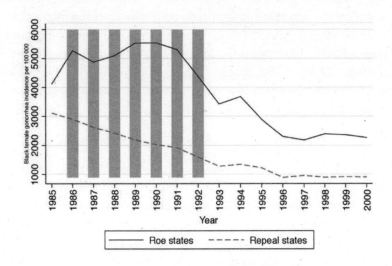

Figure 63. Differences in gonorrhea incidence among black females between repeal and *Roe* cohorts expressed as coefficient plots. Reprinted from Cunningham, S. and Cornwell, C. (2013). "The Long-Run Effect of Abortion on Sexually Transmitted Infections." *American Law and Economics Review*, 15(1):381–407. Permission from Oxford University Press.

1992, beyond which there should be no difference between repeal and *Roe* states.

It is interesting that something so simple as a staggered policy roll-out should provide two testable hypotheses that together can provide some insight into whether there is credibility to the negative selection in abortion legalization story. If we cannot find evidence for a negative parabola during this specific, narrow window, then the abortion legalization hypothesis has one more nail in its coffin.

A simple graphic for gonorrhea incidence among black 15- to 19-year-olds can help illustrate our findings. Remember, a picture is worth a thousand words, and whether it's RDD or DD, it's helpful to show pictures like these to prepare the reader for the table after table of regression coefficients. So notice what the raw data looks like in Figure 63.

First let's look at the raw data. I have shaded the years corresponding to the window where we expect to find effects. In Figure 63, we see the dynamics that will ultimately be picked up in the regression coefficients—the *Roe* states experienced a large and sustained gonorrhea epidemic that only waned once the treated cohorts emerged and overtook the entire data series.

Now let's look at regression coefficients. Our estimating equation is as follows:

$$Y_{st} = \beta_1 Repeal_s + \beta_2 DT_t + \beta_{3t} Repeal_s \times DT_t + X_{st}\psi + \alpha_s DS_s + \varepsilon_{st}$$

where Y is the log number of new gonorrhea cases for 15- to 19-year-olds (per 100,000 of the population); $Repeal_s$ equals 1 if the state legalized abortion prior to *Roe*; DT_t is a year dummy; DS_s is a state dummy; t is a time trend; X is a matrix of covariates. In the paper, I sometimes included state-specific linear trends, but for this analysis, I present the simpler model. Finally, ε_{st} is a structural error term assumed to be conditionally independent of the regressors. All standard errors, furthermore, were clustered at the state level allowing for arbitrary serial correlation.

I present the plotted coefficients from this regression for simplicity (and because pictures can be so powerful) in Figure 64. As can be seen in Figure 64, there is a negative effect during the window where *Roe* has not *fully* caught up, and that negative effect forms a parabola—just as our theory predicted.

Now, a lot of people might be done, but if you are reading this book, you have revealed that you are not like a lot of people. *Credibly* identified causal effects requires both finding effects, and ruling out alternative explanations. This is necessary because the fundamental problem of causal inference keeps us blind to the truth. But one way to alleviate some of that doubt is through rigorous placebo analysis. Here I present evidence from a triple difference in which an untreated cohort is used as a within-state control.

We chose the 25- to 29-year-olds in the same states as within-state comparison groups instead of 20- to 24-year-olds after a lot of thought. Our reasoning was that we needed an age group that was close enough to capture common trends but far enough so as not to violate SUTVA. Since 15- to 19-year-olds were more likely than 25- to 29-year-olds to have sex with 20- to 24-year-olds, we chose the

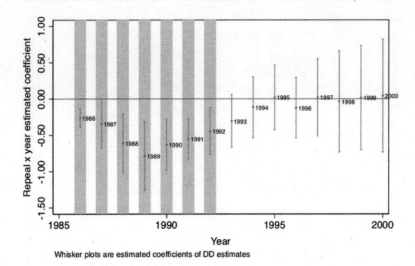

Whisker plots are estimated coefficients of DD estimates

Figure 64. Coefficients and standard errors from DD regression equation.

slightly older group as the within-stage control. But there's a trade-off here. Choose a group too close and you get SUTVA violations. Choose a group too far and they no longer can credibly soak up the heterogeneities you're worried about. The estimating equation for this regression is:

$$Y_{ast} = \beta_1 \text{ Repeal}_s + \beta_2 DT_t + \beta_{3t} \text{ Repeal}_s \cdot DT_t + \delta_1 DA + \delta_2 \text{ Repeal}_s \cdot DA$$
$$+ \delta_{3t} DA \cdot DT_t + \delta_{4t} \text{ Repeal}_s \cdot DA \cdot DT_t + X_{st}\xi + \alpha_{1s} DS_s + \alpha_{2s} DS_s \cdot DA$$
$$+ \gamma_1 t + \gamma_{2s} DS_s \cdot t + \gamma_3 DA \cdot t + \gamma_{4s} DS_s \cdot DA \cdot t + \epsilon_{ast},$$

where the DDD parameter we are estimating is δ_{4t}—the full interaction. In case this wasn't obvious, there are 7 separate dummies because our DDD parameter has all three interactions. Thus since there are eight combinations, we had to drop one as the omitted group, and control separately for the other seven. Here we present the table of coefficients. Note that the effect should be concentrated only among the treatment years as before, and second, it should form a parabola. The results are presented in Figure 65.

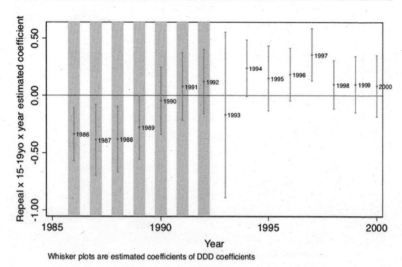

Whisker plots are estimated coefficients of DDD coefficients

Figure 65. DDD coefficients of abortion legalization on 15- to 19-year-old Black female log gonorrhea rates.

Here we see the prediction start to break down. Though there are negative effects for years 1986 to 1990, the 1991 and 1992 coefficients are positive, which is not consistent with our hypothesis. Furthermore, only the first four coefficients are statistically significant. Nevertheless, given the demanding nature of DDD, perhaps this is a small victory in favor of Gruber et al. [1999] and Donohue and Levitt [2001]. Perhaps the theory that abortion legalization had strong selection effects on cohorts has some validity.

Putting aside whether you believe the results, it is still valuable to replicate the results based on this staggered design. Recall that I said the DDD design requires stacking the data, which may seem like a bit of a black box, so I'd like to examine these data now.[7]

---

7 In the original Cunningham and Cornwell [2013], we estimated models with multiway clustering correction, but the package for this in Stata is no longer supported. Therefore, we will estimate the same models as in Cunningham and Cornwell [2013] using cluster robust standard errors. In all prior analysis, I clustered the standard errors at the state level so as to maintain consistency with this code.

STATA

abortion_dd.do

```
1   * DD estimate of 15-19 year olds in repeal states vs Roe states
2   use https://github.com/scunning1975/mixtape/raw/master/abortion.dta, clear
3   xi: reg lnr i.repeal*i.year i.fip acc ir pi alcohol crack poverty income ur if bf15==1
    ↪  [aweight=totpop], cluster(fip)
4
5   * ssc install parmest, replace
6
7   parmest, label for(estimate min95 max95 %8.2f) li(parm label estimate min95
    ↪  max95) saving(bf15_DD.dta, replace)
8
9   use ./bf15_DD.dta, replace
10
11  keep in 17/31
12
13  gen        year=1986 in 1
14  replace year=1987 in 2
15  replace year=1988 in 3
16  replace year=1989 in 4
17  replace year=1990 in 5
18  replace year=1991 in 6
19  replace year=1992 in 7
20  replace year=1993 in 8
21  replace year=1994 in 9
22  replace year=1995 in 10
23  replace year=1996 in 11
24  replace year=1997 in 12
25  replace year=1998 in 13
26  replace year=1999 in 14
27  replace year=2000 in 15
28
29  sort year
30
31  twoway (scatter estimate year, mlabel(year) mlabsize(vsmall) msize(tiny)) (rcap
    ↪  min95 max95 year, msize(vsmall)), ytitle(Repeal x year estimated
    ↪  coefficient) yscale(titlegap(2)) yline(0, lwidth(vvvthin) lcolor(black))
    ↪  xtitle(Year) xline(1986 1987 1988 1989 1990 1991 1992, lwidth(vvvthick)
    ↪  lpattern(solid) lcolor(ltblue)) xscale(titlegap(2)) title(Estimated effect of
    ↪  abortion legalization on gonorrhea) subtitle(Black females 15-19 year-olds)
    ↪  note(Whisker plots are estimated coefficients of DD estimator from Column
    ↪  b of Table 2.) legend(off)
```

R
abortion_dd.R

```r
1   #-- DD estimate of 15-19 year olds in repeal states vs Roe states
2   library(tidyverse)
3   library(haven)
4   library(estimatr)
5
6   read_data <- function(df)
7   {
8     full_path <- paste("https://raw.github.com/scunning1975/mixtape/master/",
9               df, sep = "")
10    df <- read_dta(full_path)
11    return(df)
12  }
13
14  abortion <- read_data("abortion.dta") %>%
15    mutate(
16      repeal = as_factor(repeal),
17      year   = as_factor(year),
18      fip    = as_factor(fip),
19      fa     = as_factor(fa),
20    )
21
22  reg <- abortion %>%
23    filter(bf15 == 1) %>%
24    lm_robust(lnr ~ repeal*year + fip + acc + ir + pi + alcohol+ crack + poverty+
    ↪  income+ ur,
25          data = ., weights = totpop, clusters = fip)
26
27  abortion_plot <- tibble(
28    sd = reg$std.error[-1:-75],
29    mean = reg$coefficients[-1:-75],
30    year = c(1986:2000))
31
32  abortion_plot %>%
33    ggplot(aes(x = year, y = mean)) +
34    geom_rect(aes(xmin=1986, xmax=1992, ymin=-Inf, ymax=Inf), fill = "cyan", alpha
    ↪  = 0.01)+
35    geom_point()+
36    geom_text(aes(label = year), hjust=-0.002, vjust = -0.03)+
37    geom_hline(yintercept = 0) +
38    geom_errorbar(aes(ymin = mean - sd*1.96, ymax = mean + sd*1.96), width = 0.2,
39          position = position_dodge(0.05))
```

The second line estimates the regression equation. The dynamic DD coefficients are captured by the repeal-year interactions. These are the coefficients we used to create box plots in Figure 64. You can check these yourself.

Note, for simplicity, I only estimated this for the black females (bf15==1) but you could estimate for the black males (bm15==1), white females (wf15==1), or white males (wm15==1). We do all four in the paper, but here we only focus on the black females aged 15–19 because the purpose of this section is to help you understand the estimation. I encourage you to play around with this model to see how robust the effects are in your mind using only this linear estimation.

But now I want to show you the code for estimating a triple difference model. Some reshaping had to be done behind the scenes for this data structure, but it would take too long to post that here. For now, I will simply produce the commands that produce the black female result, and I encourage you to explore the panel data structure so as to familiarize yourself with the way in which the data are organized.

Notice that some of these were already interactions (e.g., yr), which was my way to compactly include all of the interactions. I did this primarily to give myself more control over what variables I was using. But I encourage you to study the data structure itself so that when you need to estimate your own DDD, you'll have a good handle on what form the data must be in in order to execute so many interactions.

STATA
abortion_ddd.do

```
1   use https://github.com/scunning1975/mixtape/raw/master/abortion.dta, clear
2
3   * DDD estimate for 15-19 year olds vs. 20-24 year olds in repeal vs Roe states
4   gen yr=(repeal) & (younger==1)
5   gen wm=(wht==1) & (male==1)
6   gen wf=(wht==1) & (male==0)
7   gen bm=(wht==0) & (male==1)
8   gen bf=(wht==0) & (male==0)
9   char year[omit] 1985
10  char repeal[omit] 0
11  char younger[omit] 0
```

(continued)

STATA *(continued)*

```
11   char fip[omit] 1
12   char fa[omit] 0
13   char yr[omit] 0
14   xi: reg lnr i.repeal*i.year i.younger*i.repeal i.younger*i.year i.yr*i.year i.fip*t acc
  ↪    pi ir alcohol crack  poverty income ur if bf==1 & (age==15 | age==25)
  ↪    [aweight=totpop], cluster(fip)
15
16   parmest, label for(estimate min95 max95 %8.2f) li(parm label estimate min95
  ↪    max95) saving(bf15_DDD.dta, replace)
17
18   use ./bf15_DDD.dta, replace
19
20   keep in 82/96
21
22   gen           year=1986 in 1
23   replace year=1987 in 2
24   replace year=1988 in 3
25   replace year=1989 in 4
26   replace year=1990 in 5
27   replace year=1991 in 6
28   replace year=1992 in 7
29   replace year=1993 in 8
30   replace year=1994 in 9
31   replace year=1995 in 10
32   replace year=1996 in 11
33   replace year=1997 in 12
34   replace year=1998 in 13
35   replace year=1999 in 14
36   replace year=2000 in 15
37
38   sort year
39
40   twoway (scatter estimate year, mlabel(year) mlabsize(vsmall) msize(tiny)) (rcap
  ↪    min95 max95 year, msize(vsmall)), ytitle(Repeal x 20-24yo x year estimated
  ↪    coefficient) yscale(titlegap(2)) yline(0, lwidth(vvvthin) lcolor(black))
  ↪    xtitle(Year) xline(1986 1987 1988 1989 1990 1991 1992, lwidth(vvvthick)
  ↪    lpattern(solid) lcolor(ltblue)) xscale(titlegap(2)) title(Estimated effect of
  ↪    abortion legalization on gonorrhea) subtitle(Black females 15-19 year-olds)
  ↪    note(Whisker plots are estimated coefficients of DDD estimator from
  ↪    Column b of Table 2.) legend(off)
41
```

R

abortion_ddd.R

```
1  library(tidyverse)
2  library(haven)
3  library(estimatr)
4
5  read_data <- function(df)
6  {
7  full_path <- paste("https://raw.github.com/scunning1975/mixtape/master/",
8              df, sep = "")
9   df <- read_dta(full_path)
10   return(df)
11  }
12
13  abortion <- read_data("abortion.dta") %>%
14   mutate(
15    repeal = as_factor(repeal),
16    year   = as_factor(year),
17    fip    = as_factor(fip),
18    fa     = as_factor(fa),
19    younger = as_factor(younger),
20    yr     = as_factor(case_when(repeal == 1 & younger == 1 ~ 1, TRUE ~ 0)),
21    wm     = as_factor(case_when(wht == 1 & male == 1 ~ 1, TRUE ~ 0)),
22    wf     = as_factor(case_when(wht == 1 & male == 0 ~ 1, TRUE ~ 0)),
23    bm     = as_factor(case_when(wht == 0 & male == 1 ~ 1, TRUE ~ 0)),
24    bf     = as_factor(case_when(wht == 0 & male == 0 ~ 1, TRUE ~ 0))
25   ) %>%
26   filter(bf == 1 & (age == 15 | age == 25))
27
28  regddd <- lm_robust(lnr ~ repeal*year + younger*repeal + younger*year + yr*year
    ↪ + fip*t + acc + ir + pi + alcohol + crack + poverty + income + ur,
29             data = abortion, weights = totpop, clusters = fip)
30
31  abortion_plot <- tibble(
32   sd = regddd$std.error[110:124],
33   mean = regddd$coefficients[110:124],
34   year = c(1986:2000))
35
```

(continued)

R *(continued)*

```
36   abortion_plot %>%
37     ggplot(aes(x = year, y = mean)) +
38     geom_rect(aes(xmin=1986, xmax=1992, ymin=-Inf, ymax=Inf), fill = "cyan", alpha
    ↪  = 0.01)+
39     geom_point()+
40     geom_text(aes(label = year), hjust=-0.002, vjust = -0.03)+
41     geom_hline(yintercept = 0) +
42     geom_errorbar(aes(ymin = mean-sd*1.96, ymax = mean+sd*1.96), width = 0.2,
43             position = position_dodge(0.05))
```

Going beyond Cunningham and Cornwell [2013]. The US experience with abortion legalization predicted a parabola from 1986 to 1992 for 15- to 19-year-olds, and using a DD design, that's what I found. I also estimated the effect using a DDD design, and while the effects weren't as pretty as what I found with DD, there appeared to be something going on in the general vicinity of where the model predicted. So boom goes the dynamite, right? Can't we be done finally? Not quite.

Whereas my original study stopped there, I would like to go a little farther. The reason can be seen in the following Figure 66. This is a modified version of Figure 62, with the main difference being I have created a new parabola for the 20- to 24-year-olds.

Look carefully at Figure 66. Insofar as the early 1970s cohorts were treated in utero with abortion legalization, then we should see not just a parabola for the 15- to 19-year-olds for 1986 to 1992 but also for the 20- to 24-year-olds for years 1991 to 1997 as the cohorts continued to age.[8]

I did not examine the 20- to 24-year-old cohort when I first wrote this paper because at that time I doubted that the selection effects for risky sex would persist into adulthood given that youth display considerable risk-taking behavior. But with time come new perspectives, and these days I don't have strong priors that the selection effects would necessarily vanish after teenage years. So I'd like to conduct that additional analysis here and now for the first time. Let's estimate the same DD model as before, only for Black females aged 20–24.

---

8 There is a third prediction on the 25- to 29-year-olds, but for the sake of space, I only focus on the 20- to 24-year-olds.

STATA

abortion_dd2.do

```
1   use https://github.com/scunning1975/mixtape/raw/master/abortion.dta, clear
2
3   * Second DD model for 20-24 year old black females
4   char year[omit] 1985
5   xi: reg lnr i.repeal*i.year i.fip acc ir pi alcohol crack poverty income ur if (race==2
    ↪   & sex==2 & age==20) [aweight=totpop], cluster(fip)
```

R

abortion_dd2.R

```
1   library(tidyverse)
2   library(haven)
3   library(estimatr)
4
5   read_data <- function(df)
6   {
7     full_path <- paste("https://raw.github.com/scunning1975/mixtape/master/",
8                 df, sep = "")
9     df <- read_dta(full_path)
10    return(df)
11  }
12
13  abortion <- read_data("abortion.dta") %>%
14    mutate(
15      repeal = as_factor(repeal),
16      year   = as_factor(year),
17      fip    = as_factor(fip),
18      fa     = as_factor(fa),
19    )
20
21  reg <- abortion %>%
22    filter(race == 2 & sex == 2 & age == 20) %>%
23    lm_robust(lnr ~ repeal*year + fip + acc + ir + pi + alcohol+ crack + poverty+
          ↪   income+ ur,
24            data = ., weights = totpop, clusters = fip)
```

As before, we will focus just on the coefficient plots. We show that in
Figure 67. There are a couple of things about this regression output
that are troubling. First, there is a negative parabola showing up where
there wasn't necessarily one predicted—the 1986–1992 period. Note
that is the period where only the 15- to 19-year-olds were the treated

		CDC Surveillance Data in Calendar Year														
Age in calendar year		1985	1986	1987	1988	1989	1990	1991	1992	1993	1994	1995	1996	1997	1998	1999	2000
15		70	71	72	73	74	75	76	77	78	79	80	81	82	83	84	85
16		69	70	71	72	73	74	75	76	77	78	79	80	81	82	83	84
17		68	69	70	71	72	73	74	75	76	77	78	79	80	81	82	83
18		67	68	69	70	71	72	73	74	75	76	77	78	79	80	81	82
19		66	67	68	69	70	71	72	73	74	75	76	77	78	79	80	81
20		65	66	67	68	69	70	71	72	73	74	75	76	77	78	79	80
21		64	65	66	67	68	69	70	71	72	73	74	75	76	77	78	79
22		63	64	65	66	67	68	69	70	71	72	73	74	75	76	77	78
23		62	63	64	65	66	67	68	69	70	71	72	73	74	75	76	77
24		61	62	63	64	65	66	67	68	69	70	71	72	73	74	75	76
25		60	61	62	63	64	65	66	67	68	69	70	71	72	73	74	75
26		59	60	61	62	63	64	65	66	67	68	69	70	71	72	73	74
27		58	59	60	61	62	63	64	65	66	67	68	69	70	71	72	73
28		57	58	59	60	61	62	63	64	65	66	67	68	69	70	71	72
29		56	57	58	59	60	61	62	63	64	65	66	67	68	69	70	71
	Repeal (1)	0	0	0	0	0	0	1	2	3	4	5	5	5	5	5	5
	No Repeal (2)	0	0	0	0	0	0	0	0	0	1	2	3	4	5	5	5
	Difference (3)	0	0	0	0	0	0	1	2	3	3	3	2	1	0	0	0

Left axis label: Number of cohorts (age 20–24) exposed, reforms in 71, 74

Figure 66. Theoretical predictions of abortion legalization on age profiles of gonorrhea incidence for 20–24-year-olds.

cohorts, suggesting that our 15- to 19-year-old analysis was picking up something other than abortion legalization. But that was also the justification for using DDD, as clearly something else is going on in the repeal versus *Roe* states during those years that we cannot adequately control for with our controls and fixed effects.

The second thing to notice is that there is *no* parabola in the treatment window for the treatment cohort. The effect sizes are negative in the beginning, but shrink in absolute value when they should be growing. In fact, the 1991 to 1997 period is one of convergence to zero, not divergence between these two sets of states.

But as before, maybe there are strong trending unobservables for all groups masking the abortion legalization effect. To check, let's use my DDD strategy with the 25- to 29-year-olds as the within-state control group. We can implement this by using the Stata code, abortion_ddd2.do and abortion_ddd2.R.

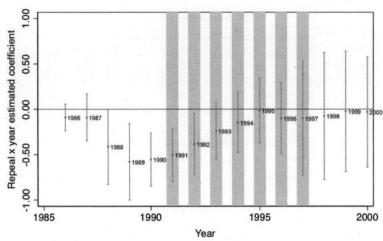

Whisker plots are estimated coefficients of DD estimates

Figure 67. Coefficients and standard errors from DD regression equation for the 20- to 24-year-olds.

STATA
abortion_ddd2.do

```
1   use https://github.com/scunning1975/mixtape/raw/master/abortion.dta, clear
2
3   * Second DDD model for 20-24 year olds vs 25-29 year olds black females in
    ↪ repeal vs Roe states
4   gen younger2 = 0
5   replace younger2 = 1 if age == 20
6   gen yr2=(repeal==1) & (younger2==1)
7   gen wm=(wht==1) & (male==1)
8   gen wf=(wht==1) & (male==0)
9   gen bm=(wht==0) & (male==1)
10  gen bf=(wht==0) & (male==0)
11  char year[omit] 1985
12  char repeal[omit] 0
13  char younger2[omit] 0
14  char fip[omit] 1
15  char fa[omit] 0
16  char yr2[omit] 0
17  xi: reg lnr i.repeal*i.year i.younger2*i.repeal i.younger2*i.year i.yr2*i.year i.fip*t
    ↪ acc pi ir alcohol crack  poverty income ur if bf==1 & (age==20 | age==25)
    ↪ [aweight=totpop], cluster(fip)
```

R
abortion_ddd2.R

```
1   library(tidyverse)
2   library(haven)
3   library(estimatr)
4
5   read_data <- function(df)
6   {
7    full_path <- paste("https://raw.github.com/scunning1975/mixtape/master/",
8              df, sep = "")
9    df <- read_dta(full_path)
10   return(df)
11  }
12
13  abortion <- read_data("abortion.dta") %>%
14   mutate(
15    repeal  = as_factor(repeal),
16    year   = as_factor(year),
17    fip    = as_factor(fip),
18    fa    = as_factor(fa),
19    younger2 = case_when(age == 20 ~ 1, TRUE ~ 0),
20    yr2    = as_factor(case_when(repeal == 1 & younger2 == 1 ~ 1, TRUE ~ 0)),
21    wm    = as_factor(case_when(wht == 1 & male == 1 ~ 1, TRUE ~ 0)),
22    wf    = as_factor(case_when(wht == 1 & male == 0 ~ 1, TRUE ~ 0)),
23    bm    = as_factor(case_when(wht == 0 & male == 1 ~ 1, TRUE ~ 0)),
24    bf    = as_factor(case_when(wht == 0 & male == 0 ~ 1, TRUE ~ 0))
25   )
26
27  regddd <- abortion %>%
28   filter(bf == 1 & (age == 20 | age ==25)) %>%
29   lm_robust(lnr ~ repeal*year + acc + ir + pi + alcohol + crack + poverty + income
       ↪ + ur,
30        data = ., weights = totpop, clusters = fip)
```

Figure 68 shows the DDD estimated coefficients for the treated
cohort relative to a slightly older 25- to 29-year-old cohort. It's possible
that the 25- to 29-year-old cohort is too close in age to function as
a satisfactory within-state control; if those age 20−24 have sex with
those who are age 25−29, for instance, then SUTVA is violated. There
are other age groups, though, that you can try in place of the 25- to 29-
year-olds, and I encourage you to do it for both the experience and the
insights you might gleam.

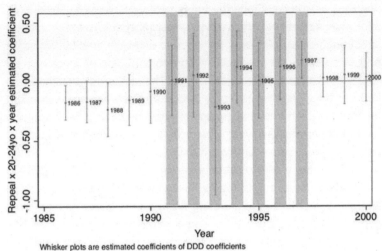

Figure 68. Coefficients and standard errors from DDD regression equation for the 20-
to 24-year-olds vs. 25- to 29-year-olds.

But let's back up and remember the big picture. The abortion legalization hypothesis made a series of predictions about where negative
parabolic treatment effects should appear in the data. And while we
found some initial support, when we exploited more of those predictions, the results fell apart. A fair interpretation of this exercise is that
our analysis does *not* support the abortion legalization hypothesis.
Figure 68 shows several point estimates at nearly zero, and standard
errors so large as to include both positive and negative values for these
interactions.

I included this analysis because I wanted to show you the power of
a theory with numerous unusual yet testable predictions. Imagine for
a moment if a parabola had showed up for all age groups in precisely
the years predicted by the theory. Wouldn't we *have* to update our priors
about the abortion legalization selection hypothesis? With predictions
so narrow, what else could be causing it? It's precisely because the predictions are so specific, though, that we are able to reject the abortion
legalization hypothesis, at least for gonorrhea.

Placebos as critique. Since the fundamental problem of causal inference blocks our direct observation of causal effects, we rely on many direct and indirect pieces of evidence to establish credible causality. And as I said in the previous section on DDD, one of those indirect pieces of evidence is placebo analysis. The reasoning goes that if we find, using our preferred research design, effects where there shouldn't be, then maybe our original findings weren't credible in the first place. Using placebo analysis within your own work has become an essential part of empirical work for this reason.

But another use of placebo analysis is to evaluate the credibility of popular estimation strategies themselves. This kind of use helps improve a literature by uncovering flaws in a research design which can then help stimulate the creation of stronger methods and models. Let's take two exemplary studies that accomplished this well: Auld and Grootendorst [2004] and Cohen-Cole and Fletcher [2008].

To say that the Becker and Murphy [1988] "rational addiction" model has been influential would be an understatement. It has over 4,000 cites and has become one of the most common frameworks in health economics. It created a cottage industry of empirical studies that persists to this day. Alcohol, tobacco, gambling, even sports, have all been found to be "rationally addictive" commodities and activities using various empirical approaches.

But some researchers cautioned the research community about these empirical studies. Rogeberg [2004] critiqued the theory on its own grounds, but I'd like to focus on the empirical studies based on the theory. Rather than talk about any specific paper, I'd like to provide a quote from Melberg [2008], who surveyed researchers who had written on rational addiction:

> A majority of [our] respondents believe the literature is a success story that demonstrates the power of economic reasoning. At the same time, they also believe the empirical evidence is weak, and they disagree both on the type of evidence that would validate the theory and the policy implications. Taken together, this points to an interesting gap. On the one hand, most of the respondents claim that the theory has valuable real world implications. On the other hand, they do not believe the theory has received empirical support. [1]

Rational addiction should be held to the same empirical standards as in theory. The strength of the model has always been based on the economic reasoning, which economists obviously find compelling. But were the empirical designs flawed? How could we know?

Auld and Grootendorst [2004] is not a test of the rational addiction model. On the contrary, it is an "anti-test" of the empirical rational addiction models common at the time. Their goal was not to evaluate the theoretical rational addiction model, in other words, but rather the empirical rational addiction models themselves. How do they do this? Auld and Grootendorst [2004] used the empirical rational addiction model to evaluate commodities that could not plausibly be considered addictive, such as eggs, milk, orange, and apples. They found that the empirical rational addiction model implied milk was extremely addictive, perhaps one of the most addictive commodities studied.[9] Is it credible to believe that eggs and milk are "rationally addictive" or is it more likely the research designs used to evaluate the rational addiction model were flawed? Auld and Grootendorst [2004] study cast doubt on the empirical rational addiction model, not the theory.

Another problematic literature was the peer-effects literature. Estimating peer effects is notoriously hard. Manski [1993] said that the deep endogeneity of social interactions made the identification of peer effects difficult and possibly even impossible. He called this problem the "mirroring" problem. If "birds of a feather flock together," then identifying peer effects in observational settings may just be impossible due to the profound endogeneities at play.

Several studies found significant network effects on outcomes like obesity, smoking, alcohol use, and happiness. This led many researchers to conclude that these kinds of risk behaviors were "contagious" through peer effects [Christakis and Fowler, 2007]. But these studies did not exploit randomized social groups. The peer groups were purely endogenous. Cohen-Cole and Fletcher [2008] showed using similar models and data that even attributes that *couldn't* be transmitted between peers—acne, height, and headaches—appeared "contagious" in observational data using the Christakis and Fowler

---

9 Milk is ironically my favorite drink, even over IPAs, so I am not persuaded by this anti-test.

[2007] model for estimation. Note, Cohen-Cole and Fletcher [2008] does not reject the idea of theoretical contagions. Rather, they point out that the Manski critique should guide peer effect analysis if social interactions are endogenous. They provide evidence for this indirectly using placebo analysis.[10]

Compositional change within repeated cross-sections. DD can be applied to repeated cross-sections, as well as panel data. But one of the risks of working with the repeated cross-sections is that unlike panel data (e.g., individual-level panel data), repeated cross-sections run the risk of compositional changes. Hong [2013] used repeated cross-sectional data from the Consumer Expenditure Survey (CEX) containing music expenditure and internet use for a random sample of households. The author's study exploited the emergence and immense popularity of Napster, the first file-sharing software widely used by Internet users, in June 1999 as a natural experiment. The study compared Internet users and Internet non-users before and after the emergence of Napster. At first glance, they found that as Internet diffusion increased from 1996 to 2001, spending on music for Internet users fell faster than that for non-Internet users. This was initially evidence that Napster was responsible for the decline, until this was investigated more carefully.

But when we look at Table 76, we see evidence of compositional changes. While music expenditure fell over the treatment period, the demographics of the two groups also changed over this period. For instance, the age of Internet users grew while income fell. If older people are less likely to buy music in the first place, then this could independently explain some of the decline. This kind of compositional change is a like an omitted variable bias built into the sample itself caused by time-variant unobservables. Diffusion of the Internet appears to be related to changing samples as younger music fans are early adopters.

10 Breakthroughs in identifying peer effects eventually emerged, but only from studies that serendipitously had randomized peer groups such as Sacerdote [2001], Lyle [2009], Carrell et al. [2019], Kofoed and McGovney [2019], and several others. Many of these papers either used randomized roommates or randomized companies at military academies. Such natural experiments are rare opportunities for studying peer effects for their ability to overcome the mirror problem.

Identification of causal effects would need for the treatment itself to be exogenous to such changes in the composition.

Final thoughts. There are a few other caveats I'd like to make before moving on. First, it is important to remember the concepts we learned in the early DAG chapter. In choosing covariates in a DD design, you must resist the temptation to simply load the regression up with a kitchen sink of regressors. You should resist if only because in so doing, you may inadvertently include a collider, and if a collider is conditioned on, it introduces strange patterns that may mislead you and your audience. There is unfortunately no way forward except, again, deep institutional familiarity with both the factors that determined treatment assignment on the ground, as well as economic theory itself. Second, another issue I skipped over entirely is the question of how the outcome is modeled. Very little thought if any is given to how exactly we should model some outcome. Just to take one example, should we use the log or the levels themselves? Should we use the quartic root? Should we use rates? These, it turns, out are critically important because for many of them, the parallel trends assumption needed for identification will not be achieved—even though it will be achieved under some other unknown transformation. It is for this reason that you can think of many DD designs as having a parametric element because you must make strong commitments about the functional form itself. I cannot provide guidance to you on this, except that maybe using the pre-treatment leads as a way of finding parallelism could be a useful guide.

Twoway Fixed Effects with Differential Timing

I have a bumper sticker on my car that says "I love Federalism (for the natural experiments)" (Figure 69). I made these bumper stickers for my students to be funny, and to illustrate that the United States is a never-ending laboratory. Because of state federalism, each US state has been given considerable discretion to govern itself with policies and reforms. Yet, because it is a union of states, US researchers have

Table 76. Changes between Internet and non-Internet users over time.

Year	1997		1998		1999	
	Internet user	Non-user	Internet user	Non-user	Internet user	Non-user
Average expenditure						
Recorded music	$25.73	$10.90	$24.18	$9.97	$20.92	$9.37
Entertainment	$195.03	$96.71	$193.38	$84.92	$182.42	$80.19
Zero expenditure						
Recorded music	0.56	0.79	0.60	0.80	0.64	0.81
Entertainment	0.08	0.32	0.09	0.35	0.14	0.39
Demographics						
Age	40.2	49.0	42.3	49.0	44.1	49.4
Income	$52,887	$30,459	$51,995	$26,189	$49,970	$26,649
High school graduate	0.18	0.31	0.17	0.32	0.21	0.32
Some college	0.37	0.28	0.35	0.27	0.34	0.27
College grad	0.43	0.21	0.45	0.21	0.42	0.20
Manager	0.16	0.08	0.16	0.08	0.14	0.08

Note: Sample means from the Consumer Expenditure Survey.

Figure 69. A bumper sticker for nerds.

access to many data sets that have been harmonized across states, making it even more useful for causal inference.

Goodman-Bacon [2019] calls the staggered assignment of treatments across geographic units over time the "differential timing" of treatment. What he means is unlike the simple 2×2 that we discussed earlier (e.g., New Jersey and Pennsylvania), where treatment units were all treated at the same time, the more common situation is one where geographic units receive treatments at different points in time. And this happens in the United States because each area (state, municipality) will adopt a policy when it wants to, for its own reasons. As a result, the adoption of some treatment will tend to be differentially timed across units.

This introduction of differential timing means there are basically two types of DD designs. There is the 2×2 DD we've been discussing wherein a single unit or a group of units all receive some treatment at the same point in time, like Snow's cholera study or Card and Krueger [1994]. And then there is the DD with differential timing in which groups receive treatment at different points in time, like Cheng and Hoekstra [2013]. We have a very good understanding of the 2×2 design, how it works, why it works, when it works, and when it does not work. But we did not until Goodman-Bacon [2019] have as good an understanding of the DD design with differential timing. So let's get down to business and discuss that now by reminding ourselves of the 2×2 DD that we introduced earlier.

$$\widehat{\delta}_{kU}^{2 \times 2} = \left(\overline{y}_k^{\text{post}(k)} - \overline{y}_k^{\text{pre}(k)} \right) - \left(\overline{y}_U^{\text{post}(k)} - \overline{y}_U^{\text{pre}(k)} \right)$$

where k is the treatment group, U is the never-treated group, and everything else is self-explanatory. Since this involves sample means, we can calculate the differences manually. Or we can estimate it with the following regression:

$$y_{it} = \beta D_i + \tau \, \text{Post}_t + \delta(D_i \times \text{Post}_t) + X_{it} + \varepsilon_{it}$$

But a more common situation you'll encounter will be a DD design with differential timing. And while the decomposition is a bit complicated, the regression equation itself is straightforward:

$$y_{it} = \alpha_0 + \delta D_{it} + X_{it} + \alpha_i + \alpha_t + \epsilon_{it}$$

When researchers estimate this regression these days, they usually use the linear fixed-effects model that I discussed in the previous panel chapter. These linear panel models have gotten the nickname "twoway fixed effects" because they include both time fixed effects and unit fixed effects. Since this is such a popular estimator, it's important we understand exactly what it is doing and what is it not.

Bacon Decomposition theorem. Goodman-Bacon [2019] provides a helpful decomposition of the twoway fixed effects estimate of $\widehat{\delta}$. Given this is the go-to model for implementing differential timing designs, I have found his decomposition useful. But as there are some other decompositions of twoway fixed effects estimators, such as another important paper by de Chaisemartin and D'Haultfœuille [2019], I'll call it the Bacon decomposition for the sake of branding.

The punchline of the Bacon decomposition theorem is that the twoway fixed effects estimator is a weighted average of all potential 2×2 DD estimates where weights are both based on group sizes and variance in treatment. Under the assumption of variance weighted common trends (VWCT) and time invariant treatment effects, the variance weighted ATT is a weighted average of all possible ATTs. And under more restrictive assumptions, that estimate perfectly matches the ATT. But that is not true when there are time-varying treatment effects, as time-varying treatment effects in a differential timing design estimated with twoway fixed effects can generate a bias. As such, twoway fixed-effects models may be severely biased, which is echoed in de Chaisemartin and D'Haultfœuille [2019].

To make this concrete, let's start with a simple example. Assume in this design that there are three groups: an early treatment group (k), a group treated later (*l*), and a group that is never treated (*U*). Groups *k* and *l* are similar in that they are both treated but they differ in that *k* is treated earlier than *l*.

Let's say there are 5 periods, and *k* is treated in period 2. Then it spends 40% of its time under treatment, or 0.4. But let's say *l* is treated in period 4. Then it spends 80% of its time treated, or 0.8. I represent this time spent in treatment for a group as $\overline{D}_k = 0.4$ and $\overline{D}_l = 0.8$. This is important, because the length of time a group spends in treatment determines its treatment variance, which in turn affects the weight that 2×2 plays in the final adding up of the DD parameter itself. And rather than write out 2×2 DD estimator every time, we will just represent each 2×2 as $\widehat{\delta}_{ab}^{2 \times 2,j}$ where *a* and *b* are the treatment groups, and *j* is the index notation for any treatment group. Thus if we wanted to know the 2×2 for group *k* compared to group *U*, we would write $\widehat{\delta}_{kU}^{2 \times 2,k}$ or, maybe to save space, just $\widehat{\delta}_{kU}^{k}$.

So, let's get started. First, in a single differential timing design, how many 2×2s are there anyway? Turns out there are a lot. To see, let's make a toy example. Let's say there are three timing groups (*a*, *b*, and *c*) and one untreated group (*U*). Then there are 9 2×2 DDs. They are:

a to b	b to a	c to a
a to c	b to c	c to b
a to U	b to U	c to U

See how it works? Okay, then let's return to our simpler example where there are two timing groups *k* and *l* and one never-treated group. Groups *k* and *l* will get treated at time periods t_k^* and t_l^*. The earlier period before anyone is treated will be called the "pre" period, the period between *k* and *l* treated is called the "mid" period, and the period after *l* is treated is called the "post" period. This will be *much* easier to understand with some simple graphs. Let's look at Figure 70. Recall the definition of a 2×2 DD is

$$\widehat{\delta}_{kU}^{2 \times 2} = \left(\overline{y}_k^{\text{post}(k)} - \overline{y}_k^{\text{pre}(k)} \right) - \left(\overline{y}_U^{\text{post}(k)} - \overline{y}_U^{\text{pre}(k)} \right)$$

where *k* and *U* are just place-holders for any of the groups used in a 2×2.

Substituting the information in each of the four panels of Figure 70 into the equation will enable you to calculate what each specific 2×2

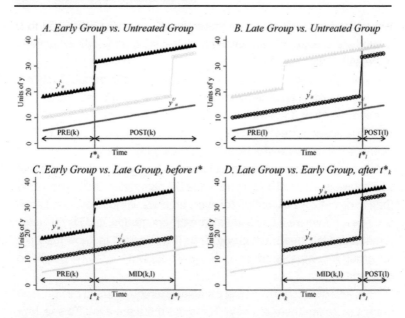

Figure 70. Four 2 × 2 DDs [Goodman-Bacon, 2019]. Reprinted with permission from authors.

is. But we can really just summarize these into three really important 2 × 2s, which are:

$$\widehat{\delta}_{kU}^{2\times2} = \left(\overline{y}_k^{post(k)} - \overline{y}_k^{pre(k)}\right) - \left(\overline{y}_U^{post(k)} - \overline{y}_U^{pre(k)}\right)$$

$$\widehat{\delta}_{kl}^{2\times2} = \left(\overline{y}_k^{mid(k,l)} - \overline{y}_k^{pre(k)}\right) - \left(\overline{y}_l^{mid(k,l)} - \overline{y}_l^{pre(k)}\right)$$

$$\widehat{\delta}_{lk}^{2\times2} = \left(\overline{y}_l^{post(l)} - \overline{y}_l^{mid(k,l)}\right) - \left(\overline{y}_k^{post(l)} - \overline{y}_k^{mid(k,l)}\right)$$

where the first 2 × 2 is any timing group compared to the untreated group (k or l), the second is a group compared to the yet-to-be-treated timing group, and the last is the eventually-treated group compared to the already-treated controls.

With this notation in mind, the DD parameter estimate can be decomposed as follows: $\widehat{\delta}^{DD}$

$$\widehat{\delta}^{DD} = \sum_{k\neq U} s_{kU}\widehat{\delta}_{kU}^{2\times2} + \sum_{k\neq U}\sum_{l>k} s_{kl}\left[\mu_{kl}\widehat{\delta}_{kl}^{2\times2,k} + (1-\mu_{kl})\widehat{\delta}_{kl}^{2\times2,l}\right] \quad (9.1)$$

where the first 2×2 is the k compared to U and the l compared to U (combined to make the equation shorter).[11] So what are these weights exactly?

$$s_{ku} = \frac{n_k n_u \overline{D}_k (1 - \overline{D}_k)}{\widehat{Var}(\tilde{D}_{it})}$$

$$s_{kl} = \frac{n_k n_l (\overline{D}_k - \overline{D}_l)(1 - (\overline{D}_k - \overline{D}_l))}{\widehat{Var}(\tilde{D}_{it})}$$

$$\mu_{kl} = \frac{1 - \overline{D}_k}{1 - (\overline{D}_k - \overline{D}_l)}$$

where n refers to sample sizes, $\overline{D}_k(1 - \overline{D}_k)$ $(\overline{D}_k - \overline{D}_l)(1 - (\overline{D}_k - \overline{D}_l))$ expressions refer to variance of treatment, and the final equation is the same for two timing groups.[12]

Two things immediately pop out of these weights that I'd like to bring to your attention. First, notice how "group" variation matters, as opposed to unit-level variation. The Bacon decomposition shows that it's group variation that twoway fixed effects is using to calculate that parameter you're seeking. The more states that adopted a law at the same time, the bigger they influence that final aggregate estimate itself.

The other thing that matters in these weights is *within-group* treatment variance. To appreciate the subtlety of what's implied, ask yourself—how long does a group have to be treated in order to maximize its treatment variance? Define $X = D(1 - D) = D - D^2$, take the derivative of V with respect to $\overline{D}$, set $\frac{dV}{d\overline{D}}$ equal to zero, and solve for $\overline{D}_*$. Treatment variance is maximized when $\overline{D} = 0.5$. Let's look at three values of $\overline{D}$ to illustrate this.

---

11 All of this decomposition comes from applying the Frisch-Waugh theorem to the underlying twoway fixed effects estimator.

12 A more recent version of Goodman-Bacon [2019] rewrites this weighting but they are numerically the same, and for these purposes, I prefer the weighting scheme discussed in an earlier version of the paper. See Goodman-Bacon [2019] for the equivalence between his two weighting descriptions.

$$\overline{D} = 0.1; 0.1 \times 0.9 = 0.09$$
$$\overline{D} = 0.4; 0.4 \times 0.6 = 0.24$$
$$\overline{D} = 0.5; 0.5 \times 0.5 = 0.25$$

So what are we learning from this, exactly? Well, what we are learning is that being treated in the *middle* of the panel actually directly influences the numerical value you get when twoway fixed effects are used to estimate the ATT. That therefore means lengthening or shortening the panel can actually change the point estimate purely by changing group treatment variance and nothing more. Isn't that kind of strange though? What criteria would we even use to determine the best length?

But what about the "treated on treated weights," or the s_{kl} weight. That doesn't have a $\overline{D}(1 - \overline{D})$ expression. Rather, it has a $(\overline{D}_k - \overline{D}_l)(1 - (\overline{D}_k - \overline{D}_l))$ expression. So the "middle" isn't super clear. That's because it isn't the middle of treatment for a single group, but rather it's the middle of the panel for the *difference* in treatment variance. For instance, let's say k spends 67% of time treated and l spends 15% of time treated. Then $\overline{D}_k - \overline{D}_l = 0.52$ and therefore $0.52 \times 0.48 = 0.2496$, which as we showed is very nearly the max value of the variance as is possible (e.g., 0.25). Think about this for a moment—twoway fixed effects with differential timing weights the 2×2s comparing the two ultimate treatment groups more if the gap in treatment time is close to 0.5.

Expressing the decomposition in potential outcomes. Up to now, we just showed what was inside the DD parameter estimate when using twoway fixed effects: it was nothing more than an "adding up" of all possible 2×2s weighted by group shares and treatment variance. But that only tells us what DD is numerically; it does not tell us whether the parameter estimate maps onto a meaningful average treatment effect. To do that, we need to take those sample averages and then use the switching equations replace them with potential outcomes. This is key to moving from numbers to estimates of causal effects.

Bacon's decomposition theorem expresses the DD coefficient in terms of sample average, making it straightforward to substitute with potential outcomes using a modified switching equation. With a little creative manipulation, this will be revelatory. First, let's define any year-specific ATT as

$$ATT_k(\tau) = E[Y_{it}^1 - Y_{it}^0 \mid k, t = \tau]$$

Next, let's define it over a time window W (e.g., a post-treatment window)

$$ATT_k(\tau) = E[Y_{it}^1 - Y_{it}^0 \mid k, \tau \in W]$$

Finally, let's define differences in average potential outcomes over time as:

$$\Delta Y_k^h(W_1, W_0) = E[Y_{it}^h \mid k, W_1] - E[Y_{it}^h \mid k, W_0]$$

for $h = 0$ (i.e., Y^0) or $h = 1$ (i.e., Y^1)

With trends, differences in mean potential outcomes is non-zero. You can see that in Figure 71.

We'll return to this, but I just wanted to point it out to you so that it would be concrete in your mind when we return to it later.

We can move now from the 2×2s that we decomposed earlier directly into the ATT, which is ultimately the main thing we want to know. We covered this earlier in the chapter, but review it again here to maintain progress on my argument. I will first write down the 2×2 expression, use the switching equation to introduce potential outcome notation, and through a little manipulation, find some ATT expression.

$$\hat{\delta}_{kU}^{2 \times 2} = \Big(E[Y_j \mid \text{Post}] - E[Y_j \mid \text{Pre}] \Big) - \Big(E[Y_u \mid \text{Post}] - E[Y_u \mid \text{Pre}] \Big)$$

$$= \underbrace{\Big(E[Y_j^1 \mid \text{Post}] - E[Y_j^0] \mid \text{Pre}] \Big) - \Big(E[Y_u^0 \mid \text{Post}] - E[Y_u^0 \mid \text{Pre}] \Big)}_{\text{Switching equation}}$$

$$+ \underbrace{E[Y_j^0 \mid \text{Post}] - E[Y_j^0 \mid \text{Post}]}_{\text{Adding zero}}$$

$$= \underbrace{E[Y_j^1 \mid \text{Post}] - E[Y_j^0 \mid \text{Post}]}_{\text{ATT}}$$

$$+ \underbrace{\Big[E[Y_j^0 \mid \text{Post}] - E[Y_j^0 \mid \text{Pre}] \Big] - \Big[E[Y_U^0 \mid \text{Post}] - E[Y_U^0 \mid \text{Pre}] \Big]}_{\text{Non-parallel trends bias in } 2 \times 2 \text{ case}}$$

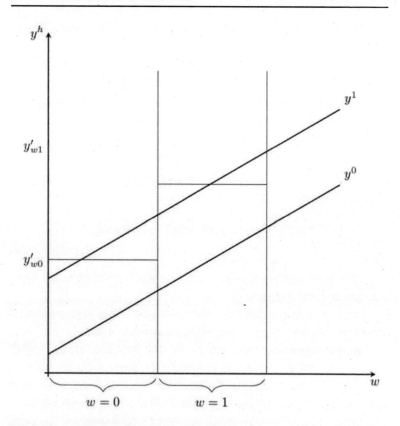

Figure 71. Changing average potential outcomes.

This can be rewritten even more compactly as:

$$\widehat{\delta}_{kU}^{2\times2} = ATT_{\text{Post},j} + \underbrace{\Delta Y^0_{\text{Post,Pre},j} - \Delta Y^0_{\text{Post,Pre},U}}_{\text{Selection bias!}}$$

The 2×2 DD can be expressed as the sum of the ATT itself *plus* a parallel trends assumption, and without parallel trends, the estimator is biased. Ask yourself—which of these two differences in the parallel trends assumption is counterfactual, $\Delta Y^0_{\text{Post,Pre},j}$ or $\Delta Y^0_{\text{Post,Pre},U}$? Which one is observed, in other words, and which one is not observed? Look and see if you can figure it out from this drawing in Figure 72.

Only if these are parallel—the counterfactual trend *and* the observable trend—does the selection bias term zero out and ATT is identified.

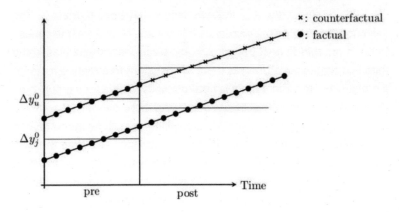

Figure 72. Visualization of parallel trends.

But let's keep looking within the decomposition, as we aren't done. The other two 2×2s need to be defined since they appear in Bacon's decomposition also. And they are:

$$\hat{\delta}_{kU}^{2\times2} = ATT_k\text{Post} + \Delta Y_k^0(\text{Post}(k),\text{Pre}(k)) - \Delta Y_U^0(\text{Post}(k),\text{Pre}) \quad (9.2)$$

$$\hat{\delta}_{kl}^{2\times2} = ATT_k(MID) + \Delta Y_k^0(MID,\text{Pre}) - \Delta Y_l^0(MID,\text{Pre}) \quad (9.3)$$

These look the same because you're always comparing the treated unit with an untreated unit (though in the second case it's just that they haven't been treated *yet*).

But what about the 2×2 that compared the late groups to the already-treated earlier groups? With a lot of substitutions like we did we get:

$$\hat{\delta}_{lk}^{2\times2} = ATT_{l,\text{Post}(l)}$$
$$+ \underbrace{\Delta Y_l^0(\text{Post}(l),MID) - \Delta Y_k^0(\text{Post}(l),MID)}_{\text{Parallel-trends bias}}$$
$$- \underbrace{(ATT_k(\text{Post}) - ATT_k(Mid))}_{\text{Heterogeneity in time bias!}} \quad (9.4)$$

I find it interesting our earlier decomposition of the simple difference in means into ATE + selection bias + heterogeneity treatment effects bias resembles the decomposition of the late to early 2×2 DD.

The first line is the *ATT* that we desperately hope to identify. The selection bias zeroes out insofar as Y^0 for k and l has the same parallel trends from *mid* to *post* period. And the treatment effects bias in the third line zeroes out *so long as* there are constant treatment effects for a group *over time*. But if there is heterogeneity in time for a group, then the two *ATT* terms will not be the same, and therefore will not zero out.

But we can sign the bias if we are willing to assume monotonicity, which means the *mid* term is smaller in absolute value than the *post* term. Under monotonicity, the interior of the parentheses in the third line is positive, and therefore the bias is negative. For positive ATT, this will bias the effects towards zero, and for negative ATT, it will cause the estimated ATT to become even more negative.

Let's pause and collect these terms. The decomposition formula for DD is:

$$\widehat{\delta}^{DD} = \sum_{k \neq U} s_{kU} \widehat{\delta}_{kU}^{2 \times 2} + \sum_{k \neq U} \sum_{l > k} s_{kl} \left[\mu_{kl} \widehat{\delta}_{kl}^{2 \times 2, k} + (1 - \mu_{kl}) \widehat{\delta}_{kl}^{2 \times 2, l} \right]$$

We will substitute the following three expressions into that formula.

$$\widehat{\delta}_{kU}^{2 \times 2} = ATT_k(\text{Post}) + \Delta Y_l^0(\text{Post}, \text{Pre}) - \Delta Y_U^0(\text{Post}, \text{Pre})$$

$$\widehat{\delta}_{kl}^{2 \times 2, k} = ATT_k(\text{Mid}) + \Delta Y_l^0(\text{Mid}, \text{Pre}) - \Delta Y_l^0(\text{Mid}, \text{Pre})$$

$$\widehat{\delta}_{lk}^{2 \times 2, l} = ATT_l \text{Post}(l) + \Delta Y_l^0(\text{Post}(l), \text{MID}) - \Delta Y_k^0(\text{Post}(l), \text{MID})$$
$$- (ATT_k(\text{Post}) - ATT_k(\text{Mid}))$$

Substituting all three terms into the decomposition formula is a bit overwhelming, so let's simplify the notation. The estimated DD parameter is equal to:

$$p \lim \widehat{\delta}_{n \to \infty}^{DD} = VWATT + VWCT - \Delta ATT \tag{9.5}$$

In the next few sections, I discuss each individual element of this expression.

Variance weighted ATT. We begin by discussing the variance weighted average treatment effect on the treatment group, or *VWATT*. Its unpacked expression is:

$$VWATT = \sum_{k \neq U} \sigma_{kU} ATT_k(\text{Post}(k)) \qquad (9.6)$$

$$+ \sum_{k \neq U} \sum_{l > k} \sigma_{kl} \left[\mu_{kl} ATT_k(\text{MID}) + (1 - \mu_{kl}) ATT_l(\text{POST}(l)) \right] \qquad (9.7)$$

where σ is like s, only population terms not samples. Notice that the VWATT simply contains the three ATTs identified above, each of which was weighted by the weights contained in the decomposition formula. While these weights sum to one, that weighting is irrelevant if the ATT are identical.[13]

When I learned that the DD coefficient was a weighted average of all individual 2×2s, I was not terribly surprised. I may not have intuitively known that the weights were based on group shares and treatment variance, but I figured it was probably a weighted average nonetheless. I did not have that same experience, though, when I worked through the other two terms. I now turn to the other two terms: the VWCT and the ΔATT.

Variance weighted common trends. VWCT stands for variance weighted common trends. This is just the collection of non-parallel-trends biases we previously wrote out, but notice—identification requires *variance weighted* common trends to hold, which is actually a bit weaker than we thought before with identical trends. You get this with identical trends, but what Goodman-Bacon [2019] shows us is that *technically* you don't need identical trends because the weights can make it hold even if we don't have exact parallel trends. Unfortunately, this is a bit of a pain to write out, but since it's important, I will.

$$VWCT = \sum_{k \neq U} \sigma_{kU} \left[\Delta Y_k^0(\text{Post}(k), \text{Pre}) - \Delta Y_U^0(\text{Post}(k), \text{Pre}) \right]$$

$$+ \sum_{k \neq U} \sum_{l > k} \sigma_{kl} \left[\mu_{kl} \{ \Delta Y_k^0(\text{Mid}, \text{Pre}(k)) - \Delta Y_l^0(\text{Mid}, \text{Pre}(k)) \} \right.$$

$$\left. + (1 - \mu_{kl}) \{ \Delta Y_l^0(\text{Post}(l), \text{Mid}) - \Delta Y_k^0(\text{Post}(l), \text{Mid}) \} \right] \qquad (9.8)$$

---

13 Heterogeneity in ATT across k and l is not the source of any biases. Only heterogeneity over time for k or l's ATT introduces bias. We will discuss this in more detail later.

Notice that the VWCT term simply collects all the non-parallel-trend biases from the three 2×2s. One of the novelties, though, is that the non-parallel-trend biases are also weighted by the same weights used in the VATT.

This is actually a new insight. On the one hand, there are a lot of terms we need to be zero. On the other hand, it's ironically a *weaker* identifying assumption strictly identical common trends as the weights can technically correct for unequal trends. VWCT will zero out with exact parallel trends and in those situations where the weights adjust the trends to zero out. This is good news (sort of).

ATT heterogeneity within time bias. When we decomposed the simple difference in mean outcomes into the sum of the ATE, selection bias, and heterogeneous treatment effects bias, it really wasn't a huge headache. That was because if the ATT differed from the ATU, then the simple difference in mean outcomes became the sum of ATT and selection bias, which was still an interesting parameter. But in the Bacon decomposition, ATT heterogeneity over time introduces bias that is not so benign. Let's look at what happens when there is time-variant within-group treatment effects.

$$\Delta ATT = \sum_{k \neq U} \sum_{l > k} (1 - \mu_{kl}) \Big[ATT_k(\text{Post}(l) - ATT_k(\text{Mid})) \Big] \quad (9.9)$$

Heterogeneity in the ATT has two interpretations: you can have heterogeneous treatment effects *across* groups, and you can have heterogeneous treatment effects *within* groups over time. The ΔATT is concerned with the latter only. The first case would be heterogeneity *across units* but not within groups. When there is heterogeneity across groups, then the VWATT is simply the average over group-specific ATTs weighted by a function of sample shares and treatment variance. There is no bias from this kind of heterogeneity.[14]

But it's the second case—when ATT is constant across units but heterogeneous within groups over time—that things get a little worrisome. Time-varying treatment effects, even if they are identical across units, generate cross-group heterogeneity because of the differing

---

14 Scattering the weights against the individual 2×2s can help reveal if the overall coefficient is driven by a few different 2×2s with large weights.

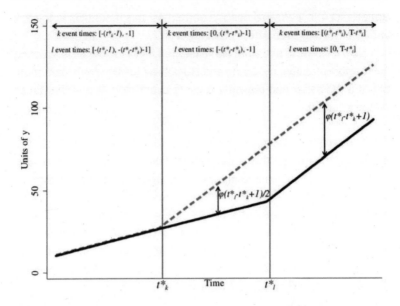

Figure 73. Within-group heterogeneity in the *ATT*. Goodman-Bacon, A. (2019). "Difference-in-Differences with Variation in Treatment Timing." Unpublished Manuscript. Permission from author.

post-treatment windows, and the fact that earlier-treated groups are serving as controls for later-treated groups. Let's consider a case where the counterfactual outcomes are identical, but the treatment effect is a linear break in the trend (Figure 73). For instance, $Y_{it}^1 = Y_{it}^0 + \theta(t - t_1^* + 1)$ similar to Meer and West [2016].

Notice how the first 2×2 uses the later group as its control in the middle period, *but* in the late period, the later-treated group is using the earlier treated as its control. When is this a problem?

It's a problem if there are a lot of those 2×2s or if their weights are large. If they are negligible portions of the estimate, then even if it exists, then given their weights are small (as group shares are also an important piece of the weighting not just the variance in treatment) the bias may be small. But let's say that doesn't hold. Then what is going on? The effect is biased because the control group is experiencing a trend in outcomes (e.g., heterogeneous treatment effects), and this bias feeds through to the later 2×2 according to the size of the weights, $(1 - \mu_{kl})$. We will need to correct for this if our plan is to stick with the twoway fixed effects estimator.

Now it's time to use what we've learned. Let's look at an interesting and important paper by Cheng and Hoekstra [2013] to both learn more about a DD paper and replicate it using event studies and the Bacon decomposition.

Castle-doctrine statutes and homicides. Cheng and Hoekstra [2013] evaluated the impact that a gun reform had on violence and to illustrate various principles and practices regarding differential timing. I'd like to discuss those principles in the context of this paper. This next section will discuss, extend, and replicate various parts of this study.

Trayvon Benjamin Martin was a 17-year-old African-American young man when George Zimmerman shot and killed him in Sanford, Florida, on February 26, 2012. Martin was walking home alone from a convenience store when Zimmerman spotted him, followed him from a distance, and reported him to the police. He said he found Martin's behavior "suspicious," and though police officers urged Zimmerman to stay back, Zimmerman stalked and eventually provoked Martin. An altercation occurred and Zimmerman fatally shot Martin. Zimmerman claimed self-defense and was nonetheless charged with Martin's death. A jury acquitted him of second-degree murder and of manslaughter.

Zimmerman's actions were interpreted by the jury to be legal because in 2005, Florida reformed when and where lethal self-defense could be used. Whereas once lethal self-defense was only legal inside the home, a new law, "Stand Your Ground," had extended that right to other public places. Between 2000 and 2010, twenty-one states explicitly expanded the castle-doctrine statute by extending the places outside the home where lethal force could be legally used.[15] These states had removed a long-standing tradition in the common law that placed the duty to retreat from danger on the victim. After these reforms, though, victims no longer had a duty to retreat in public places if they felt threatened; they could retaliate in lethal self-defense.

---

15 These laws are called castle-doctrine statutes because the home—where lethal self-defense had been protected—is considered one's castle.

Other changes were also made. In some states, individuals who used lethal force outside the home were *assumed* to be reasonably afraid. Thus, a prosecutor would have to prove fear was not reasonable, allegedly an almost impossible task. Civil liability for those acting under these expansions was also removed. As civil liability is a lower threshold of guilt than criminal guilt, this effectively removed the remaining constraint that might keep someone from using lethal force outside the home.

From an economic perspective, these reforms lowered the cost of killing someone. One could use lethal self-defense in situations from which they had previously been barred. And as there was no civil liability, the expected cost of killing someone was now lower. Thus, insofar as people are sensitive to incentives, then depending on the elasticities of lethal self-defense with respect to cost, we expect an increase in lethal violence for the marginal victim. The reforms may have, in other words, caused homicides to rise.

One can divide lethal force into true and false positives. The true positive use of lethal force would be those situations in which, had the person not used lethal force, he or she would have been murdered. Thus, the true positive case of lethal force is simply a transfer of one life (the offender) for another (the defender). This is tragic, but official statistics would not record an net increase in homicides relative to the counterfactual—only which person had been killed. But a false positive causes a net increase in homicides relative to the counterfactual. Some arguments can escalate unnecessarily, and yet under common law, the duty to retreat would have defused the situation before it spilled over into lethal force. Now, though, under these castle-doctrine reforms, that safety valve is removed, and thus a killing occurs that would not have in counterfactual, leading to a net increase in homicides.

But that is not the only possible impact of the reforms—deterrence of violence is also a possibility under these reforms. In Lott and Mustard [1997], the authors found that concealed-carry laws reduced violence. They suggested this was caused by deterrence—thinking someone may be carrying a concealed weapon, the rational criminal is deterred from committing a crime. Deterrence dates back to Becker [1968] and Jeremy Bentham before him. Expanding the arenas where lethal force could be used could also deter crime. Since this theoretical possibility depends crucially on key elasticities, which may in fact be zero, deterrence from expanding where guns can be used to kill someone is ultimately an empirical question.

Cheng and Hoekstra [2013] chose a difference-in-differences design for their project where the castle doctrine law was the treatment and timing was differential across states. Their estimating equation was

$$Y_{it} = \alpha + \delta D_{it} + \gamma X_{it} + \sigma_i + \tau_t + \varepsilon_{it}$$

where D_{it} is the treatment parameter. They estimated this equation using a standard twoway fixed effects model as well as count models. Ordinarily, the treatment parameter will be a 0 or 1, but in Cheng and Hoekstra [2013], it's a variable ranging from 0 to 1, because some states get the law change mid-year. So if they got the law in July, then D_{it} equals 0 before the year of adoption, 0.5 in the year of adoption and 1 thereafter. The X_{it} variable included a particular kind of control that they called "region-by-year fixed effects," which was a vector of dummies for the census region to which the state belonged interacted with each year fixed effect. This was done so that explicit counterfactuals were forced to come from within the same census region.[16] As the results are not dramatically different between their twoway fixed effects and count models, I will tend to emphasize results from the twoway fixed effects.

The data they used is somewhat standard in crime studies. They used the FBI Uniform Crime Reports Summary Part I files from 2000 to 2010. The FBI Uniform Crime Reports is a harmonized data set on eight "index" crimes collected from voluntarily participating police agencies across the country. Participation is high and the data goes back many decades, making it attractive for many contemporary questions regarding the crime policy. Crimes were converted into rates, or "offenses per 100,000 population."

Cheng and Hoekstra [2013] rhetorically open their study with a series of simple placebos to check whether the reforms were spuriously correlated with crime trends more generally. Since oftentimes

---

16 This would violate SUTVA insofar as gun violence spills over to a neighboring state when the own state passes a reform.

many crimes are correlated because of unobserved factors, this has some appeal, as it rules out the possibility that the laws were simply being adopted in areas where crime rates were already rising. For their falsifications they chose motor vehicle thefts and larcenies, neither of which, they reasoned, should be credibly connected to lowering the cost of using lethal force in public.

There are so many regression coefficients in Table 77 because applied microeconomists like to report results under increasingly restrictive models. In this case, each column is a new regression with additional controls such as additional fixed-effects specifications, time-varying controls, a one-year lead to check on the pre-treatment differences in outcomes, and state-specific trends. As you can see, many of these coefficients are very small, and because they are small, even large standard errors yield a range of estimates that are still not very large.

Next they look at what they consider to be crimes that might be deterred if policy created a credible threat of lethal retaliation in public: burglary, robbery, and aggravated assault.

Insofar as castle doctrine has a deterrence effect, then we would expect a *negative* effect of the law on offenses. But all of the regressions shown in Table 78 are actually positive, and very few are significant even still. So the authors conclude they cannot detect any deterrence—which does not mean it didn't happen; just that they cannot reject the null for large effects.

Now they move to their main results, which is interesting because it's much more common for authors to lead with their main results. But the rhetoric of this paper is somewhat original in that respect. By this point, the reader has seen a lot of null effects from the laws and may be wondering, "What's going on? This law isn't spurious and isn't causing deterrence. Why am I reading this paper?"

The first thing the authors did was show a series of figures showing the *raw data* on homicides for treatment and control states. This is always a challenge when working with differential timing, though. For instance, approximately twenty states adopted a castle-doctrine law from 2005 to 2010, but *not at the same time*. So how are you going to show this visually? What is the pre-treatment period, for instance, for the *control group* when there is differential timing? If one state adopts

Table 77. Falsification Tests: The effect of castle doctrine laws on larceny and motor vehicle theft.

	OLS – Weighted by State Population					
	1	2	3	4	5	6
Panel A. Larceny	**Log(Larceny Rate)**					
Castle Doctrine Law	0.00300	−0.00600	−0.00910	−0.0858	−0.00401	−0.00284
	(0.0161)	(0.0147)	(0.0139)	(0.0139)	(0.0128)	(0.0180)
0 to 2 years before adoption of castle doctrine law				0.00112 (0.0105)		
Observation	550	550	550	550	550	550
Panel B. Motor Vehicle Theft	**Log(Motor Vehicle Theft Rate)**					
Castle Doctrine Law	0.0517	−0.0389	−0.0252	−0.0294	−0.0165	−0.00708
	(0.0563)	(0.448)	(0.0396)	(0.0469)	(0.0354)	(0.0372)
0 to 2 years before adoption of castle doctrine law				−0.00896 (0.0216)		
Observation	550	550	550	550	550	550
State and year fixed effects	Yes	Yes	Yes	Yes	Yes	Yes
Region-by-year fixed effects		Yes	Yes	Yes	Yes	Yes
Time-varying controls			Yes	Yes	Yes	Yes
Controls for larceny or motor theft					Yes	
State-specific linear time trends						Yes

Notes: Each column in each panel represents a separate regression. The unit of observation is state-year. Robust standard errors are clustered at the state level. Time-varying controls include policing and incarceration rates, welfare and public assistance spending, median income, poverty rate, unemployment rate, and demographics. *Significant at the 10 percent level. **Significant at the 5 percent level. ***Significant at the 1 percent level.

Table 78. The deterrence effects of castle-doctrine laws: Burglary, robbery, and aggravated assault.

	OLS – Weighted by State Population					
	1	2	3	4	5	6
Panel A. Burglary	Log(Burglary Rate)					
Castle-doctrine law	0.0780***	0.0290	0.0223	0.0181	0.0327*	0.0237
0 to 2 years before adoption of castle-doctrine law	(0.0255)	(0.0236)	(0.0223)	(0.0265) −0.009606 (0.0133)	(0.0165)	(0.0207)
Panel B. Robbery	Log(Robbery Rate)					
Castle-doctrine law	0.0408	0.0344	0.0262	0.0197	0.0376**	0.0515*
0 to 2 years before adoption of castle-doctrine law	(0.0254)	(0.0224)	(0.0229)	(0.0257) −0.0138 (0.0153)	(0.0181)	(0.0274)
Panel C. Aggravated Assault	Log(Aggravated Assault Rate)					
Castle-doctrine law	0.0434	0.0397	0.0372	0.0330	0.0424	0.0414
0 to 2 years before adoption of castle-doctrine law	(0.0387)	(0.0407)	(0.0319)	(0.0367) −0.00897 (0.0147)	(0.0291)	(0.0285)
Observation	550	550	550	550	550	550
State and year fixed effects	Yes	Yes	Yes	Yes	Yes	Yes
Region-by-year fixed effects		Yes	Yes	Yes	Yes	Yes

(Continued)

Table 78. (*Continued*)

	OLS – Weighted by State Population			
Time-varying controls	Yes	Yes	Yes	Yes
Controls for larceny or motor theft			Yes	
State-specific linear time trends				Yes

Notes: Each column in each panel represents a separate regression. The unit of observation is state-year. Robust standard errors are clustered at the state level. Time-varying controls include policing and incarceration rates, welfare and public assistance spending, median income, poverty rate, unemployment rate, and demographics. *Significant at the 10 percent level. **Significant at the 5 percent level. ***Significant at the 1 percent level.

in 2005, but another in 2006, then what precisely is the pre- and post-treatment for the control group? So that's a bit of a challenge, and yet if you stick with our guiding principle that causal inference studies desperately need data visualization of the main effects, your job is to solve it with creativity and honesty to make beautiful figures. Cheng and Hoekstra [2013] could've presented regression coefficients on leads and lags, as that is very commonly done, but knowing these authors firsthand, their preference is to give the reader pictures of the raw data to be as transparent as possible. Therefore, they showed multiple figures where each figure was a "treatment group" compared to all the "never-treated" units. Figure 74 shows the Florida case.

Notice that before the passage of the law, the offenses are fairly flat for treatment and control. Obviously, as I've emphasized, this is not a direct *test* of the parallel-trends assumption. Parallel trends in the pre-treatment period are neither necessary nor sufficient. The identifying assumption, recall, is that of variance-weighted common trends, which are entirely based on parallel counterfactual trends, not pre-treatment trends. But researchers use parallel pre-treatment trends like a hunch that the counterfactual trends would have been parallel. In one sense, parallel pre-treatment rules out some obvious spurious factors that we should be worried about, such as the law adoption

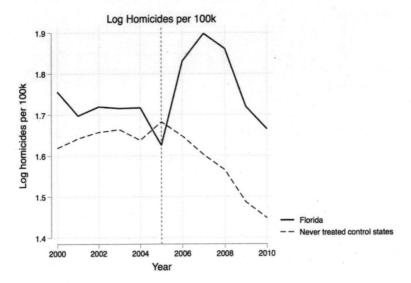

Figure 74. Raw data of log homicides per 100,000 for Florida versus never-treated control states.

happening around the timing of a change, even if that's simply nothing more than seemingly spurious factors like rising homicides. But that's clearly not happening here—homicides weren't diverging from controls pre-treatment. They were following a similar trajectory before Florida passed its law and *only then* did the trends converge. Notice that after 2005, which is when the law occurs, there's a sizable jump in homicides. There are additional figures like this, but they all have this set up—they show a treatment *group* over time compared to the same "never-treated" group.

Insofar as the cost of committing lethal force has fallen, then we expect to see more of it, which implies a positive coefficient on the estimated δ term assuming the heterogeneity bias we discussed earlier doesn't cause the twoway fixed effects estimated coefficient to flip signs. It should be different from zero both statistically and in a meaningful magnitude. They present four separate types of specifications—three using OLS, one using negative binomial. But I will only report the weighted OLS regressions for the sake of space.

There's a lot of information in Table 79, so let's be sure not to get lost. First, all coefficients are positive and similar in magnitude—between 8% and 10% increases in homicides. Second, three of the four

Table 79. The effect of castle-doctrine laws on homicide.

Panel A. Homicide OLS—Weights	1	2	Log(Homicide rate) 3	4	5	6
Castle-doctrine law	0.0801**	0.0946***	0.0937***	0.0955**	0.0985***	0.100**
	(0.0342)	(0.0279)	(0.0290)	(0.0367)	(0.0299)	(0.0388)
0 to 2 years before adoption of castle-doctrine law					0.00398	
					(0.0222)	
Observation	550	550	550	550	550	550
State and year fixed effects	Yes	Yes	Yes	Yes	Yes	Yes
Region-by-year fixed effects		Yes	Yes	Yes	Yes	Yes
Time-varying controls			Yes	Yes	Yes	Yes
Controls for larceny or motor theft					Yes	Yes
State-specific linear time trends						Yes

Notes: Each column in each panel represents a separate regression. The unit of observation is state-year. Robust standard errors are clustered at the state level. Time-varying controls include policing and incarceration rates, welfare and public assistance spending, median income, poverty rate, unemployment rate, and demographics. *Significant at the 10 percent level. **Significant at the 5 percent level. ***Significant at the 1 percent level.

Table 80. Randomization inference averages [Cheng and Hoekstra, 2013].

Method	Average estimate	Estimates larger than actual estimate
Weighted OLS	−0.003	0/40
Unweighted OLS	0.001	1/40
Negative binomial	0.001	0/40

panels are almost entirely significant. It appears that the bulk of their evidence suggests the castle-doctrine statute caused an increase in homicides around 8%.

Not satisfied, the authors implemented a kind of randomization inference-based test. Specifically, they moved the eleven-year panel back in time covering 1960–2009 and estimated forty placebo "effects" of passing castle doctrine one to forty years earlier. When they did this, they found that the average effect from this exercise was essentially zero. Those results are summarized here. It appears there is something statistically unusual about the actual treatment profile compared to the placebo profiles, because the actual profile yields effect sizes larger than all but one case in any of the placebo regressions run.

Cheng and Hoekstra [2013] found no evidence that castle-doctrine laws deter violent offenses, but they did find that it increased homicides. An 8% net increase in homicide rates translates to around six hundred additional homicides per year across the twenty-one adopting states. Thinking back to to the killing of Trayvon Martin by George Zimmerman, one is left to wonder whether Trayvon might still be alive had Florida not passed Stand Your Ground. This kind of counterfactual reasoning can drive you crazy, because it is unanswerable—we simply don't know, cannot know, and never will know the answer to counterfactual questions. The fundamental problem of causal inference states that we need to know what would have happened that fateful night without Stand Your Ground and compare that with what happened with Stand Your Ground to know what can and cannot be placed at the feet of that law. What we do know is that under certain assumptions related to the DD design, homicides were on net around 8%–10% higher than they would've been when compared against explicit counterfactuals. And while that doesn't answer every question, it suggests that a nontrivial number of deaths can be blamed on laws similar to Stand Your Ground.

Replicating Cheng and Hoekstra [2013], sort of. Now that we've discussed Cheng and Hoekstra [2013], I want to replicate it, or at least do some work on their data set to illustrate certain things that we've discussed, like event studies and the Bacon decomposition. This analysis will be slightly different from what they did, though, because their policy variable was on the interval [0,1] rather than being a pure dummy. That's because they carefully defined their policy variable according to the month in which the law was passed (e.g., June) divided by a total of 12 months. So if a state passed the last in June, then they would assign a 0.5 in the first year, and a 1 thereafter. While there's nothing wrong with that approach, I am going to use a dummy because it makes the event studies a bit easier to visualize, and the Bacon decomposition only works with dummy policy variables.

First, I will replicate his main homicide results from Panel A, column 6, of Figure 74.

STATA
castle_1.do

```
1   use https://github.com/scunning1975/mixtape/raw/master/castle.dta, clear
2   set scheme cleanplots
3   * ssc install bacondecomp
4
5   * define global macros
6   global crime1 jhcitizen_c jhpolice_c murder homicide  robbery assault burglary
     ↪   larceny motor robbery_gun_r
7   global demo blackm_15_24 whitem_15_24 blackm_25_44 whitem_25_44
     ↪   //demographics
8   global lintrend trend_1-trend_51 //state linear trend
9   global region r20001-r20104  //region-quarter fixed effects
10  global exocrime l_larceny l_motor // exogenous crime rates
11  global spending l_exp_subsidy l_exp_pubwelfare
12  global xvar l_police unemployrt poverty l_income l_prisoner l_lagprisoner $demo
     ↪   $spending
13
14  label variable post "Year of treatment"
15  xi: xtreg l_homicide i.year $region $xvar $lintrend post [aweight=popwt], fe
     ↪   vce(cluster sid)
16
```

```
                                    R
                              castle_1.R
1   library(bacondecomp)
2   library(tidyverse)
3   library(haven)
4   library(lfe)
5
6   read_data <- function(df)
7   {
8     full_path <- paste("https://raw.github.com/scunning1975/mixtape/master/",
9               df, sep = "")
10    df <- read_dta(full_path)
11    return(df)
12  }
13
14  castle <- read_data("castle.dta")
15
16  #--- global variables
17  crime1 <- c("jhcitizen_c", "jhpolice_c",
18          "murder", "homicide",
19          "robbery", "assault", "burglary",
20          "larceny", "motor", "robbery_gun_r")
21
22  demo <- c("emo", "blackm_15_24", "whitem_15_24",
23        "blackm_25_44", "whitem_25_44")
24
25  # variables dropped to prevent colinearity
26  dropped_vars <- c("r20004", "r20014",
27            "r20024", "r20034",
28            "r20044", "r20054",
29            "r20064", "r20074",
30            "r20084", "r20094",
31            "r20101", "r20102", "r20103",
32            "r20104", "trend_9", "trend_46",
33            "trend_49", "trend_50", "trend_51"
34  )
35
36  lintrend <- castle %>%
37    select(starts_with("trend")) %>%
38    colnames %>%
39    # remove due to colinearity
40    subset(.,! . %in% dropped_vars)
```

(continued)

R *(continued)*

```
41
42   region <- castle %>%
43     select(starts_with("r20")) %>%
44     colnames %>%
45     # remove due to colinearity
46     subset(.,! . %in% dropped_vars)
47
48
49   exocrime <- c("l_lacerny", "l_motor")
50   spending <- c("l_exp_subsidy", "l_exp_pubwelfare")
51
52
53   xvar <- c(
54     "blackm_15_24", "whitem_15_24", "blackm_25_44", "whitem_25_44",
55     "l_exp_subsidy", "l_exp_pubwelfare",
56     "l_police", "unemployrt", "poverty",
57     "l_income", "l_prisoner", "l_lagprisoner"
58   )
59
60   law <- c("cdl")
61
62   dd_formula <- as.formula(
63     paste("l_homicide ~ ",
64       paste(
65         paste(xvar, collapse = " + "),
66         paste(region, collapse = " + "),
67         paste(lintrend, collapse = " + "),
68         paste("post", collapse = " + "), sep = " + "),
69       "| year + sid | 0 | sid"
70     )
71   )
72
73   #Fixed effect regression using post as treatment variable
74   dd_reg <- felm(dd_formula, weights = castle$popwt, data = castle)
75   summary(dd_reg)
76
77
```

Here we see the main result that castle doctrine expansions led to an approximately 10% increase in homicides. And if we use the post-dummy, which is essentially equal to 0 unless the state had fully covered castle doctrine expansions, then the effect is more like 7.6%.

But now, I'd like to go beyond their study to implement an event study. First, we need to define pre-treatment leads and lags. To do this, we use a "time_til" variable, which is the number of years until or after the state received the treatment. Using this variable, we then create the leads (which will be the years prior to treatment) and lags (the years post-treatment).

STATA
castle_2.do

```
1   * Event study regression with the year of treatment (lag0) as the omitted
    ↪   category.
2   xi: xtreg l_homicide  i.year $region lead9 lead8 lead7 lead6 lead5 lead4 lead3
    ↪   lead2 lead1 lag1-lag5 [aweight=popwt], fe vce(cluster sid)
```

R
castle_2.R

```
1   castle <- castle %>%
2   mutate(
3     time_til = year - treatment_date,
4     lead1 = case_when(time_til == -1 ~ 1, TRUE ~ 0),
5     lead2 = case_when(time_til == -2 ~ 1, TRUE ~ 0),
6     lead3 = case_when(time_til == -3 ~ 1, TRUE ~ 0),
7     lead4 = case_when(time_til == -4 ~ 1, TRUE ~ 0),
8     lead5 = case_when(time_til == -5 ~ 1, TRUE ~ 0),
9     lead6 = case_when(time_til == -6 ~ 1, TRUE ~ 0),
10    lead7 = case_when(time_til == -7 ~ 1, TRUE ~ 0),
11    lead8 = case_when(time_til == -8 ~ 1, TRUE ~ 0),
12    lead9 = case_when(time_til == -9 ~ 1, TRUE ~ 0),
13
14    lag0 = case_when(time_til == 0 ~ 1, TRUE ~ 0),
15    lag1 = case_when(time_til == 1 ~ 1, TRUE ~ 0),
16    lag2 = case_when(time_til == 2 ~ 1, TRUE ~ 0),
17    lag3 = case_when(time_til == 3 ~ 1, TRUE ~ 0),
18    lag4 = case_when(time_til == 4 ~ 1, TRUE ~ 0),
19    lag5 = case_when(time_til == 5 ~ 1, TRUE ~ 0)
20  )
```

(continued)

R *(continued)*

```
21  event_study_formula <- as.formula(
22    paste("l_homicide ~ + ",
23      paste(
24        paste(region, collapse = " + "),
25        paste(paste("lead", 1:9, sep = ""), collapse = " + "),
26        paste(paste("lag", 1:5, sep = ""), collapse = " + "), sep = " + "),
27        "| year + state | 0 | sid"
28    ),
29  )
30
31  event_study_reg <- felm(event_study_formula, weights = castle$popwt, data =
    ↪ castle)
32  summary(event_study_reg)
```

Our omitted category is the year of treatment, so all coefficients are with respect to that year. You can see from the coefficients on the leads that they are not statistically different from zero prior to treatment, except for leads 8 and 9, which may be because there are only three states with eight years prior to treatment, and one state with nine years prior to treatment. But in the years prior to treatment, leads 1 to 6 are equal to zero and statistically insignificant, although they do technically have large confidence intervals. The lags, on the other hand, are all positive and not too dissimilar from one another except for lag 5, which is around 17%.

Now it is customary to plot these event studies, so let's do that now. I am going to show you an easy way and a longer way to do this. The longer way gives you ultimately more control over what exactly you want the event study to look like, but for a fast and dirty method, the easier way will suffice. For the easier way, you will need to install a program in Stata called coefplot, written by Ben Jann, author of estout.[17]

---

17 Ben Jann is a valuable contributor to the Stata community for creating several community ado packages, such as -estout- for making tables and -coefplot- for making pictures of regression coefficients.

STATA
castle_3.do

```
1   * Plot the coefficients using coefplot
2   * ssc install coefplot
3
4   coefplot, keep(lead9 lead8 lead7 lead6 lead5 lead4 lead3 lead2 lead1 lag1 lag2
    ↪   lag3 lag4 lag5) xlabel(, angle(vertical)) yline(0) xline(9.5) vertical
    ↪   msymbol(D) mfcolor(white) ciopts(lwidth(*3) lcolor(*.6)) mlabel
    ↪   format(%9.3f) mlabposition(12) mlabgap(*2) title(Log Murder Rate)
```

R
castle_3.R

```
1
2   # order of the coefficients for the plot
3   plot_order <- c("lead9", "lead8", "lead7",
4           "lead6", "lead5", "lead4", "lead3",
5           "lead2", "lead1", "lag1",
6           "lag2", "lag3", "lag4", "lag5")
7
8   # grab the clustered standard errors
9   # and average coefficient estimates
10  # from the regression, label them accordingly
11  # add a zero'th lag for plotting purposes
12  leadslags_plot <- tibble(
13    sd = c(event_study_reg$cse[plot_order], 0),
14    mean = c(coef(event_study_reg)[plot_order], 0),
15    label = c(-9,-8,-7,-6, -5, -4, -3, -2, -1, 1,2,3,4,5, 0)
16  )
17
18  # This version has a point-range at each
19  # estimated lead or lag
20  # comes down to stylistic preference at the
21  # end of the day!
22  leadslags_plot %>%
23    ggplot(aes(x = label, y = mean,
24          ymin = mean-1.96*sd,
25          ymax = mean+1.96*sd)) +
26    geom_hline(yintercept = 0.035169444, color = "red") +
27    geom_pointrange() +
```

(continued)

R *(continued)*
28   theme_minimal() +
29   xlab("Years before and after castle doctrine expansion") +
30   ylab("log(Homicide Rate)") +
31   geom_hline(yintercept = 0,
32   linetype = "dashed") +
33   geom_vline(xintercept = 0,
34   linetype = "dashed")
35
36
37

Let's look now at what this command created. As you can see in Figure 75, eight to nine years prior to treatment, treatment states have significantly lower levels of homicides, but as there are so few states that even have these values (one with −9 and three with −8), we may want to disregard the relevance of these negative effects if for no other reason than that there are so few units in the dummy and we know from earlier that that can lead to very high overrejection rates [MacKinnon and Webb, 2017]. Instead, notice that for the six years prior to treatment, there is virtually no difference between the treatment states and the control states.

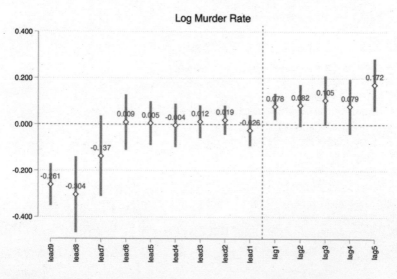

Figure 75. Homicide event-study plots using coefplot. Cheng and Hoekstra [2013].

But, after the year of treatment, that changes. Log murders begin rising, which is consistent with our post dummy that imposed zeros on all pre-treatment leads and required that the average effect post-treatment be a constant.

I promised to show you how to make this graph in a way that gave more flexibility, but you should be warned, this is a bit more cumbersome.

STATA

castle_4.do

```
1  xi: xtreg l_homicide  i.year $region $xvar $lintrend post [aweight=popwt], fe
   ↪  vce(cluster sid)
2
3  local DDL = _b[post]
4  local DD : display %03.2f _b[post]
5  local DDSE : display %03.2f _se[post]
6  local DD1 = -0.10
7
8  xi: xtreg l_homicide  i.year $region lead9 lead8 lead7 lead6 lead5 lead4 lead3
   ↪  lead2 lead1 lag1-lag5 [aweight=popwt], fe vce(cluster sid)
9
10 outreg2 using "./eventstudy_levels.xls", replace keep(lead9 lead8 lead7 lead6
   ↪  lead5 lead4 lead3 lead2 lead1 lag1-lag5) noparen noaster addstat(DD, `DD',
   ↪  DDSE, `DDSE')
11
12
13 *Pull in the ES Coefs
14 xmluse "./eventstudy_levels.xls", clear cells(A3:B32) first
15 replace VARIABLES = subinstr(VARIABLES,"lead","",.)
16 replace VARIABLES = subinstr(VARIABLES,"lag","",.)
17 quietly destring _all, replace ignore(",")
18 replace VARIABLES = -9 in 2
19 replace VARIABLES = -8 in 4
20 replace VARIABLES = -7 in 6
21 replace VARIABLES = -6 in 8
22 replace VARIABLES = -5 in 10
23 replace VARIABLES = -4 in 12
24 replace VARIABLES = -3 in 14
25 replace VARIABLES = -2 in 16
26
```

(continued)

	STATA *(continued)*
27	
28	
29	replace VARIABLES = -1 in 18
30	replace VARIABLES = 1 in 20
31	replace VARIABLES = 2 in 22
32	replace VARIABLES = 3 in 24
33	replace VARIABLES = 4 in 26
34	replace VARIABLES = 5 in 28
35	drop in 1
36	compress
37	quietly destring _all, replace ignore(",")
38	compress
39	
40	
41	
42	ren VARIABLES exp
43	gen b = exp<.
44	replace exp = -9 in 2
45	replace exp = -8 in 4
46	replace exp = -7 in 6
47	replace exp = -6 in 8
48	replace exp = -5 in 10
49	replace exp = -4 in 12
50	replace exp = -3 in 14
51	replace exp = -2 in 16
52	replace exp = -1 in 18
53	replace exp = 1 in 20
54	replace exp = 2 in 22
55	replace exp = 3 in 24
56	replace exp = 4 in 26
57	replace exp = 5 in 28
58	
59	* Expand the dataset by one more observation so as to include the comparison ↪ year
60	local obs =_N+1
61	set obs `obs'
62	for var _all: replace X = 0 in `obs'
63	replace b = 1 in `obs'
64	replace exp = 0 in `obs'
65	keep exp l_homicide b

(continued)

STATA *(continued)*

```
66    set obs 30
67    foreach x of varlist exp l_homicide b {
68        replace `x'=0 in 30
69        }
70    reshape wide l_homicide, i(exp) j(b)
71
72
73    * Create the confidence intervals
74    cap drop *lb* *ub*
75    gen lb = l_homicide1 - 1.96*l_homicide0
76    gen ub = l_homicide1 + 1.96*l_homicide0
77
78
79    * Create the picture
80    set scheme s2color
81    #delimit ;
82    twoway (scatter l_homicide1 ub lb exp ,
83              lpattern(solid dash dash dot dot solid solid)
84              lcolor(gray gray gray red blue)
85              lwidth(thick medium medium medium medium thick thick)
86              msymbol(i i i i i i i i i i i i i i) msize(medlarge medlarge)
87              mcolor(gray black gray gray red blue)
88              c(l l l l l l l l l l l l l l)
89              cmissing(n n n n n n n n n n n n n n)
90              xline(0, lcolor(black) lpattern(solid))
91              yline(0, lcolor(black))
92              xlabel(-9 -8 -7 -6 -5 -4 -3 -2 -1 0 1 2 3 4 5 , labsize(medium))
93              ylabel(, nogrid labsize(medium))
94              xsize(7.5) ysize(5.5)
95              legend(off)
96              xtitle("Years before and after castle doctrine expansion",
                 ↪   size(medium))
97              ytitle("Log Murders ", size(medium))
98              graphregion(fcolor(white) color(white) icolor(white) margin(zero))
99              yline(`DDL', lcolor(red) lwidth(thick)) text(`DD1' -0.10 "DD
                 ↪   Coefficient = `DD' (s.e. = `DDSE')")
100           )
101   ;
102
103   #delimit cr;
```

R
castle_4.R

```
1
2    # This version includes
3    # an interval that traces the confidence intervals
4    # of your coefficients
5    leadslags_plot %>%
6    ggplot(aes(x = label, y = mean,
7            ymin = mean-1.96*sd,
8            ymax = mean+1.96*sd)) +
9    # this creates a red horizontal line
10   geom_hline(yintercept = 0.035169444, color = "red") +
11   geom_line() +
12   geom_point() +
13   geom_ribbon(alpha = 0.2) +
14   theme_minimal() +
15   # Important to have informative axes labels!
16   xlab("Years before and after castle doctrine expansion") +
17   ylab("log(Homicide Rate)") +
18   geom_hline(yintercept = 0) +
19   geom_vline(xintercept = 0)
```

You can see the figure that this creates in Figure 76. The difference between coefplot and this twoway command connects the event-study coefficients with lines, whereas coefplot displayed them as coefficients hanging in the air. Neither is right or wrong; I merely wanted you to see the differences for your own sake and to have code that you might experiment with and adapt to your own needs.

But the thing about this graph is that the leads are imbalanced. There's only one state, for instance, in the ninth lead, and there's only three in the eighth lead. So I'd like you to do two modifications to this. First, I'd like you to replace the sixth lead so that it is now equal to leads 6−9. In other words, we will force these late adopters to have the same coefficient as those with six years until treatment. When you do that, you should get Figure 77.

Next, let's balance the event study by dropping the states who only show up in the seventh, eighth, and ninth leads.[18] When you do this, you should get Figure 78.

---

18 Alex Bartik once recommended this to me.

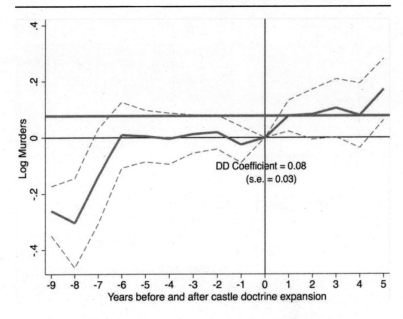

Figure 76. Homicide event study plots created manually with twoway. Cheng and Hoekstra [2013].

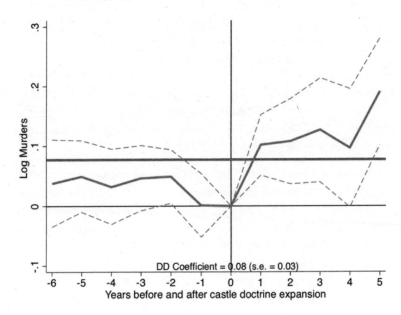

Figure 77. Homicide event-study plots using twoway. Cheng and Hoekstra [2013].

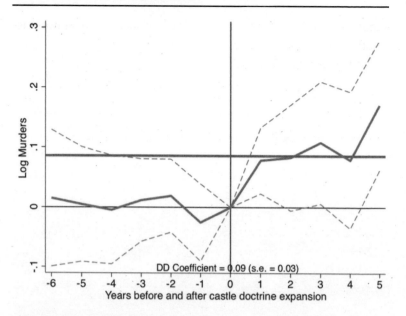

Figure 78. Homicide event-study plots using twoway. Cheng and Hoekstra [2013].

If nothing else, exploring these different specifications and cuts of the data can help you understand just how confident you should be that prior to treatment, treatment and control states genuinely were pretty similar. And if they weren't similar, it behooves the researcher to at minimum provide some insight to others as to why the treatment and control groups were dissimilar in levels. Because after all—if they were different in levels, then it's entirely plausible they would be different in their counterfactual trends too because why else are they different in the first place [Kahn-Lang and Lang, 2019].

Bacon decomposition. Recall that we run into trouble using the twoway fixed-effects model in a DD framework insofar as there are heterogeneous treatment effects over time. But the problem here only occurs with those $2 \times 2s$ that use late-treated units compared to early-treated units. If there are few such cases, then the issue is much less problematic depending on the magnitudes of the weights and the size of the DD coefficients themselves. What we are now going to do is simply evaluate the frequency with which this issue occurs using the Bacon decomposition. Recall that the Bacon decomposition decomposes the twoway fixed effects estimator of the DD parameter into weighted averages of individual $2 \times 2s$ across the four types of $2 \times 2s$ possible. The

Bacon decomposition uses a binary treatment variable, so we will reestimate the effect of castle-doctrine statutes on logged homicide rates by coding a state as "treated" if any portion of the year it had a castle-doctrine amendment. We will work with the special case of no covariates for simplicity, though note that the decomposition works with the inclusion of covariates as well [Goodman-Bacon, 2019]. Stata users will need to download -ddtiming- from Thomas Goldring's website, which I've included in the first line.

First, let's estimate the actual model itself using a post dummy equaling one if the state was covered by a castle-doctrine statute that year. Here we find a smaller effect than many of Cheng and Hoekstra's estimates because we do not include their state-year interaction fixed effects strategy among other things. But this is just for illustrative purposes, so let's move to the Bacon decomposition itself. We can decompose the parameter estimate into the three different types of 2×2s, which I've reproduced in Table 81.

STATA
castle_5.do

```
1   use https://github.com/scunning1975/mixtape/raw/master/castle.dta, clear
2   * ssc install bacondecomp
3
4   * define global macros
5   global crime1 jhcitizen_c jhpolice_c murder homicide  robbery assault burglary
      ↪   larceny motor robbery_gun_r
6   global demo blackm_15_24 whitem_15_24 blackm_25_44 whitem_25_44
      ↪   //demographics
7   global lintrend trend_1-trend_51 //state linear trend
8   global region r20001-r20104  //region-quarter fixed effects
9   global exocrime l_larceny l_motor // exogenous crime rates
10  global spending l_exp_subsidy l_exp_pubwelfare
11  global xvar l_police unemployrt poverty l_income l_prisoner l_lagprisoner $demo
      ↪   $spending
12  global law cdl
13
14  * Bacon decomposition
15  net install ddtiming, from(https://tgoldring.com/code/)
16  areg l_homicide post i.year, a(sid) robust
17  ddtiming l_homicide post, i(sid) t(year)
18
```

R

castle_5.R

```
1   library(bacondecomp)
2   library(lfe)
3
4   df_bacon <- bacon(l_homicide ~ post,
5            data = castle, id_var = "state",
6            time_var = "year")
7
8   # Diff-in-diff estimate is the weighted average of
9   # individual 2x2 estimates
10  dd_estimate <- sum(df_bacon$estimate*df_bacon$weight)
11
12  # 2x2 Decomposition Plot
13  bacon_plot <- ggplot(data = df_bacon) +
14    geom_point(aes(x = weight, y = estimate,
15            color = type, shape = type), size = 2) +
16    xlab("Weight") +
17    ylab("2x2 DD Estimate") +
18    geom_hline(yintercept = dd_estimate, color = "red") +
19    theme_minimal() +
20    theme(
21      legend.title = element_blank(),
22      legend.background = element_rect(
23        fill="white", linetype="solid"),
24      legend.justification=c(1,1),
25      legend.position=c(1,1)
26    )
27
28  bacon_plot
29
30  # create formula
31  bacon_dd_formula <- as.formula(
32    'l_homicide ~ post | year + sid | 0 | sid')
33
34  # Simple diff-in-diff regression
35  bacon_dd_reg <- felm(formula = bacon_dd_formula, data = castle)
36  summary(bacon_dd_reg)
37
38  # Note that the estimate from earlier equals the
39  # coefficient on post
40  dd_estimate
41
```

Table 81. Bacon decomposition example.

DD Comparison	Weight	Avg DD Est
Earlier T vs. Later C	0.077	−0.029
Later T vs. Earlier C	0.024	0.046
T vs. Never treated	0.899	0.078

Dep var	Log(homicide rate)
Castle-doctrine law	0.069
	(0.034)

Taking these weights, let's just double check that they do indeed add up to the regression estimate we just obtained using our twoway fixed-effects estimator.[19]

$$di(0.077 * -0.029) + (0.024 * 0.046) + (0.899 * 0.078) = 0.069$$

That is our main estimate, and thus confirms what we've been building on, which is that the DD parameter estimate from a twoway fixed-effects estimator is simply a weighted average over the different types of 2×2s in any differential design. Furthermore, we can see in the Bacon decomposition that most of the 0.069 parameter estimate is coming from comparing the treatment states to a group of never-treated states. The average DD estimate for that group is 0.078 with a weight of 0.899. So even though there is a later to early 2×2 in the mix, as there always will be with any differential timing, it is small in terms of influence and ultimately pulls down the estimate.

But let's now visualize this as distributing the weights against the DD estimates, which is a useful exercise. The horizontal line in Figure 79 shows the average DD estimate we obtained from our fixed-effects regression of 0.069. But then what are these other graphics? Let's review.

Each icon in the graphic represents a single 2×2 DD. The horizontal axis shows the weight itself, whereas the vertical axis shows the

19 This result is different from Cheng and Hoekstra's because it does not include the region by year fixed effects. I exclude them for simplicity.

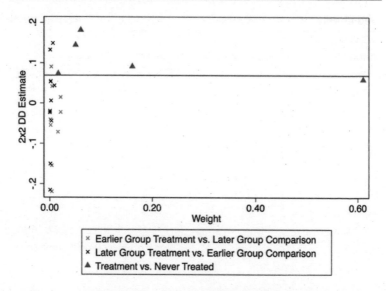

Figure 79. Bacon decomposition of DD into weights and single 2×2s.

magnitude of that particular 2×2. Icons further to the right therefore will be more influential in the final average DD than those closer to zero.

There are three kinds of icons here: an early to late group comparison (represented with a light $\times$), a late to early (dark $\times$), and a treatment compared to a never-treated (dark triangle). You can see that the dark triangles are all above zero, meaning that each of these 2×2s (which correspond to a particular set of states getting the treatment in the same year) is positive. Now they are spread out somewhat—two groups are on the horizontal line, but the rest are higher. What appears to be the case, though, is that the group with the largest weight is really pulling down the parameter estimate and bringing it closer to the 0.069 that we find in the regression.

The future of DD. The Bacon decomposition is an important phase in our understanding of the DD design when implemented using the twoway fixed effects linear model. Prior to this decomposition, we had only a metaphorical understanding of the necessary conditions for identifying causal effects using differential timing with a twoway fixed-effects estimator. We thought that since the 2×2 required parallel trends, that that "sort of" must be what's going on with differential timing too. And we weren't too far off—there is a version of parallel trends in the identifying assumptions of DD using twoway fixed effects with differential timing. But what Goodman-Bacon [2019] also showed is that the weights themselves drove the numerical estimates too, and that while some of it was intuitive (e.g., group shares being influential) others were not (e.g., variance in treatment being influential).

The Bacon decomposition also highlighted some of the unique challenges we face with differential timing. Perhaps no other problem is better highlighted in the diagnostics of the Bacon decomposition than the problematic "late to early" 2×2 for instance. Given any heterogeneity bias, the late to early 2×2 introduces biases *even with variance weighted common trends holding*! So, where to now?

From 2018 to 2020, there has been an explosion of work on the DD design. Much of it is unpublished, and there has yet to appear any real consensus among applied people as to how to handle it. Here I would like to outline what I believe could be a map as you attempt to navigate the future of DD. I have attempted to divide this new work into three categories: weighting, selective choice of "good" 2×2s, and matrix completion.

What we know now is that there are two fundamental problems with the DD design. First, there is the issue of weighting itself. The twoway fixed-effects estimator weights the individual 2×2s in ways that do not make a ton of theoretical sense. For instance, why do we think that groups at the middle of the panel should be weighted more than those at the end? There's no theoretical reason we should believe that. But as Goodman-Bacon [2019] revealed, that's precisely what twoway fixed effects does. And this is weird because you can change your results simply by adding or subtracting years to the panel—not just because this changes the 2×2, but also because it changes the variance in treatment itself! So that's weird.[20]

But this is not really the fatal problem, you might say, with twoway fixed-effects estimates of a DD design. The bigger issue was what we

---

20 This is less an issue with event study designs because the variance of treatment indicator is the same for everyone.

saw in the Bacon decomposition—you will inevitably use past treated units as controls for future treated units, or what I called the "late to early 2×2." This happens both in the event study and in the designs modeling the average treatment effect with a dummy variable. Insofar as it takes more than one period for the treatment to be fully incorporated, then insofar as there's substantial weight given to the late to early 2×2s, the existence of heterogeneous treatment effects skews the parameter away from the ATT—maybe even flipping signs![21]

Whereas the weird weighting associated with twoway fixed effects is an issue, it's something you can at least check into because the Bacon decomposition allows you to separate out the 2×2 average DD values from their weights. Thus, if your results are changing by adding years because your underlying 2×2s are changing, you simply need to investigate it in the Bacon decomposition. The weights and the 2×2s, in other words, are things that can be directly calculated, which can be a source of insight into why twoway fixed effects estimator is finding what it finds.

But the second issue is a different beast altogether. And one way to think of the emerging literature is that many authors are attempting to solve the problem that some of these 2×2s (e.g., the late to early 2×2) are problematic. Insofar as they are problematic, can we improve over our static twoway fixed-effects model? Let's take a few of these issues up with examples from the growing literature.

Another solution to the weird weighting twoway fixed-effects problem has been provided by Callaway and Sant'Anna [2019].[22] Callaway and Sant'Anna [2019] approach the DD framework very differently than Goodman-Bacon [2019]. Callaway and Sant'Anna [2019] use an

---

[21] In all seriousness, it is practically modal in applied papers that utilize a DD design to imagine that dynamic treatment effects are at least plausible ex ante, if not expected. This kind of "dynamic treatment effects" is usually believed as a realistic description of what we think could happen in any policy environment. As such, the biases associated with panel fixed effects model with twoway fixed effects is, to be blunt, scary. Rarely have I seen a study wherein the treatment was merely a one-period shift in size. Even in the Miller et al. [2019] paper, the effect of ACA-led Medicaid expansions was a gradual reduction in annual mortality over time. Figure 60 really is probably a *typical* kind of event study, not an exceptional one.

[22] Sant'Anna has been particularly active in this area in producing elegant econometric solutions to some of these DD problems.

approach that allows them to estimate what they call the group-time average treatment effect, which is just the ATT for a given group at any point in time. Assuming parallel trends conditional on time-invariant covariates and overlap in a propensity score, which I'll discuss below, you can calculate group ATT by time (relative time like in an event study or absolute time). One unique part of these authors' approach is that it is non-parametric as opposed to regression-based. For instance, under their identifying assumptions, their nonparametric estimator for a group ATT by time is:

$$ATT(g,t) = E\left[\left(\frac{G_g}{E[G_g]} - \frac{\dfrac{p_g(X)C}{1-p_g(X)}}{E\left[\dfrac{p_g(X)C}{1-p_g(X)}\right]}\right)(Y_t - Y_{g-1})\right]$$

where the weights, p, are propensity scores, G is a binary variable that is equal to 1 if an individual is first treated in period g, and C is a binary variable equal to one for individuals in the control group. Notice there is no time index, so these C units are the never-treated group. If you're still with me, you should find the weights straightforward. Take observations from the control group as well as group g, and omit the other groups. Then weight up those observations from the control group that have characteristics similar to those frequently found in group g and weight down observations from the control group that have characteristics rarely found in group g. This kind of reweighting procedure guarantees that the covariates of group g and the control group are balanced. You can see principles from earlier chapters making their way into this DD estimation—namely, balance on covariates to create exchangeable units on observables.

But because we are calculating group-specific ATT by time, you end up with a lot of treatment effect parameters. The authors address this by showing how one can take all of these treatment effects and collapse them into more interpretable parameters, such as a larger ATT. All of this is done without running a regression, and therefore avoids some of the unique issues created in doing so.

One simple solution might be to estimate your event-study model and simply take the mean over all lags using a linear combination of all point estimates [Borusyak and Jaravel, 2018]. Using this method, we in fact find considerably larger effects or nearly twice the size as we get from the simpler static twoway fixed-effects model. This is perhaps an improvement because weights can be large on the long-run effects due to large effects from group shares. So if you want a summary measure, it's better to estimate the event study and then average them after the fact.

Another great example of a paper wrestling with the biases brought up by heterogeneous treatment effects is Sun and Abraham [2020]. This paper is primarily motivated by problems created in event studies, but you can see some of the issues brought up in Goodman-Bacon [2019]. In an event study with differential timing, as we discussed earlier, leads and lags are often used to measure dynamics in the treatment itself. But these can produce causally uninterpretable results because they will assign non-convex weights to cohort-specific treatment effects. Similar to Callaway and Sant'Anna [2019], they propose estimating a group-specific dynamic effect and from those calculate a group specific estimate.

The way I organize these papers in my mind is around the idea of heterogeneity in time, the use of twoway fixed effects, and differential timing. The theoretical insight from all these papers is the coefficients on the static twoway fixed-effects leads and lags will be unintelligible if there is heterogeneity in treatment effects over time. In this sense, we are back in the world that Goodman-Bacon [2019] revealed, in which heterogeneity treatment effect biases create real challenges for the DD design using twoway fixed effects.[23]

Their alternative is estimate a "saturated" model to ensure that the heterogeneous problem never occurs in the first place. The proposed alternative estimation technique is to use an interacted specification that is saturated in relative time indicators as well as cohort indicators. The treatment effect associated with this design is called the interaction-weighted estimator, and using it, the DD parameter is equivalent to the difference between the average change in outcomes for a given cohort in those periods prior to treatment and the average

---

23 One improvement over the binary treatment approach to estimating the treatment effect is when using an event study, the variance of treatment issues are moot.

changes for those units that had not been treated at the time interval. Additionally, this method uses the never-treated units as controls, and thereby avoids the hairy problems noted in Goodman-Bacon [2019] when computing later to early $2 \times 2s$.[24]

Another paper that attempts to circumvent the weirdness of the regression-based method when there are numerous late to early $2 \times 2s$ is Cengiz et al. [2019]. This is bound to be a classic study in labor for its exhaustive search for detectable repercussions of the minimum wage on low-paying jobs. The authors ultimately find little evidence to support any concern, but how do they come to this conclusion?

Cengiz et al. [2019] take a careful approach by creating separate samples. The authors want to know the impact of minimum-wage changes on low-wage jobs across 138 state-level minimum-wage changes from 1979 to 2016. The authors in an appendix note the problems with aggregating individual DD estimates into a single parameter, and so tackle the problem incrementally by creating 138 separate data sets associated with a minimum-wage event. Each sample has both treatment groups and control groups, but not all units are used as controls. Rather, only units that were not treated within the sample window are allowed to be controls. Insofar as a control is not treated during the sample window associated with a treatment unit, it can be by this criteria used as a control. These 138 estimates are then stacked to calculate average treatment effects. This is an alternative method to the twoway fixed-effects DD estimator because it uses a more stringent criteria for whether a unit can be considered a control. This in turn circumvents the heterogeneity problems that Goodman-Bacon [2019] notes because Cengiz et al. [2019] essentially create 138 DD situations in which controls are always "never-treated" for the duration of time under consideration.

But the last methodology I will discuss that has emerged in the last couple of years is a radical departure from the regression-based methodology altogether. Rather than use a twoway fixed-effects estimator to estimate treatment effects with differential timing, Athey et al. [2018] propose a machine-learning-based methodology called "matrix

---

24 But in selecting only the never-treated as controls, the approach may have limited value for those situations where the number of units in the never-treated pool is extremely small.

completion" for panel data. The estimator is exotic and bears some resemblance to matching imputation and synthetic control. Given the growing popularity of placing machine learning at the service of causal inference, I suspect that once Stata code for matrix completion is introduced, we will see this procedure used more broadly.

Matrix completion for panel data is a machine-learning-based approach to causal inference when one is working explicitly with panel data and differential timing. The application of matrix completion to causal inference has some intuitive appeal given one of the ways that Rubin has framed causality is as a missing data problem. Thus, if we are missing the matrix of counterfactuals, we might explore whether this method from computer science could assist us in recovering it. Imagine we could create two matrices of potential outcomes: a matrix of Y^0 potential outcomes for all panel units over time and Y^1. Once treatment occurs, a unit switches from Y^0 to Y^1 under the switching equation, and therefore the missing data problem occurs. Missingness is simply another way of describing the fundamental problem of causal inference for there will never be a complete set of matrices enabling calculation of interesting treatment parameters given the switching equation only assigns one of them to reality.

Say we are interested in this treatment effect parameter:

$$\widehat{\delta_{ATT}} = \frac{1}{N_T} \sum \left(Y_{it}^1 - Z_{it}^0 \right)$$

where Y^1 are the observed outcomes in a panel unit at some post-treatment period, Z^0 is the estimated missing elements of the Y^0 matrix for the post-treatment period, and N_T is the number of treatment units. Matrix completion uses the observed elements of the matrix's realized values to predict the missing elements of the Y^0 matrix (missing due to being in the post-treatment period and therefore having switched from Y^0 to Y^1).

Analytically, this imputation is done via something called regularization-based prediction. The objective in this approach is to optimally predict the missing elements by minimizing a convex function of the difference between the observed matrix of Y^0 and the unknown complete matrix Z^0 using nuclear norm regularization. Let Ω denote the row and column indices (i,j) of the observed entries of the outcomes, then the objective function can be written as

$$\widehat{Z^0} = \arg\min_{Z^0} \sum_{(i,j)\in\Omega} \frac{(Y_{it}^0 - Z_{it}^0)^2}{|\Omega|} + \Lambda||Z^0||$$

where $||Z^0||$ is the nuclear norm (sum of singular values of $Z0$). The regularization parameter Λ is chosen using tenfold cross validation. Athey et al. [2018] show that this procedure outperforms other methods in terms of root mean squared prediction error.

Unfortunately, at present estimation using matrix completion is not available in Stata. R packages for it do exist, such as the gsynth package, but it has to be adapted for Stata users. And until it is created, I suspect adoption will lag.

Conclusion

America's institutionalized state federalism provides a constantly evolving laboratory for applied researchers seeking to evaluate the causal effects of laws and other interventions. It has for this reason probably become one of the most popular forms of identification among American researchers, if not the most common. A Google search of the phrase "difference-in-differences" brought up 45,000 hits. It is arguably the most common methodology you will use—more than IV or matching or even RDD, despite RDD's greater perceived credibility. There is simply a never-ending flow of quasi-experiments being created by our decentralized data-generating process in the United States made even more advantageous by so many federal agencies being responsible for data collection, thus ensuring improved data quality and consistency.

But, what we have learned in this chapter is that while there is a current set of identifying assumptions and practices associated with the DD design, differential timing does introduce some thorny challenges that have long been misunderstood. Much of the future of DD appears to be mounting solutions to problems we are coming to understand better, such as the odd weighting of regression itself and problematic 2×2 DDs that bias the aggregate ATT when heterogeneity in the treatment effects over time exists. Nevertheless, DD—and specifically, regression-based DD—is not going away. It is the single most popular design in the applied researcher's toolkit and likely will be for many years to come. Thus it behooves the researcher to study this literature carefully so that they can better protect against various forms of bias.

Synthetic Control

Introducing the Comparative Case Study

The first appearance of the synthetic control estimator was a 2003 article where it was used to estimate the impact of terrorism on economic activity [Abadie and Gardeazabal, 2003]. Since that publication, it has become very popular—particularly after the release of an R and Stata package coinciding with Abadie et al. [2010]. A Google Scholar search for the words "synthetic control" and "Abadie" yielded over 3,500 hits at the time of writing. The estimator has been so influential that Athey and Imbens [2017b] said it was "arguably the most important innovation in the policy evaluation literature in the last 15 years" [3].

To understand the reasons you might use synthetic control, let's back up to the broader idea of the comparative case study. In qualitative case studies, such as Alexis de Tocqueville's classic *Democracy in America*, the goal is to reason inductively about the causal effect of events or characteristics of a single unit on some outcome using logic and historical analysis. But it may not give a very satisfactory answer to these causal questions because sometimes qualitative comparative case studies lack an explicit counterfactual. As such, we are usually left with description and speculation about the causal pathways connecting various events to outcomes.

Quantitative comparative case studies are more explicitly causal designs. They usually are natural experiments and are applied to only a single unit, such as a single school, firm, state, or country. These kinds of quantitative comparative case studies compare the evolution of an aggregate outcome with either some single other outcome, or as is more often the case, a chosen set of similar units that serve as a control group.

As Athey and Imbens [2017b] point out, one of the most important contributions to quantitative comparative case studies is the synthetic control model. The synthetic control model was developed in Abadie and Gardeazabal [2003] in a study of terrorism's effect on aggregate income, which was then elaborated on in a more exhaustive treatment [Abadie et al., 2010]. Synthetic controls models optimally choose a set of weights which when applied to a group of corresponding units produce an optimally estimated counterfactual to the unit that received the treatment. This counterfactual, called the "synthetic unit," serves to outline what would have happened to the aggregate treated unit had the treatment never occurred. It is a powerful, yet surprisingly simple, generalization of the difference-in-differences strategy. We will discuss it with a motivating example—the paper on the famous Mariel boatlift by Card [1990].

Cuba, Miami, and the Mariel Boatlift. Labor economists have debated the effect of immigration on local labor-market conditions for many years [Card and Peri, 2016]. Do inflows of immigrants depress wages and the employment of natives in local labor-markets? For Card [1990], this was an empirical question, and he used a natural experiment to evaluate it.

In 1980, Fidel Castro announced that anyone wishing to leave Cuba could do so. With Castro's support, Cuban Americans helped arrange the Mariel boatlift, a mass exodus from Cuba's Mariel Harbor to the United States (primarily Miami) between April and October 1980. Approximately 125,000 Cubans emigrated to Florida over six months. The emigration stopped only because Cuba and the US mutually agreed to end it. The event increased the Miami labor force by 7%, largely by depositing a record number of low-skilled workers into a relatively small geographic area.

Card saw this as an ideal natural experiment. It was arguably an exogenous shift in the labor-supply curve, which would allow him to determine if wages fell and employment increased, consistent with a simple competitive labor-market model. He used individual-level data on unemployment from the Current Population Survey for Miami and chose four comparison cities (Atlanta, Los Angeles, Houston, and Tampa−St. Petersburg). The choice of these four cities is delegated to a footnote in which Card argues that they were similar based on demographics and economic conditions. Card estimated a simple DD model and found, surprisingly, no effect on wages or native unemployment. He argued that Miami's labor-market was capable of absorbing the surge in labor supply because of similar surges two decades earlier.

The paper was very controversial, probably not so much because he attempted to answer empirically an important question in labor economics using a natural experiment, but rather because the result violated conventional wisdom. It would not be the last word on the subject, and I don't take a stand on this question; rather, I introduce it to highlight a few characteristics of the study. Notably, a recent study replicated Card's paper using synthetic control and found similar results [Peri and Yasenov, 2018].

Card's study was a comparative case study that had strengths and weaknesses. The policy intervention occurred at an aggregate level, for which aggregate data was available. But the problems with the study were that the selection of the control group was *ad hoc* and subjective. Second, the standard errors reflect sampling variance as opposed to uncertainty about the ability of the control group to reproduce the counterfactual of interest. Abadie and Gardeazabal [2003] and Abadie et al. [2010] introduced the synthetic control estimator as a way of addressing both issues simultaneously.

Abadie and Gardeazabal [2003] method uses a weighted average of units in the donor pool to model the counterfactual. The method is based on the observation that, when the units of analysis are a few aggregate units, a combination of comparison units (the "synthetic control") often does a better job of reproducing characteristics of a treated unit than using a single comparison unit alone. The comparison unit, therefore, in this method is selected to be the weighted average of all comparison units that best resemble the characteristics of the treated unit(s) in the pre-treatment period.

Abadie et al. [2010] argue that this method has many distinct advantages over regression-based methods. For one, the method precludes extrapolation. It uses instead interpolation, because the estimated causal effect is always based on a comparison between some outcome in a given year and a counterfactual in the same year. That is, it uses as its counterfactual a convex hull of control group units, and thus the counterfactual is based on where data actually is, as opposed to extrapolating beyond the support of the data, which can occur in extreme situations with regression [King and Zeng, 2006].

A second advantage has to do with processing of the data. The construction of the counterfactual does not require access to the post-treatment outcomes during the design phase of the study, unlike regression. The advantage here is that it helps the researcher avoid "peeking" at the results while specifying the model. Care and honesty must still be used, as it's just as easy to look at the outcomes during the design phase as it is to not, but the point is that it is hypothetically possible to focus just on design, and not estimation, with this method [Rubin, 2007, 2008].

Another advantage, which is oftentimes a reason people will object to a study, is that the weights that are chosen make explicit what each unit is contributing to the counterfactual. Now this is in many ways a strict advantage, except when it comes to defending those weights in a seminar. Because someone can see that Idaho is contributing 0.3 to your modeling of Florida, they are now able to argue that it's absurd to think Idaho is anything like Florida. But contrast this with regression, which also weights the data, but does so blindly. The only reason no one objects to what regression produces as a weight is that they *cannot see the weights*. They are implicit rather than explicit. So I see this explicit production of weights as a distinct advantage because it makes synthetic control more transparent than regression-based designs (even if will likely require fights with the audience and reader that you wouldn't have had otherwise).

A fourth advantage, which I think is often unappreciated, is that it bridges a gap between qualitative and quantitative types. Qualitative researchers are often the very ones focused on describing a single unit, such as a country or a prison [Perkinson, 2010], in great detail.

They are usually the experts on the histories surrounding those institutions. They are usually the ones doing comparative case studies in the first place. Synthetic control places a valuable tool into their hands which enables them to choose counterfactuals—a process that in principle can improve their work insofar as they are interested in evaluating some particular intervention.

Picking synthetic controls. Abadie et al. [2010] argue that synthetic control removes subjective researcher bias, but it turns out it is somewhat more complicated. The frontier of this method has grown considerably in recent years, along different margins, one of which is via the model-fitting exercise itself. Some new ways of trying to choose more principled models have appeared, particularly when efforts to fit the data with the synthetic control in the pre-treatment period are imperfect. Ferman and Pinto [2019] and Powell [2017], for instance, propose alternative solutions to this problem. Ferman and Pinto [2019] examine the properties of using de-trended data. They find that it can have advantages, and even dominate DD, in terms of bias and variance.

When there are transitory shocks, which is common in practice, the fit deteriorates, thus introducing bias. Powell [2017] provides a parametric solution, however, which cleverly exploits information in the procedure that may help reconstruct the treatment unit. Assume that Georgia receives some treatment but for whatever reason the convex hull assumption does not hold in the data (i.e., Georgia is unusual). Powell [2017] shows that if Georgia appears in the synthetic control for some other state, then it is possible to recover the treatment effect through a kind of backdoor procedure. Using the appearance of Georgia as a control in all the placebos can then be used to reconstruct the counterfactual.

But still there remain questions regarding the selection of the covariates that will be used for any matching. Through repeated iterations and changes to the matching formula, a person can potentially reintroduce bias through the endogenous selection of covariates used in a specification search. While the weights are optimally chosen to minimize some distance function, through the choice of the covariates themselves, the researcher can in principle select different weights.

She just doesn't have a lot of control over it, because ultimately the weights are optimal for a given set of covariates, but selecting models that suit one's priors is still possible.

Ferman et al. [2020] addressed this gap in the literature by providing guidance on principled covariate selection. They consider a variety of commonly used synthetic control specifications (e.g., all pre-treatment outcome values, the first three quarters of the pre-treatment outcome values). They then run the randomization inference test to calculate empirical p-values. They find that the probability of falsely rejecting the null in at least one specification for a 5% significance test can be as high as 14% when there are pre-treatment periods. The possibilities for specification searching remain high even when the number of pre-treatment periods is large, too. They consider a sample with four hundred pre-treatment periods and still find a false-positive probability of around 13% that at least one specification was significant at the 5% level. Thus, even with a large number of pre-treatment periods, it is theoretically possible to "hack" the analysis in order to find statistical significance that suits one's priors.

Given the broad discretion available to a researcher to search over countless specifications using covariates and pretreatment combinations, one might conclude that fewer time periods are better if for no other reason than that this limits the ability to conduct an endogenous specification search. Using Monte Carlo simulations, though, they find that models which use more pre-treatment outcome lags as predictors—consistent with statements made by Abadie et al. [2010] originally—do a better job controlling for unobserved confounders, whereas those which limit the number of pre-treatment outcome lags substantially misallocate more weights and should not be considered in synthetic control applications.

Thus, one of the main takeaways of Ferman et al. [2020] is that, despite the hope that synthetic control would remove subjective researcher bias by creating weights based on a data-driven optimization algorithm, this may be somewhat overstated in practice. Whereas it remains true that the weights are optimal in that they uniquely minimize the distance function, the point of Ferman et al. [2020] is to note that the distance function is still, at the end of the day, *endogenously chosen by the researcher*.

So, given this risk of presenting results that may be cherry picked, what should we do? Ferman et al. [2020] suggest presenting multiple results under a variety of commonly specified specifications. If it is regularly robust, the reader may have sufficient information to check this, as opposed to only seeing one specification which may be the cherry-picked result.

Formalization. Let Y_{jt} be the outcome of interest for unit j of $J+1$ aggregate units at time t, and treatment group be $j=1$. The synthetic control estimator models the effect of the intervention at time T_0 on the treatment group using a linear combination of optimally chosen units as a synthetic control. For the post-intervention period, the synthetic control estimator measures the causal effect as $Y_{1t} - \sum_{j=2}^{J+1} w_j^* Y_{jt}$ where w_j^* is a vector of optimally chosen weights.

Matching variables, X_1 and X_0, are chosen as predictors of post-intervention outcomes and must be unaffected by the intervention. The weights are chosen so as to minimize the norm, $||X_1 - X_0 W||$ subject to weight constraints. There are two weight constraints. First, let $W = (w_2, \ldots, w_{J+1})'$ with $w_j \geq 0$ for $j = 2, \ldots, J+1$. Second, let $w_2 + \cdots + w_{J+1} = 1$. In words, no unit receives a negative weight, but can receive a zero weight.[1] And the sum of all weights must equal one.

As I said, Abadie et al. [2010] consider

$$||X_1 - X_0 W|| = \sqrt{(X_1 - X_0 W)' V (X_1 - X_0 W)}$$

where V is some $(k \times k)$ symmetric and positive semidefinite matrix. Let X_{jm} be the value of the mth covariates for unit j. Typically, V is diagonal with main diagonal $v_1, \ldots, v_k$. Then the synthetic control weights minimize:

$$\sum_{m=1}^{k} v_m \left(X_{1m} - \sum_{j=2}^{J+1} w_j X_{jm} \right)^2$$

where v_m is a weight that reflects the relative importance that we assign to the mth variable when we measure the discrepancy between the treated unit and the synthetic control.

---

1 See Doudchenko and Imbens [2016] for work relaxing the non-negativity constraint.

The choice of V, as should be seen by now, is important because W^* depends on one's choice of V. The synthetic control $W^*(V)$ is meant to reproduce the behavior of the outcome variable for the treated unit in the absence of the treatment. Therefore, the weights $v_1, \ldots, v_k$ should reflect the predictive value of the covariates.

Abadie et al. [2010] suggests different choices of V, but ultimately it appears from practice that most people choose a V that minimizes the mean squared prediction error:

$$\sum_{t=1}^{T_0} \left(Y_{1t} - \sum_{j=2}^{J+1} w_j^*(V) Y_{jt} \right)^2$$

What about unobserved factors? Comparative case studies are complicated by unmeasured factors affecting the outcome of interest as well as heterogeneity in the effect of observed and unobserved factors. Abadie et al. [2010] note that if the number of pre-intervention periods in the data is "large," then matching on pre-intervention outcomes can allow us to control for the heterogeneous responses to multiple unobserved factors. The intuition here is that only units that are alike on unobservables and observables would follow a similar trajectory pre-treatment.

California's Proposition 99. Abadie and Gardeazabal [2003] developed the synthetic control estimator so as to evaluate the impact that terrorism had on the Basque region in Spain. But Abadie et al. [2010] expound on the method by using a cigarette tax in California called Proposition 99. Their example uses a placebo-based method for inference, so let's look more closely at their paper.

In 1988, California passed comprehensive tobacco control legislation called Proposition 99. Proposition 99 increased cigarette taxes by \$0.25 a pack, spurred clean-air ordinances throughout the state, funded anti-smoking media campaigns, earmarked tax revenues to health and anti-smoking budgets, and produced more than \$100 million a year in anti-tobacco projects. Other states had similar control programs, and they were dropped from their analysis.

Figure 80 shows changes in cigarette sales for California and the rest of the United States annually from 1970 to 2000. As can be seen, cigarette sales fell after Proposition 99, but as they were already falling,

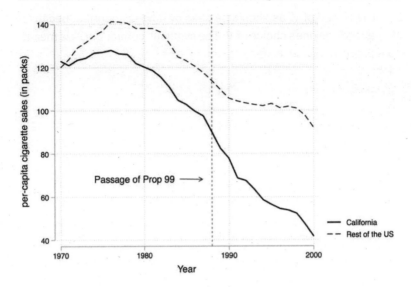

Figure 80. Cigarettes sales in California vs. the rest of the country.

it's not clear if there was any effect—particularly since they were falling in the rest of the country at the same time.

Using their method, though, they select an optimal set of weights that when applied to the rest of the country produces the figure shown in Figure 81. Notice that pre-treatment, this set of weights produces a nearly identical time path for California as the real California itself, but post-treatment the two series diverge. There appears at first glance to have been an effect of the program on cigarette sales.

The variables they used for their distance minimization are listed in Table 82. Notice that this analysis produces values for the treatment group and control group that facilitate a simple investigation of balance. This is not a technical test, as there is only one value per variable per treatment category, but it's the best we can do with this method. And it appears that the variables used for matching are similar across the two groups, particularly for the lagged values.

Like RDD, synthetic control is a picture-intensive estimator. Your estimator is basically a picture of two series which, if there is a causal effect, diverge from another post-treatment, but resemble each other

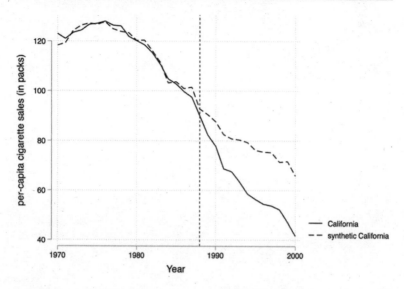

Figure 81. Cigarette sales in California vs. synthetic California.

Table 82. Balance table.

| Variables | California | | Average of |
	Real	Synthetic	38 Control States
Ln(GDP per capita)	10.08	9.86	9.86
Percent aged 15–24	17.40	17.40	17.29
Retail price	89.42	89.41	87.27
Beer consumption per capita	24.28	24.20	23.75
Cigarette sales per capita 1988	90.10	91.62	114.20
Cigarette sales per capita 1980	120.20	120.43	136.58
Cigarette sales per capita 1975	127.10	126.99	132.81

Note: All variables except lagged cigarette sales are averaged for the 1980–1988 period. Beer consumption is averaged 1984–1988.

pre-treatment. It is common to therefore see a picture just showing the difference between the two series (Figure 82).

But so far, we have only covered estimation. How do we determine whether the observed difference between the two series is a *statistically significant* difference? After all, we only have two observations

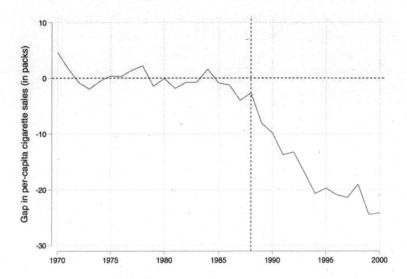

Figure 82. Gap in cigarette sales for estimation pre- and post-treatment.

per year. Maybe the divergence between the two series is nothing more than prediction error, and any model chosen would've done that, even if there was no treatment effect. Abadie et al. [2010] suggest that we use an old-fashioned method to construct exact *p*-values based on Fisher [1935]. Firpo and Possebom [2018] call the null hypothesis used in this test the "no treatment effect whatsoever," which is the most common null used in the literature. Whereas they propose an alternative null for inference, I will focus on the original null proposed by Abadie et al. [2010] for this exercise. As discussed in an earlier chapter, randomization inference assigns the treatment to every untreated unit, recalculates the model's key coefficients, and collects them into a distribution which are then used for inference. Abadie et al. [2010] recommend calculating a set of root mean squared prediction error (RMSPE) values for the pre- and post-treatment period as the test statistic used for inference.[2] We proceed as follows:

---

2 What we will do is simply reassign the treatment to each unit, putting California back into the donor pool each time, estimate the model for that "placebo," and record information from each iteration.

1. Iteratively apply the synthetic control method to each country/state in the donor pool and obtain a distribution of placebo effects.
2. Calculate the RMSPE for each placebo for the pre-treatment period:

$$RMSPE = \left(\frac{1}{T-T_0} \sum_{t=T_0+t}^{T} \left(Y_{1t} - \sum_{j=2}^{J+1} w_j^* Y_{jt} \right)^2 \right)^{\frac{1}{2}}$$

3. Calculate the RMSPE for each placebo for the post-treatment period (similar equation but for the post-treatment period).
4. Compute the ratio of the post- to pre-treatment RMSPE.
5. Sort this ratio in descending order from greatest to highest.
6. Calculate the treatment unit's ratio in the distribution as $p = \frac{RANK}{TOTAL}$.

In other words, what we want to know is whether California's treatment effect is extreme, which is a relative concept compared to the donor pool's own placebo ratios.

There are several different ways to represent this. The first is to overlay California with all the placebos using Stata twoway command, which I'll show later. Figure 83 shows what this looks like. And I think you'll agree, it tells a nice story. Clearly, California is in the tails of some distribution of treatment effects.

Abadie et al. [2010] recommend iteratively dropping the states whose pre-treatment RMSPE is considerably different than California's because as you can see, they're kind of blowing up the scale and making it hard to see what's going on. They do this in several steps, but I'll just skip to the last step (Figure 84). In this, they've dropped any state unit from the graph whose pre-treatment RMSPE is more than two times that of California's. This therefore limits the picture to just units whose model fit, pre-treatment, was pretty good, like California's.

But, ultimately, inference is based on those exact p-values. So the way we do this is we simply create a histogram of the ratios, and more or less mark the treatment group in the distribution so that the reader can see the exact p-value associated with the model. I produce that here in Figure 85.

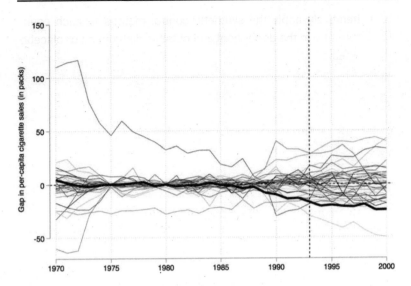

Figure 83. Placebo distribution using all units as donor pool.

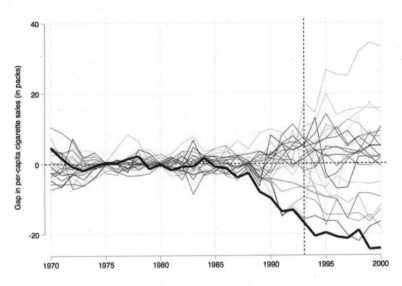

Figure 84. Placebo graphs using only units whose p-treatment RMSPE was no more than two times the size of California's.

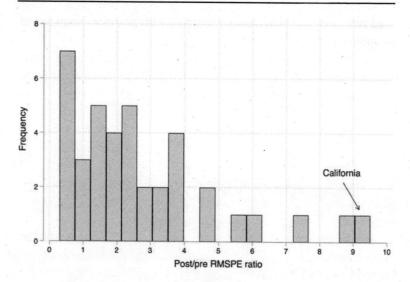

Figure 85. Histogram of post/pre RMSPE of all units.

As can be seen, California is ranked first out of thirty-eight state units.[3] This gives an exact *p*-value of 0.026, which is less than the conventional 5% most journals want to (arbitrarily) see for statistical significance.

Falsifications. In Abadie et al. [2015], the authors studied the effect of the reunification of Germany on gross domestic product. One of the contributions this paper makes, though, is a recommendation for testing the validity of the estimator through a falsification exercise. To illustrate this, let me review the study. The 1990 reunification of Germany brought together East Germany and West Germany, which after years of separation had developed vastly different cultures, economies, and political systems. The authors were interested in evaluating the effect that that reunification had on economic output, but as with the smoking study, they thought that the countries were simply too dissimilar from any one country to make a compelling comparison group, so they

---

3 Recall, they dropped several states who had similar legislation passed over this time period.

used synthetic control to create a composite comparison group based on optimally chosen countries.

One of the things the authors do in this study is provide some guidance as to how to check whether the model you chose is a reasonable one. The authors specifically recommend rewinding time from the date of the treatment itself and estimating their model on an earlier (placebo) date. Since placebo dates should have no effect on output, that provides some assurances that any deviations found in 1990 might be due to structural breaks caused by the reunification itself. And in fact they don't find any effect when using 1975 as a placebo date, suggesting that their model has good in and out of sample predictive properties.

We include this second paper primarily to illustrate that synthetic control methods are increasingly expected to pursue numerous falsification exercises in addition to simply estimating the causal effect itself. In this sense, researchers have pushed others to hold it to the same level of scrutiny and skepticism as they have with other methodologies such as RDD and IV. Authors using synthetic control must do more than merely run the synth command when doing comparative case studies. They must find the exact p-values through placebo-based inference, check for the quality of the pre-treatment fit, investigate the balance of the covariates used for matching, and check for the validity of the model through placebo estimation (e.g., rolling back the treatment date).

Prison Construction and Black Male Incarceration

The project that you'll be replicating here is a project I have been working on with several coauthors over the last few years.[4] Here's the backdrop.

In 1980, the Texas Department of Corrections (TDC) lost a major civil action lawsuit, *Ruiz v. Estelle*; Ruiz was the prisoner who brought the case, and Estelle was the warden. The case argued that TDC was engaging in unconstitutional practices related to overcrowding and

---

4 You can find one example of an unpublished manuscript here coauthored with Sam Kang: http://scunning.com/prison_booms_and_drugs_20.eps.

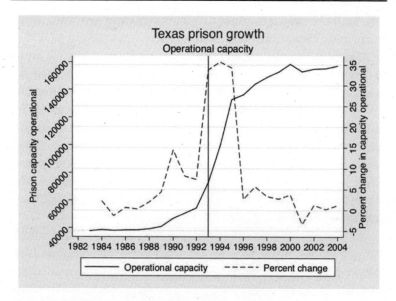

Figure 86. Prison capacity (operational capacity) expansion.

other prison conditions. Texas lost the case, and as a result, was forced to enter into a series of settlements. To amend the issue of overcrowding, the courts placed constraints on the number of inmates who could be placed in cells. To ensure compliance, TDC was put under court supervision until 2003.

Given these constraints, the construction of new prisons was the only way that Texas could keep arresting as many people as its police departments wanted to without having to release those whom the TDC had already imprisoned. If it didn't build more prisons, the state would be forced to increase the number of people to whom it granted parole. That is precisely what happened; following *Ruiz v. Estelle*, Texas used parole more intensively.

But then, in the late 1980s, Texas Governor Bill Clements began building prisons. Later, in 1993, Texas Governor Ann Richards began building even more prisons. Under Richards, state legislators approved $1 billion for prison construction, which would double the state's ability to imprison people within three years. This can be seen in Figure 86.

As can be seen, Clements' prison capacity expansion projects were relatively small compared those by Richards. But Richards's

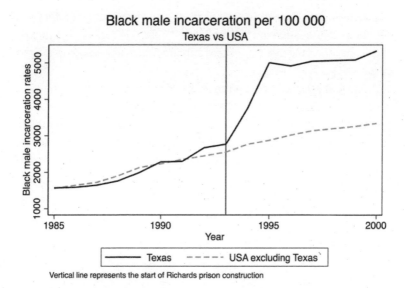

Figure 87. Rates of incarceration of African American men under the State of Texas's prison-building boom.

investments in "operational capacity," or the ability to house prisoners, were gigantic. The number of prison beds available for the state to keep filling with prisoners grew more than 30% for three years, meaning that the number of prison beds more than doubled in just a short period of time.

What was the effect of building so many prisons? Just because prison capacity expands doesn't mean incarceration will grow. But, among other reasons, because the state was intensively using paroles to handle the flow, that's precisely what did happen. The analysis that follows will show the effect of this prison-building boom by the state of Texas on the incarceration of African American men.

As you can see from Figure 87, the incarceration rate of black men doubled in only three years. Texas basically went from being a typical, modal state when it came to incarceration to one of the most severe in only a short period.

What we will now do is analyze the effect that the prison construction under Governor Ann Richards had on the incarceration of black men using synthetic control. The R file is much more straightforward than the synthetic control file, which is broken up into several parts.

I have therefore posted to Github both a texassynth.do file that will run all of this seamlessly, as well as a "Read Me" document to help you understand the directories and subdirectories needed do this. Let's begin.

The first step is to create the figure showing the effect of the 1993 prison construction on the incarceration of black men. I've chosen a set of covariates and pre-treatment outcome variables for the matching; I encourage you, though, to play around with different models. We can already see from Figure 87 that prior to 1993, Texas Black male incarceration were pretty similar to the rest of the country. What this is going to mean for our analysis is that we have every reason to believe that the convex hull likely exists in this application.

STATA
synth_1.do

```
1   cd /users/scott\_cunningham/downloads/texas/do
2   * Estimation 1: Texas model of black male prisoners (per capita)
3   use https://github.com/scunning1975/mixtape/raw/master/texas.dta, clear
4   ssc install synth
5   ssc install mat2txt
6   #delimit;
7   synth      bmprison
8              bmprison(1990) bmprison(1992) bmprison(1991) bmprison(1988)
9              alcohol(1990) aidscapita(1990) aidscapita(1991)
10             income ur poverty black(1990) black(1991) black(1992)
11             perc1519(1990)
12                    ,
13         trunit(48) trperiod(1993) unitnames(state)
14         mspeperiod(1985(1)1993) resultsperiod(1985(1)2000)
15         keep(../data/synth/synth\_bmprate.dta) replace fig;
16         mat list e(V_matrix);
17         #delimit cr
18         graph save Graph ../Figures/synth\_tx.gph, replace}
```

R
synth_1.R

```
1   library(tidyverse)
2   library(haven)
3   library(Synth)
4   library(devtools)
```

(continued)

R *(continued)*

```
5   if(!require(SCtools)) devtools::install_github("bcastanho/SCtools")
6   library(SCtools)
7
8   read_data <- function(df)
9   {
10    full_path <- paste("https://raw.github.com/scunning1975/mixtape/master/",
11              df, sep = "")
12    df <- read_dta(full_path)
13    return(df)
14  }
15
16  texas <- read_data("texas.dta") %>%
17    as.data.frame(.)
18
19  dataprep_out <- dataprep(
20    foo = texas,
21    predictors = c("poverty", "income"),
22    predictors.op = "mean",
23    time.predictors.prior = 1985:1993,
24    special.predictors = list(
25      list("bmprison", c(1988, 1990:1992), "mean"),
26      list("alcohol", 1990, "mean"),
27      list("aidscapita", 1990:1991, "mean"),
28      list("black", 1990:1992, "mean"),
29      list("perc1519", 1990, "mean")),
30    dependent = "bmprison",
31    unit.variable = "statefip",
32    unit.names.variable = "state",
33    time.variable = "year",
34    treatment.identifier = 48,
35    controls.identifier = c(1,2,4:6,8:13,15:42,44:47,49:51,53:56),
36    time.optimize.ssr = 1985:1993,
37    time.plot = 1985:2000
38  )
39
40  synth_out <- synth(data.prep.obj = dataprep_out)
41
42  path.plot(synth_out, dataprep_out)
43
44
45
46
```

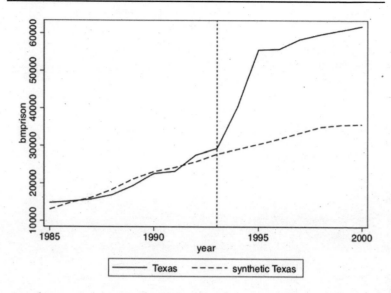

Figure 88. African-American male incarceration.

Regarding the Stata syntax of this file: I personally prefer to make the delimiter a semicolon because I want to have all syntax for synth on the same screen. I'm more of a visual person, so that helps me. Next the synth syntax. The syntax goes like this: call synth, then call the outcome variable (bmprison), then the variables you want to match on. Notice that you can choose either to match on the entire pre-treatment average, or you can choose particular years. I choose both. Also recall that Abadie et al. [2010] notes the importance of controlling for pre-treatment outcomes to soak up the heterogeneity; I do that here as well. Once you've listed your covariates, you use a comma to move to Stata options. You first have to specify the treatment unit. The FIPS code for Texas is a 48, hence the 48. You then specify the treatment period, which is 1993. You list the period of time which will be used to minimize the mean squared prediction error, as well as what years to display. Stata will produce both a figure as well as a data set with information used to create the figure. It will also list the V matrix. Finally, I change the delimiter back to carriage return, and save the figure in the /Figures subdirectory. Let's look at what these lines made (Figure 88).

STATA
synth_2.do

```
1   * Plot the gap in predicted error
2   use ../data/synth/synth_bmprate.dta, clear
3   keep _Y_treated _Y_synthetic _time
4   drop if _time==.
5   rename _time year
6   rename _Y_treated treat
7   rename _Y_synthetic counterfact
8   gen gap48=treat-counterfact
9   sort year
10  #delimit ;
11  twoway (line gap48 year,lp(solid)lw(vthin)lcolor(black)), yline(0,
    ↪  lpattern(shortdash) lcolor(black))
12      xline(1993, lpattern(shortdash) lcolor(black)) xtitle("",si(medsmall))
    ↪      xlabel(#10)
13      ytitle("Gap in black male prisoner prediction error", size(medsmall))
    ↪      legend(off);
14      #delimit cr
15      save ../data/synth/synth_bmprate_48.dta, replace}
```

R
synth_2.R

```
1   gaps.plot(synth_out, dataprep_out)
```

This is the kind of outcome that you ideally want to have specifically, a very similar pre-treatment trend in the synthetic Texas group compared to the actual Texas group, and a divergence in the post-treatment period. We will now plot the gap between these two lines using our programming commands in the accompanying code.

The figure that this makes is shown in Figure 89. It is essentially nothing more than the "gap" between Texas and synthetic Texas values, per year, in Figure 88.

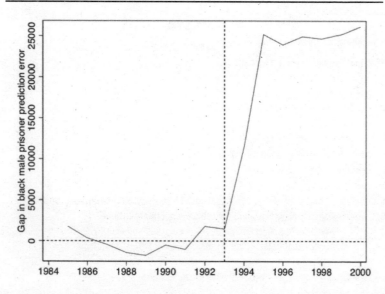

Figure 89. Gap between actual Texas and synthetic Texas.

Table 83. Synthetic control weights.

State name	Weight
California	0.408
Florida	0.109
Illinois	0.36
Louisiana	0.122

And finally, we will show the weights used to construct the synthetic Texas.

Now that we have our estimates of the causal effect, we move into the calculation of the exact *p*-value, which will be based on assigning the treatment to every state and reestimating our model. Texas will always be thrown back into the donor pool each time. This next part will contain multiple Stata programs, but because of the efficiency of the R package, I will only produce one R program. So all exposition henceforth will focus on the Stata commands.

STATA
synth_3.do

```
1   * Inference 1 placebo test
2   #delimit;
3   set more off;
4   use ../data/texas.dta, replace;
5   local statelist  1 2 4 5 6 8 9 10 11 12 13 15 16 17 18 20 21 22 23 24 25 26 27 28
      ↪   29 30 31 32
6       33 34 35 36 37 38 39 40 41 42 45 46 47 48 49 51 53 55;
7   foreach i of local statelist {;
8   synth       bmprison
9             bmprison(1990) bmprison(1992) bmprison(1991) bmprison(1988)
10            alcohol(1990) aidscapita(1990) aidscapita(1991)
11            income ur poverty black(1990) black(1991) black(1992)
12            perc1519(1990)
13               ,
14               trunit(`i') trperiod(1993) unitnames(state)
15               mspeperiod(1985(1)1993) resultsperiod(1985(1)2000)
16               keep(../data/synth/synth\_bmprate\_`i'.dta) replace;
17               matrix state`i' = e(RMSPE); /* check the V matrix*/
18
19  foreach i of local statelist {;
20  matrix rownames state`i'=`i';
21  matlist state`i', names(rows);
22  };
23  #delimit cr
```

R
synth_3_7.R

```
1
2   placebos <- generate.placebos(dataprep_out, synth_out, Sigf.ipop = 3)
3
4   plot_placebos(placebos)
5
6   mspe.plot(placebos, discard.extreme = TRUE, mspe.limit = 1, plot.hist = TRUE)
```

This is a loop in which it will cycle through every state and estimate the model. It will then save data associated with each model into the ../data/synth/synth_bmcrate_'i'.dta data file where 'i' is one of the state FIPS codes listed after local statelist. Now that we have each of these files, we can calculate the post-to-pre RMSPE.

STATA
synth_4.do

```
1   local statelist  1 2 4 5 6 8 9 10 11 12 13 15 16 17 18 20 21 22 23 24 25 26 27 28
    ↪  29 30 31 32
2        33 34 35 36 37 38 39 40 41 42 45 46 47 48 49 51 53 55
3   foreach i of local statelist {
4        use ../data/synth/synth_bmprate_`i' ,clear
5        keep _Y_treated _Y_synthetic _time
6        drop if _time==.
7        rename _time year
8        rename _Y_treated  treat`i'
9        rename _Y_synthetic counterfact`i'
10       gen gap`i'=treat`i'-counterfact`i'
11       sort year
12       save ../data/synth/synth_gap_bmprate`i', replace
13       }
14  use ../data/synth/synth_gap_bmprate48.dta, clear
15  sort year
16  save ../data/synth/placebo_bmprate48.dta, replace
17
18  foreach i of local statelist {
19            merge year using ../data/synth/synth_gap_bmprate`i'
20            drop _merge
21            sort year
22       save ../data/synth/placebo_bmprate.dta, replace
23       }
24
```

Notice that this is going to first create the gap between the treatment state and the counterfactual state before merging each of them into a single data file.

STATA
synth_5.do

```
1   ** Inference 2: Estimate the pre- and post-RMSPE and calculate the ratio of the
2   * post-pre RMSPE
3   set more off
4   local statelist  1 2 4 5 6 8 9 10 11 12 13 15 16 17 18 20 21 22 23 24 25 26 27 28
     ↪   29 30 31 32
5       33 34 35 36 37 38 39 40 41 42 45 46 47 48 49 51 53 55
6   foreach i of local statelist {
7
8       use ../data/synth/synth_gap_bmprate`i', clear
9       gen gap3=gap`i'*gap`i'
10      egen postmean=mean(gap3) if year>1993
11      egen premean=mean(gap3) if year<=1993
12      gen rmspe=sqrt(premean) if year<=1993
13      replace rmspe=sqrt(postmean) if year>1993
14      gen ratio=rmspe/rmspe[_n-1] if 1994
15      gen rmspe_post=sqrt(postmean) if year>1993
16      gen rmspe_pre=rmspe[_n-1] if 1994
17      mkmat rmspe_pre rmspe_post ratio if 1994, matrix (state`i')
```

In this part, we are calculating the post-RMSPE, the pre-RMSPE, and the ratio of the two. Once we have this information, we can compute a histogram. The following commands do that.

STATA
synth_6.do

```
1   * show post/pre-expansion RMSPE ratio for all states, generate histogram
2       foreach i of local statelist {
3           matrix rownames state`i'=`i'
4           matlist state`i', names(rows)
5                                       }
6   #delimit ;
7   matstate=state1/state2/state4/state5/state6/state8/state9/state10/state11/
8   state12/state13/state15/state16/state17/state18/state20/state21/state22/
9   state23/state24/state25/state26/state27/state28/state29/state30/state31/
10  state32/state33/state34/state35/state36/state37/state38/state39/state40/
11  state41/state42/state45/state46/state47/state48/state49/state51/state53/
12  state55;
13  #delimit cr
```

(continued)

STATA *(continued)*
14   * ssc install mat2txt
15   mat2txt, matrix(state) saving(../inference/rmspe_bmprate.txt) replace
16   insheet using ../inference/rmspe_bmprate.txt, clear
17   ren v1 state
18   drop v5
19   gsort -ratio
20   gen rank=_n
21   gen p=rank/46
22   export excel using ../inference/rmspe_bmprate, firstrow(variables) replace
23   import excel ../inference/rmspe_bmprate.xls, sheet("Sheet1") firstrow clear
24   histogram ratio, bin(20) frequency fcolor(gs13) lcolor(black) ylabel(0(2)6)
25   xtitle(Post/pre RMSPE ratio) xlabel(0(1)5)
26   * Show the post/pre RMSPE ratio for all states, generate the histogram.
27   list rank p if state==48

All the looping will take a few moments to run, but once it is done, it will produce a histogram of the distribution of ratios of post-RMSPE to pre-RMSPE. As you can see from the *p*-value, Texas has the second-highest ratio out of forty-six state units, giving it a *p*-value of 0.04. We can see that in Figure 90.

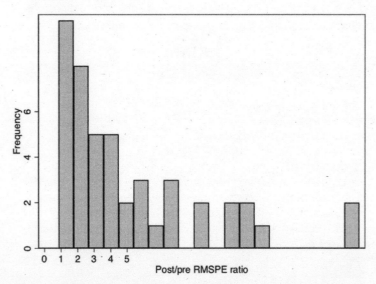

Figure 90. Histogram of the distribution of ratios of post-RMSPE to pre-RMSPE. Texas is one of the ones in the far right tail.

Notice that in addition to the figure, this created an Excel spreadsheet containing information on the pre-RMSPE, the post-RMSPE, the ratio, and the rank. We will want to use that again when we limit our display next to states whose pre-RMSPE are similar to that of Texas.

Now we want to create the characteristic placebo graph where all the state placebos are laid on top of Texas. To do that, we can use some simple syntax contained in the Stata code:

STATA

synth_7.do

```
1   * Inference 3: all the placeboes on the same picture
2   use ../data/synth/placebo_bmprate.dta, replace
3   * Picture of the full sample, including outlier RSMPE
4   #delimit;
5   twoway
6   (line gap1 year ,lp(solid)lw(vthin))
7   (line gap2 year ,lp(solid)lw(vthin))
8   (line gap4 year ,lp(solid)lw(vthin))
9   (line gap5 year ,lp(solid)lw(vthin))
10  (line gap6 year ,lp(solid)lw(vthin))
11  (line gap8 year ,lp(solid)lw(vthin))
12  (line gap9 year ,lp(solid)lw(vthin))
13  (line gap10 year ,lp(solid)lw(vthin))
14  (line gap11 year ,lp(solid)lw(vthin))
15  (line gap12 year ,lp(solid)lw(vthin))
16  (line gap13 year ,lp(solid)lw(vthin))
17  (line gap15 year ,lp(solid)lw(vthin))
18  (line gap16 year ,lp(solid)lw(vthin))
19  (line gap17 year ,lp(solid)lw(vthin))
20  (line gap18 year ,lp(solid)lw(vthin))
21  (line gap20 year ,lp(solid)lw(vthin))
22  (line gap21 year ,lp(solid)lw(vthin))
23  (line gap22 year ,lp(solid)lw(vthin))
24  (line gap23 year ,lp(solid)lw(vthin))
25  (line gap24 year ,lp(solid)lw(vthin))
26  (line gap25 year ,lp(solid)lw(vthin))
27  (line gap26 year ,lp(solid)lw(vthin))
28  (line gap27 year ,lp(solid)lw(vthin))
29  (line gap28 year ,lp(solid)lw(vthin))
30  (line gap29 year ,lp(solid)lw(vthin))
```

(continued)

	STATA (continued)
31	(line gap30 year ,lp(solid)lw(vthin))
32	(line gap31 year ,lp(solid)lw(vthin))
33	(line gap32 year ,lp(solid)lw(vthin))
34	(line gap33 year ,lp(solid)lw(vthin))
35	(line gap34 year ,lp(solid)lw(vthin))
36	(line gap35 year ,lp(solid)lw(vthin))
37	(line gap36 year ,lp(solid)lw(vthin))
38	(line gap37 year ,lp(solid)lw(vthin))
39	(line gap38 year ,lp(solid)lw(vthin))
40	(line gap39 year ,lp(solid)lw(vthin))
41	(line gap40 year ,lp(solid)lw(vthin))
42	(line gap41 year ,lp(solid)lw(vthin))
43	(line gap42 year ,lp(solid)lw(vthin))
44	(line gap45 year ,lp(solid)lw(vthin))
45	(line gap46 year ,lp(solid)lw(vthin))
46	(line gap47 year ,lp(solid)lw(vthin))
47	(line gap49 year ,lp(solid)lw(vthin))
48	(line gap51 year ,lp(solid)lw(vthin))
49	(line gap53 year ,lp(solid)lw(vthin))
50	(line gap55 year ,lp(solid)lw(vthin))
51	(line gap48 year ,lp(solid)lw(thick)lcolor(black)), /*treatment unit, Texas*/
52	yline(0, lpattern(shortdash) lcolor(black)) xline(1993, lpattern(shortdash)
	↪ lcolor(black))
53	xtitle("",si(small)) xlabel(#10) ytitle("Gap in black male prisoners prediction
	↪ error", size(small))
54	legend(off);
55	#delimit cr

Here we will only display the main picture with the placebos, though one could show several cuts of the data in which you drop states whose pre-treatment fit compared to Texas is rather poor.

Now that you have seen how to use this .do file to estimate a synthetic control model, you are ready to play around with the data yourself. All of this analysis so far has used black male (total counts) incarceration as the dependent variable, but perhaps the results would be different if we used black male incarceration. That information is contained in the data set. I would like for you to do your own analysis using the black male incarceration rate variable as the dependent variable. You will need to find a new model to fit this pattern, as it's unlikely

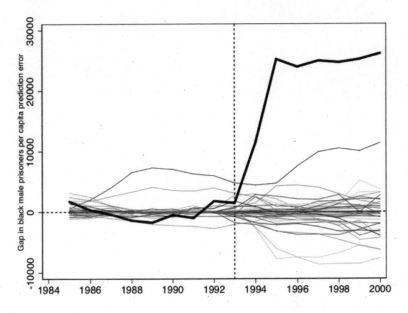

Figure 91. Placebo distribution. Texas is the black line.

that the one we used for levels will do as good a job describing rates as it did levels. In addition, you should implement the placebo-date falsification exercise that we mentioned from Abadie et al. [2015]. Choose 1989 as your treatment date and 1992 as the end of the sample, and check whether the same model shows the same treatment effect as you found when you used the correct year, 1993, as the treatment date. I encourage you to use these data and this file to learn the ins and outs of the procedure itself, as well as to think more deeply about what synthetic control is doing and how to best use it in research.

Conclusion. In conclusion, we have seen how to estimate synthetic control models in Stata. This model is currently an active area of research, and so I have decided to wait until subsequent editions to dive into the new material as many questions are unsettled. This chapter therefore is a good foundation for understanding the model and the practices around implementing it, including code in *R* and Stata to do so. I hope that you find this useful.

Conclusion

Causal inference is an important and fun area. It's fun because the potential outcomes model is both an intuitive and a philosophically stimulating way to think about causal effects. The model has also proved quite powerful at helping us better understand the assumptions needed to identify causal effects using exotic quasi-experimental research designs beyond the randomized control trial. Pearl's directed acyclic graphical models are also helpful for moving between a theoretical model and an understanding of some phenomenon and a strategy to identify the causal effect you care about. From those DAGs, you can learn whether it's even possible to design such an identification strategy with your data set. And although that can be disappointing, it is a disciplined and truthful approach to estimation. These DAGs are, in my experience, empowering, extremely useful for the design phase of a project, and adored by students.

The methods I've outlined are merely some of the most common research designs currently employed in applied microeconomics. I have tried to selectively navigate the research to bring readers as close to the frontier as possible. But I had to leave some things out. For instance, there is nothing on bounding or partial identification. But perhaps if you love this book enough, there can be a second edition that includes that important topic.

Even for the topics I did cover, these areas are constantly changing, and I encourage you to read many of the articles provided in the bibliography to learn more. I also encourage you just to use the links provided in the software code throughout this book and download the data files yourself. Play around with the programs, explore the data, and improve your own intuition on how to use R and Stata to tackle causal inference problems using these designs. I hope you found this book valuable. Good luck in your research. I wish you all the best.

Bibliography

Abadie, A., Athey, S., Imbens, G. W., and Wooldridge, J. M. (2020). Sampling-based versus design-based uncertainty in regression analysis. *Econometrica*, 88:265–296.

Abadie, A., Diamond, A., and Hainmueller, J. (2010). Synthetic control methods for comparative case studies: Estimating the effect of California's tobacco control program. *Journal of the American Statistical Association*, 105(490):493–505.

Abadie, A., Diamond, A., and Hainmueller, J. (2015). Comparative politics and the synthetic control method. *American Journal of Political Science*, 59(2):495–510.

Abadie, A. and Gardeazabal, J. (2003). The economic costs of conflict: A case study of the Basque Country. *American Economic Review*, 93(1):113–132.

Abadie, A. and Imbens, G. (2006). Large sample properties of matching estimators for average treatment effects. *Econometrica*, 74(1):235–267.

Abadie, A. and Imbens, G. (2011). Bias-corrected matching estimators for average treatment effects. *Journal of Business and Economic Statistics*, 29:1–11.

Abadie, A. and Imbens, G. W. (2008). On the failure of the bootstrap for matching estimators. *Econometrica*, 76(6):1537–1557.

Adudumilli, K. (2018). Bootstrap inference for propensity score matching. Unpublished manuscript.

Aizer, A. and Doyle, J. J. (2015). Juvenile incarceration, human capital, and future crime: Evidence from randomly assigned judges. *Quarterly Journal of Economics*, 130(2):759–803.

Almond, D., Doyle, J. J., Kowalski, A., and Williams, H. (2010). Estimating returns to medical care: Evidence from at-risk newborns. *Quarterly Journal of Economics*, 125(2):591–634.

Anderson, D. M., Hansen, B., and Rees, D. I. (2013). Medical marijuana laws, traffic fatalities, and alcohol consumption. *Journal of Law and Economics*, 56(2):333–369.

Angrist, J. D. (1990). Lifetime earnings and the Vietnam era draft lottery: Evidence from Social Security administrative records. *American Economic Review*, 80(3):313–336.

Angrist, J. D., Imbens, G. W., and Krueger, A. B. (1999). Jackknife instrumental variables estimation. *Journal of Applied Econometrics*, 14:57–67.

Angrist, J. D., Imbens, G. W., and Rubin, D. B. (1996). Identification of causal effects using instrumental variables. *Journal of the American Statistical Association*, 87:328–336.

Angrist, J. D. and Krueger, A. B. (1991). Does compulsory school attendance affect schooling and earnings? *Quarterly Journal of Economics*, 106(4):979–1014.

Angrist, J. D. and Krueger, A. B. (2001). Instrumental variables and the search for identification: From supply and demand to natural experiments. *Journal of Economic Perspectives*, 15(4):69–85.

Angrist, J. D. and Lavy, V. (1999). Using Maimonides' rule to estimate the effect of class size on scholastic achievement. *Quarterly Journal of Economics*, 114(2):533–575.

Angrist, J. D. and Pischke, J.-S. (2009). *Mostly Harmless Econometrics*. Princeton University.

Arnold, D., Dobbie, W., and Yang, C. S. (2018). Racial bias in bail decisions. *Quarterly Journal of Economics*, 133(4):1885–1932.

Ashenfelter, O. (1978). Estimating the effect of training programs on earnings. *Review of Economics and Statistics*, 60:47–57.

Athey, S., Bayati, M., Doudchenko, N., Imbens, G. W., and Khosravi, K. (2018). Matrix completion methods for causal panel data models. arXiv No. 1710.10251.

Athey, S. and Imbens, G. W. (2017a). *The Econometrics of Randomized Experiments*, 1:73–140. Elsevier.

Athey, S. and Imbens, G. W. (2017b). The state of applied econometrics: Causality and policy evaluation. *Journal of Economic Perspectives*, 31(2):3–32.

Auld, M. C. and Grootendorst, P. (2004). An empirical analysis of milk addiction. *Journal of Health Economics*, 23(6):1117–1133.

Baicker, K., Taubman, S. L., Allen, H. L., Bernstein, M., Gruber, J., Newhouse, J., Schneider, E., Wright, B., Zaslavsky, A., and Finkelstein, A. (2013). The Oregon experiment—effects of Medicaid on clinical outcomes. *New England Journal of Medicine*, 368:1713–1722.

Band, H. and Robins, J. M. (2005). Doubly robust estimation in missing data and causal inference models. *Biometrics*, 61:962–972.

Barnow, B. S., Cain, G. G., and Goldberger, A. (1981). Selection on observables. *Evaluation Studies Review Annual*, 5:43–59.

Barreca, A. I., Guldi, M., Lindo, J. M., and Waddell, G. R. (2011). Saving babies? Revisiting the effect of very low birth weight classification. *Quarterly Journal of Economics*, 126(4):2117–2123.

Barreca, A. I., Lindo, J. M., and Waddell, G. R. (2016). Heaping-induced bias in regression-discontinuity designs. *Economic Inquiry*, 54(1):268–293.

Bartik, T. J. (1991). *Who Benefits from State and Local Economic Development Policies?* W. E. Upjohn Institute for Employment Research, Kalamazoo, MI.

Becker, G. (1968). Crime and punishment: An economic approach. *Journal of Political Economy*, 76:169–217.

Becker, G. (1994). *Human Capital: A Theoretical and Empirical Analysis with Special Reference to Education*. 3rd ed. University of Chicago Press.

Becker, G. S. (1993). The economic way of looking at life. *Journal of Political Economy*, 101(3):385–409.

Becker, G. S., Grossman, M., and Murphy, K. M. (2006). The market for illegal gods: The case of drugs. *Journal of Political Economy*, 114(1):38–60.

Becker, G. S. and Murphy, K. M. (1988). A theory of rational addiction. *Journal of Political Economy*, 96(4):675–700.

Beland, L. P. (2015). Political parties and labor-market outcomes: Evidence from US states. *American Economic Journal: Applied Economics*, 7(4):198–220.

Bertrand, M., Duflo, E., and Mullainathan, S. (2004). How much should we trust differences-in-differences estimates? *Quarterly Journal of Economics*, 119(1):249–275.

Binder, J. (1998). The event study methodology since 1969. *Review of Quantitative Finance and Accounting*, 11:111–137.

Black, S. E. (1999). Do better schools matter? Parental valuation of elementary education. *Quarterly Journal of Economics*, 114(2):577–599.

Blanchard, O. J. and Katz, L. F. (1992). Regional evolutions. *Brookings Papers on Economic Activity*, 1:1–75.

Bodory, H., Camponovo, L., Huber, M., and Lechner, M. (2020). The finite sample performance of inference methods for propensity score matching and weighting estimators. *Journal of Business and Economic Statistics*, 38(1):183–200.

Borusyak, K., Hull, P., and Jaravel, X. (2019). Quasi-experimental shift-share research designs. Unpublished manuscript.

Borusyak, K. and Jaravel, X. (2018). Revisiting event study designs. Unpublished manuscript.

Bound, J., Jaeger, D. A., and Baker, R. M. (1995). Problems with instrumental variables estimation when the correlation between the instruments and the endogenous explanatory variable is weak. *Journal of the American Statistical Association*, 90(430):443–450.

Brooks, J. M. and Ohsfeldt, R. L. (2013). Squeezing the balloon: Propensity scores and unmeasured covariate balance. *Health Services Research*, 48(4):1487–1507.

Buchanan, J. (1996). Letter to the editor. *Wall Street Journal*.

Buchmueller, T. C., DiNardo, J., and Valletta, R. G. (2011). The effect of an employer health insurance mandate on health insurance coverage and the demand for labor: Evidence from Hawaii. *American Economic Journal: Economic Policy*, 3(4):25–51.

Buckles, K. and Hungerman, D. (2013). Season of birth and later outcomes: Old questions, new answers. *Review of Economics and Statistics*, 95(3):711–724.

Busso, M., DiNardo, J., and McCrary, J. (2014). New evidence on the finite sample properties of propensity score reweighting and matching estimators. *Review of Economics and Statistics*, 96(5):885–897.

Callaway, B. and Sant'Anna, P. H. C. (2019). Difference-in-differences with multiple time periods. Unpublished manuscript.

Calonico, S., Cattaneo, M. D., and Titiunik, R. (2014). Robust nonparametric confidence intervals for regression-discontinuity designs. *Econometrica*, 82(6):2295–2326.

Cameron, A. C., Gelbach, J. B., and Miller, D. L. (2008). Bootstrap-based improvements for inference with clustered errors. *Review of Economics and Statistics*, 90(3):414–427.

Cameron, A. C., Gelbach, J. B., and Miller, D. L. (2011). Robust inference with multiway clustering. *Journal of Business and Economic Statistics*, 29(2):238–249.

Card, D. (1990). The impact of the Mariel boatlift on the Miami labor-market. *Industrial and Labor Relations Review*, 43(2):245–257.

Card, D. (1995). *Aspects of Labour Economics: Essays in Honour of John Vanderkamp*. University of Toronto Press.

Card, D., Dobkin, C., and Maestas, N. (2008). The impact of nearly universal insurance coverage on health care utilization: Evidence from medicare. *American Economic Review*, 98(5):2242–2258.

Card, D., Dobkin, C., and Maestas, N. (2009). Does Medicare save lives? *Quarterly Journal of Economics*, 124(2):597–636.

Card, D. and Krueger, A. (1994). Minimum wages and employment: A case study of the fast-food industry in New Jersey and Pennsylvania. *American Economic Review*, 84:772–793.

Card, D., Lee, D. S., Pei, Z., and Weber, A. (2015). Inference on causal effects in a generalized regression kink design. *Econometrica*, 84(6):2453–2483.

Card, D. and Peri, G. (2016). Immigration economics: A review. Unpublished manuscript.

Carpenter, C. and Dobkin, C. (2009). The effect of alcohol consumption on mortality: Regression discontinuity evidence from the minimum drinking age. *American Economic Journal: Applied Economics*, 1(1):164–182.

Carrell, S. E., Hoekstra, M., and West, J. E. (2011). Does drinking impair college performance? Evidence from a regression discontinuity approach. *Journal of Public Economics*, 95:54–62.

Carrell, S. E., Hoekstra, M., and West, J. E. (2019). The impact of college diversity on behavior toward minorities. *American Economic Journal: Economic Policy*, 11(4):159–182.

Cattaneo, M. D., Frandsen, B. R., and Titiunik, R. (2015). Randomization inference in the regression discontinuity design: An application to party advantages in the U.S. Senate. *Journal of Causal Inference*, 3(1):1–24.

Cattaneo, M. D., Jansson, M., and Ma, X. (2019). Simply local polynomial density estimators. *Journal of the American Statistical Association*.

Caughey, D. and Sekhon, J. S. (2011). Elections and the regression discontinuity design: Lessons from close U.S. House races, 1942–2008. *Political Analysis*, 19:385–408.

Cengiz, D., Dube, A., Lindner, A., and Zipperer, B. (2019). The effect of minimum wages on low-wage jobs. *Quarterly Journal of Economics*, 134(3):1405–1454.

Charles, K. K. and Stephens, M. (2006). Abortion legalization and adolescent substance use. *Journal of Law and Economics*, 49:481–505.

Cheng, C. and Hoekstra, M. (2013). Does strengthening self-defense law deter crime or escalate violence? Evidence from expansions to castle doctrine. *Journal of Human Resources*, 48(3):821–854.

Christakis, N. A. and Fowler, J. H. (2007). The spread of obesity in a large social network over 32 years. *New England Journal of Medicine*, 357(4):370–379.

Cochran, W. G. (1968). The effectiveness of adjustment by subclassification in removing bias in observational studies. *Biometrics*, 24(2):295–313.

Cohen-Cole, E. and Fletcher, J. (2008). Deteching implausible social network effects in acne, height, and headaches: Longitudinal analysis. *British Medical Journal*, 337(a2533).

Coleman, T. S. (2019). Causality in the time of cholera: John Snow as a prototype for causal inference. Unpublished manuscript.

Coles, P. (2019). Einstein, Eddington, and the 1919 eclipse. *Nature*, 568(7752):306–307.

Conley, D. and Fletcher, J. (2017). *The Genome Factor: What the Social Genomics Revolution Reveals about Ourselves, Our History, and the Future*. Princeton University Press.

Cook, T. D. (2008). "Waiting for life to arrive": A history of the regression-discontinuity design in psychology, statistics, and economics. *Journal of Econometrics*, 142:636–654.

Cornwell, C. and Rupert, P. (1997). Unobservable individual effects, marriage, and the earnings of young men. *Economic Inquiry*, 35(2):1–8.

Cornwell, C. and Trumbull, W. N. (1994). Estimating the economic model of crime with panel data. *Review of Economics and Statistics*, 76(2):360–366.

Craig, M. (2006). *The Professor, the Banker, and the Suicide King: Inside the Richest Poker Game of All Time*. Grand Central Publishing.

Crump, R. K., Hotz, V. J., Imbens, G. W., and Mitnik, O. A. (2009). Dealing with limited overlap in estimation of average treatment effects. *Biometrika*, 96(1):187–199.

Cunningham, S. and Cornwell, C. (2013). The long-run effect of abortion on sexually transmitted infections. *American Law and Economics Review*, 15(1):381–407.

Cunningham, S. and Finlay, K. (2012). Parental substance abuse and foster care: Evidence from two methamphetamine supply shocks. *Economic Inquiry*, 51(1):764–782.

Cunningham, S. and Kendall, T. D. (2011). Prostitution 2.0: The changing face of sex work. *Journal of Urban Economics*, 69:273–287.

Cunningham, S. and Kendall, T. D. (2014). *Examining the Role of Client Reviews and Reputation within Online Prostitution*. Oxford University Press.

Cunningham, S. and Kendall, T. D. (2016). Prostitution labor supply and education. *Review of Economics of the Household*, Forthcoming.

Dale, S. B. and Krueger, A. B. (2002). Estimating the payoff to attending a more selective college: An application of selection on observables and unobservables. *Quarterly Journal of Economics*, 117(4):1491–1527.

de Chaisemartin, C. and D'Haultfœuille, X. (2019). Two-way fixed effects estimators with heterogeneous treatment effects. Unpublished manuscript.

Dehejia, R. H. and Wahba, S. (1999). Causal effects in nonexperimental studies: Reevaluating the evaluation of training programs. *Journal of the American Statistical Association*, 94(448):1053–1062.

Dehejia, R. H. and Wahba, S. (2002). Propensity score-matching methods for nonexperimental causal studies. *Review of Economics and Statistics*, 84(1):151–161.

Dobbie, W., Goldin, J., and Yang, C. S. (2018). The effects of pretrial detention on conviction, future crime, and employment: Evidence from randomly assigned judges. *American Economic Review*, 108(2):201–240.

Dobbie, W., Goldsmith-Pinkham, P., and Yang, C. (2017). Consumer bankruptcy and financial health. *Review of Economics and Statistics*, 99(5):853–869.

Dobkin, C. and Nicosia, N. (2009). The war on drugs: Methamphetamine, public health, and crime. *American Economic Review*, 99(1):324–349.

Donald, S. G. and Newey, W. K. (2001). Choosing the number of instruments. *Econometrica*, 69(5):1161–1191.

Donohue, J. J. and Levitt, S. D. (2001). The impact of legalized abortion on crime. *Quarterly Journal of Economics*, 116(2):379–420.

Doudchenko, N. and Imbens, G. (2016). Balancing, regression, difference-in-differences, and synthetic control methods: A synthesis. Working Paper No. 22791, National Bureau of Economic Research, Cambridge, MA.

Doyle, J. J. (2007). Child protection and adult crime: Using investigator assignment to estimate causal effects of foster care. Unpublished manuscript.

Doyle, J. J. (2008). Child protection and child outcomes: Measuring the effects of foster care. *American Economic Review*, 97(5):1583–1610.

Dube, A., Lester, T. W., and Reich, M. (2010). Minimum wage effects across state borders: Estimates using contiguous counties. *Review of Economics and Statistics*, 92(4):945–964.

Efron, B. (1979). Bootstrap methods: Another look at the jackknife. *Annals of Statistics*, 7(1):1–26.

Eggers, A. C., Fowler, A., Hainmueller, J., Hall, A. B., and Jr., J. M. S. (2014). On the validity of the regression discontinuity design for estimating electoral effects: New evidence from over 40,000 close races. *American Journal of Political Science*, 59(1):259–274.

Elwert, F. and Winship, C. (2014). Endogenous selection bias: The problem of conditioning on a collider variable. *Annual Review of Sociology*, 40:31–53.

Ferman, B. and Pinto, C. (2019). Synthetic controls with imperfect pre-treatment fit. Unpublished manuscript.

Ferman, B., Pinto, C., and Possebom, V. (2020). Cherry picking with synthetic controls. *Journal of Policy Analysis and Management*, 39(2):510–532.

Finkelstein, A., Taubman, S., Wright, B., Bernstein, M., Gruber, J., Newhouse, J. P., Allen, H., and Baicker, K. (2012). The Oregon health insurance experiment: Evidence from the first year. *Quarterly Journal of Economics*, 127(3):1057–1106.

Firpo, S. and Possebom, V. (2018). Synthetic control method: Inference, sensitivity analysis, and confidence sets. *Journal of Causal Inference*, 6(2):1–26.

Fisher, R. A. (1925). *Statistical Methods for Research Workers*. Oliver and Boyd.

Fisher, R. A. (1935). *The Design of Experiments*. Oliver and Boyd.

Foote, C. L. and Goetz, C. F. (2008). The impact of legalized abortion on crime: Comment. *Quarterly Journal of Economics*, 123(1):407–423.

Frandsen, B. R., Lefgren, L. J., and Leslie, E. C. (2019).   Judging judge fixed effects. Working Paper No. 25528, National Bureau of Economic Research, Cambridge, MA.

Freedman, D. A. (1991). Statistical models and shoe leather. *Sociological Methodology*, 21:291–313.

Freeman, R. B. (1980).   An empirical analysis of the fixed coefficient "manpower requirement" mode, 1960–1970. *Journal of Human Resources*, 15(2):176–199.

Frisch, R. and Waugh, F. V. (1933). Partial time regressions as compared with individuals trends. *Econometrica*, 1(4):387–401.

Fryer, R. (2019). An empirical analysis of racial differences in police use of force. *Journal of Political Economy*, 127(3).

Gaudet, F. J., Harris, G. S., and St. John, C. W. (1933).   Individual differences in the sentencing tendencies of judges.   *Journal of Criminal Law and Criminology*, 23(5):811–818.

Gauss, C. F. (1809). *Theoria Motus Corporum Coelestium*. Perthes et Besser, Hamburg.

Gelman, A. and Imbens, G. (2019).   Why higher-order polynomials should not be used in regression discontinuity designs. *Journal of Business and Economic Statistics*, 37(3):447–456.

Gertler, P., Shah, M., and Bertozzi, S. M. (2005).   Risky business: The market for unprotected commercial sex. *Journal of Political Economy*, 113(3):518–550.

Gilchrist, D. S. and Sands, E. G. (2016).   Something to talk about: Social spillovers in movie consumption. *Journal of Political Economy*, 124(5):1339–1382.

Goldberger, A. S. (1972).   Selection bias in evaluating treatment effects: Some formal illustrations. Unpublished manuscript.

Goldsmith-Pinkham, P. and Imbens, G. W. (2013). Social networks and the identification of peer effects. *Journal of Business and Economic Statistics*, 31(3).

Goldsmith-Pinkham, P., Sorkin, I., and Swift, H. (2020). Bartik instruments: What, when, why, and how. *American Economic Review*, Forthcoming. Working Paper No. 24408, National Bureau of Economic Research, Cambridge, MA.

Goodman-Bacon, A. (2019). Difference-in-differences with variation in treatment timing. Unpublished manuscript.

Graddy, K. (2006). The Fulton Fish Market. *Journal of Economic Perspectives*, 20(2):207–220.

Gruber, J. (1994).   The incidence of mandated maternity benefits. *American Economic Review*, 84(3):622–641.

Gruber, J., Levine, P. B., and Staiger, D. (1999).   Abortion legalization and child living circumstances: Who is the "marginal child"? *Quarterly Journal of Economics*, 114(1):263–291.

Haavelmo, T. (1943). The statistical implications of a system of simultaneous equations. *Econometrica*, 11(1):1–12.

Hahn, J., Todd, P., and van der Klaauw, W. (2001). Identification and estimation of treatment effects with a regression-discontinuity design. *Econometrica*, 69(1):201–209.

Hájek, J. (1971). *Comment on "An Essay on the Logical Foundations of Survey Sampling, Part One."* Holt, Rinehart and Winston.

Hamermesh, D. S. and Biddle, J. E. (1994). Beauty and the labor-market. *American Economic Review*, 84(5):1174–1194.

Hansen, B. (2015). Punishment and deterrence: Evidence from drunk driving. *American Economic Review*, 105(4):1581–1617.

Heckman, J. and Pinto, R. (2015). Causal analysis after Haavelmo. *Econometric Theory*, 31(1):115–151.

Heckman, J. J. (1979). Sample selection bias as a specification error. *Econometrica*, 47(1):153–161.

Heckman, J. J. and Vytlacil, E. J. (2007). *Econometric Evaluation of Social Programs, Part I: Causal Models, Structural Models, and Econometric Policy Evaluation*, 6B:4779–4874. Elsevier.

Hirano, K. and Imbens, G. W. (2001). Estimation of causal effects using propensity score weighting: An application to data on right heart catheterization. *Health Services and Outcomes Research Methodology*, 2:259–278.

Hoekstra, M. (2009). The effect of attending the flagship state university on earnings: A discontinuity-based approach. *Review of Economics and Statistics*, 91(4):717–724.

Holtzman, W. H. (1950). The unbiased estimate of the population variance and standard deviation. *American Journal of Psychology*, 63(4):615–617.

Hong, S. H. (2013). Measuring the effect of Napster on recorded music sales: Difference-in-differences estimates under compositional changes. *Journal of Applied Econometrics*, 28(2):297–324.

Hooke, R. (1983). *How to Tell the Liars from the Statisticians*. CRC Press.

Horvitz, D. G. and Thompson, D. J. (1952). A generalization of sampling without replacement from a finite universe. *Journal of the American Statistical Association*, 47(260):663–685.

Hume, D. (1993). *An Enquiry Concerning Human Understanding: With Hume's Abstract of A Treatise of Human Nature and A Letter from a Gentleman to His Friend in Edinburgh*. 2nd ed. Hackett Publishing.

Iacus, S. M., King, G., and Porro, G. (2012). Causal inference without balance checking: Coarsened exact matching. *Political Analysis*, 20(1):1–24.

Imai, K. and Kim, I. S. (2017). When should we use fixed effects regression models for causal inference with longitudinal data? Unpublished manuscript.

Imai, K. and Ratkovic, M. (2013). Covariate balancing propensity score. *Journal of the Royal Statistical Society Statistical Methodology Series B*, 76(1):243–263.

Imbens, G. and Kalyanaraman, K. (2011). Optimal bandwidth choice for the regression discontinuity estimator. *Review of Economic Studies*, 79(3):933–959.

Imbens, G. W. (2000). The role of the propensity score in estimating dose-response functions. *Biometrika*, 87(3):706–710.

Imbens, G. W. (2019). Potential outcome and directed acyclic graph approaches to causality: Relevance for empirical practices in economics. Unpublished manuscript.

Imbens, G. W. and Angrist, J. D. (1994). Identification and estimation of local average treatment effects. *Econometrica*, 62(2):467–475.

Imbens, G. W. and Lemieux, T. (2008). Regression discontinuity designs: A guide to practice. *Journal of Econometrics*, 142:615–635.

Imbens, G. W. and Rubin, D. B. (2015). *Causal Inference for Statistics, Social and Biomedical Sciences: An Introduction*. Cambridge University Press.

Jacob, B. A. and Lefgen, L. (2004). Remedial education and student achivement: A regression-discontinuity analysis. *Review of Economics and Statistics*, 86(1):226–244.

Joyce, T. (2004). Did legalized abortion lower crime? *Journal of Human Resources*, 39(1):1–28.

Joyce, T. (2009). A simple test of abortion and crime. *Review of Economics and Statistics*, 91(1):112–123.

Juhn, C., Murphy, K. M., and Pierce, B. (1993). Wage inequality and the rise in returns to skill. *Journal of Political Economy*, 101(3):410–442.

Kahn-Lang, A. and Lang, K. (2019). The promise and pitfalls of differences-in-differences: Reflections on 16 and pregnant and other applications. *Journal of Business and Economic Statistics*, DOI: 10.1080/07350015.2018.1546591.

King, G. and Nielsen, R. (2019). Why propensity scores should not be used for matching. *Political Analysis*, 27(4).

King, G. and Zeng, L. (2006). The dangers of extreme counterfactuals. *Political Analysis*, 14(2):131–159.

Kling, J. R. (2006). Incarceration length, employment, and earnings. *American Economic Review*, 96(3):863–876.

Knox, D., Lowe, W., and Mummolo, J. (2020). Administrative records mask racially biased policing. *American Political Science Review*, Forthcoming.

Kofoed, M. S. and McGovney, E. (2019). The effect of same-gender and same-race role models on occupation choice: Evidence from randomly assigned mentors at West Point. *Journal of Human Resources*, 54(2).

Kolesár, M. and Rothe, C. (2018). Inference in regression discontinuity designs with a discrete running variable. *American Economic Review*, 108(8):2277–2304.

Krueger, A. (1999). Experimental estimates of education production functions. *Quarterly Journal of Economics*, 114(2):497–532.

Lalonde, R. (1986). Evaluating the econometric evaluations of training programs with experimental data. *American Economic Review*, 76(4):604–620.

Lee, D. S. and Card, D. (2008). Regression discontinuity inference with specification error. *Journal of Econometrics*, 142(2):655–674.

Lee, D. S. and Lemieux, T. (2010). Regresion discontinuity designs in economics. *Journal of Economic Literature*, 48:281–355.

Lee, D. S., Moretti, E., and Butler, M. J. (2004). Do voters affect or elect policies? Evidence from the U.S. House. *Quarterly Journal of Economics*, 119(3):807–859.

Leslie, E. and Pope, N. G. (2018). The unintended impact of pretrial detention on case outcomes: Evidence from New York City arraignments. *Journal of Law and Economics*, 60(3):529–557.

Levine, P. B. (2004). *Sex and Consequences: Abortion, Public Policy, and the Economics of Fertility*. Princeton University Press.

Levine, P. B., Staiger, D., Kane, T. J., and Zimmerman, D. J. (1999). *Roe v. Wade* and American fertility. *American Journal of Public Health*, 89(2):199–203.

Levitt, S. D. (2004). Understanding why crime fell in the 1990s: Four factors that explain the decline and six that do not. *Journal of Economic Perspectives*, 18(1):163–190.

Lewis, D. (1973). Causation. *Journal of Philosophy*, 70(17):556–567.

Lindo, J., Myers, C., Schlosser, A., and Cunningham, S. (2019). How far is too far? New evidence on abortion clinic closures, access, and abortions. *Journal of Human Resources*, forthcoming.

Lott, J. R. and Mustard, D. B. (1997). Crime, deterrence, and the right-to-carry concealed handguns. *Journal of Legal Studies*, 26:1–68.

Lovell, M. C. (1963). Seasonal adjustment of economic time series and multiple regression analysis. *Journal of the American Statistical Association*, 58(304):991–1010.

Lovell, M. C. (2008). A simple proof of the FWL theorem. *Journal of Economic Education*, 39(1):88–91.

Lyle, D. S. (2009). The effects of peer group heterogeneity on the production of human capital at West Point. *American Economic Journal: Applied Economics*, 1(4):69–84.

MacKinnon, J. G. and Webb, M. D. (2017). Wild bootstrap inference for wildly different cluster sizes. *Journal of Applied Econometrics*, 32(2):233–254.

Manski, C. F. (1993). Identification of endogenous social effects: The reflection problem. *Review of Economic Studies*, 60:531–542.

Manski, C. F. and Pepper, J. V. (2018). How do right-to-carry laws affect crime rates? Coping with ambiguity using bounded-variation assumptions. *Review of Economics and Statistics*, 100(2):232–244.

Matsueda, R. L. (2012). *Handbook of Structural Equation Modeling*. Guilford Press.

McCrary, J. (2008). Manipulation of the running variable in the regression discontinuity design: A design test. *Journal of Econometrics*, 142:698–714.

McGrayne, S. B. (2012). *The Theory That Would Not Die: How Bayes' Rule Cracked the Enigma Code, Hunted Down Russian Submarines, and Emerged Triumphant from Two Centuries of Controversy*. Yale University Press.

Meer, J. and West, J. (2016). Effects of the minimum wage on employment dynamics. *Journal of Human Resources*, 51(2):500–522.

Melberg, H. O. (2008). Rational addiction theory: A survey of opinions. Unpublished manuscript.

Mill, J. S. (2010). *A System of Logic, Ratiocinative and Inductive*. FQ Books.

Miller, S., Altekruse, S., Johnson, N., and Wherry, L. R. (2019). Medicaid and mortality: New evidence from linked survey and administrative data. NBER Working Paper 26081.

Millimet, D. L. and Tchernis, R. (2009). On the specification of propensity scores, with applications to the analysis of trade policies. *Journal of Business and Economic Statistics*, 27(3):397–415.

Morgan, M. S. (1991). *The History of Econometric Ideas*. 2nd ed. Cambridge University Press.

Morgan, S. L. and Winship, C. (2014). *Counterfactuals and Causal Inference: Methods and Principles for Social Research*. 2nd ed. Cambridge University Press.

Mueller-Smith, M. (2015). The criminal and labor-market impacts of incarceration. Unpublished manuscript.

Needleman, M. and Needleman, C. (1969). Marx and the problem of causation. *Science and Society*, 33(3):322–339.

Neumark, D., Salas, J. I., and Wascher, W. (2014). Revisting the minimum wage-employment debate: Throwing out the baby with the bathwater? *Industrial and Labor Relations Review*, 67(2.5):608–648.

Newhouse, J. P. (1993). *Free for All? Lessons from the RAND Health Experiment*. Harvard University Press.

Norris, S., Pecenco, M., and Weaver, J. (2020). The effects of parental and sibling incarceration: Evidence from Ohio. Unpublished manuscript.

Pearl, J. (2009). *Causality*. 2nd ed. Cambridge University Press.

Peirce, C. S. and Jastrow, J. (1885). On small differences in sensation. *Memoirs of the National Academy of Sciences*, 3:73–83.

Peri, G. and Yasenov, V. (2018). The labor-market effects of a refugee wave: Synthetic control method meets the Mariel boatlift. *Journal of Human Resources*, doi: 10.3368/jhr.54.2.0217.8561R1.

Perkinson, R. (2010). *Texas Tough: The Rise of America's Prison Empire*. Picador.

Perloff, H. S. (1957). Interrelations of state income and industrial structure. *Review of Economics and Statistics*, 39(2):162–171.

Piazza, J. (2009). Megan Fox voted worst—but sexiest—actress of 2009. https://marquee.blogs.cnn.com/2009/12/30/megan-fox-voted-worst-but-sexiest-actress-of-2009/.

Powell, D. (2017). Imperfect synthetic controls: Did the Massachusetts health care reform save lives? Unpublished manuscript.

Rilke, R. M. (1929). *Letters to a Young Poet*. Merchant Books.

Rogeberg, O. (2004). Taking absurd theories seriously: Economics and the case of rational addiction theories. *Philosophy of Science*, 71:263–285.

Rosen, S. (1986). *Handbook of Labor Economics*. Vol. 1. North Holland.

Rosenbaum, P. R. and Rubin, D. B. (1983). The central role of the propensity score in observational studies for causal effects. *Biometrika*, 70(1):41–55.

Rubin, D. (1974). Estimating causal effects of treatments in randominzed and nonrandomized studies. *Journal of Educational Psychology*, 66(5):688–701.

Rubin, D. B. (1977). Assignment to treatment group on the basis of a covariate. *Journal of Educational Statistics*, 2:1–26.

Rubin, D. B. (2005). Causal inference using potential outcomes: Design, modeling, decisions. *Journal of the American Statistical Association*, 100(469):322–331.

Rubin, D. B. (2007). The design versus the analysis of observational studies for causal effects: Parallels with the design of randomized trials. *Statistics in Medicine*, 26(1):20–36.

Rubin, D. B. (2008). For objective causal inference, design trumps analysis. *Annals of Applied Statistics*, 2(3):808–840.

Sacerdote, B. (2001). Peer effects with random assignment: Results for Dartmouth roommates. *Quarterly Journal of Economics*: 681–704.

Sant'Anna, P. H. C. and Zhao, J. B. (2018). Doubly robust difference-in-differences estimators. Unpublished manuscript.

Sharpe, J. (2019). Re-evaluating the impact of immigration on the U.S. rental housing market. *Journal of Urban Economics*, 111(C):14–34.

Smith, A. (1776). *An Inquiry into the Nature and Causes of the Wealth of Nations*. Bantam Classics.

Smith, J. A. and Todd, P. E. (2001). Reconciling conflicting evidence on the performance on propensity-score matching methods. *American Economic Review*, 91(2):112–118.

Smith, J. A. and Todd, P. E. (2005). Does matching overcome LaLonde's critique of nonexperimental estimators? *Journal of Econometrics*, 125(1–2):305–353.

Snow, J. (1855). *On the Mode of Communication of Cholera*. 2nd ed. John Churchill.

Splawa-Neyman, J. (1923). On the application of probability theory to agricultural experiments: Essay on principles. *Annals of Agricultural Sciences*: 1–51.

Staiger, D. and Stock, J. H. (1997). Instrumental variables regression with weak instruments. *Econometrica*, 65(3):557–586.

Steiner, P. M., Kim, Y., Hall, C. E., and Su, D. (2017). Graphical models for quasi-experimental designs. *Sociological Methods and Research*, 46(2):155–188.

Stevenson, M. T. (2018). Distortion of justice: How the inability to pay bail affects case outcomes. *Journal of Law, Economics, and Organization*, 34(4):511–542.

Stigler, S. M. (1980). Stigler's law of eponymy. *Transactions of the New York Academy of Sciences*, 39:147–158.

Stock, J. H. and Trebbi, F. (2003). Who invented instrumental variable regression? *Journal of Economic Perspectives*, 17(3):177–194.

Sun, L. and Abraham, S. (2020). Estimating dynamic treatment effects in event studies with heterogeneous treatment effects. Unpublished manuscript.

Thistlehwaite, D. and Campbell, D. (1960). Regression-discontinuity analysis: An alternative to the ex post facto experiment. *Journal of Educational Psychology*, 51:309–317.

Thornton, R. L. (2008). The demand for, and impact of, learning HIV status. *American Economic Review*, 98(5):1829–1863.

Van der Klaauw, W. (2002). Estimating the effect of financial aid offers on college enrollment: A regression-discontinuity approach. *International Economic Review*, 43(4):1249–1287.

Waldfogel, J. (1995). The selection hypotehsis and the relationship between trial and plaintiff victory. *Journal of Political Economy*, 103(2):229–260.

White, H. (1980). A heteroskedasticity-consistent covariance matrix estimator and a direct test for heteroskedasticity. *Econometrica*, 48(4):817–838.

Wolpin, K. I. (2013). *The Limits of Inference without Theory*. MIT Press.

Wooldridge, J. (2010). *Econometric Analysis of Cross Section and Panel Data*. 2nd ed. MIT Press.

Wooldridge, J. (2015). *Introductory Econometrics: A Modern Approach*. 6th ed. South-Western College Pub.

Wright, P. G. (1928). *The Tariff on Animal and Vegetable Oils*. Macmillan.

Young, A. (2019). Chanelling Fisher: Randomization tests and the statistical insignificance of seemingly significant experimental results. *Quarterly Journal of Economics*, 134(2):557–598.

Yule, G. U. (1899). An investigation into the causes of changes in pauperism in England, chiefly during the last two intercensal decades. *Journal of Royal Statistical Society*, 62:249–295.

Zhao, Q. (2019). Covariate balancing propensity score by tailored loss functions. *Annals of Statistics*, 47(2):965–993.

Zubizarreta, J. R. (2015). Stable weights that balance covariates for estimation with incomplete outcome data. *Journal of the American Statistical Association*, 110(511):910–922.

Permissions

MATHEMATICS

Words and Music by Dante Smith, David Gates and Christopher Martin
Copyright © 1999 Sony/ATV Music Publishing LLC, EMI Black-wood Music Inc., Empire International, Medina Sound Music, EMI April Music Inc. and Gifted Pearl Music
All Rights on behalf of Sony/ATV Music Publishing LLC, EMI Blackwood Music Inc., Empire International and Medina Sound Music Sony/ATV Music Publishing LLC, 424 Church Street, Suite 1200, Nashville, TN 37219
All Rights on behalf of Gifted Pearl Music Administered by World-wide by Kobalt Songs Music Publishing
International Copyright Secured All Rights Reserved
Reprinted by Permission of Hal Leonard LLC

MINORITY REPORT

Words and Music by Shawn Carter, Shaffer Smith, Mark Batson, Andre Young and Dawaun Parker
© 2006 EMI APRIL MUSIC INC., CARTER BOYS MUSIC, UNI-VERSAL MUSIC—Z SONGS, SUPER SAYIN PUBLISHING, BAT FUTURE MUSIC, KOBALT MUSIC COPYRIGHTS SARL, WC MUSIC CORP., AIN'T NOTHING BUT FUNKIN' MUSIC, WARNER-TAMERLANE PUBLISHING CORP., PSALM 144 VERSE 1 MUSIC and ALIEN STATUS MUSIC
All Rights for CARTER BOYS MUSIC Controlled and Administered by EMI APRIL MUSIC INC.
All Rights for SUPER SAYIN PUBLISHING Controlled and Adminis-tered by UNIVERSAL MUSIC—Z SONGS

MO' MONEY, MO' PROBLEMS

JUMP AROUND

Words and Music by Larry E. Muggerud, Erik Schrody, Bob Relf, Earl Nelson, Fred Smith, Jimmy Webb, Charlie Cronander, Kal Mann, and Dave Appell

ULTRALIGHT BEAM

Words and Music by Cydel Young, Kasseem Dean, Malik Jones, Kanye West, Jerome Potter, Sam Griesemer, Kirk Franklin, Kelly Price, Noah Goldstein, Nico Segal, Mike Dean, Terius Nash, Derek Watkins and Chancelor Bennett

CHANGES

Words and Music by Tupac Shakur, Deon Evans and Bruce Hornsby

WHAT'S THE DIFFERENCE

Words and Music by Melvin Bradford, Alvin Joiner, Marshall Mathers, Richard Bembery, Charles Aznavour and Stephon Cunningham

   © 1999 MELVIN BRADFORD MUSIC, HARD WORKING BLACK FOLKS, INC., WC MUSIC CORP. and COPUBLISHERS

   All Rights for MELVIN BRADFORD MUSIC and HARD WORKING BLACK FOLKS, INC. Administered by WC MUSIC CORP.

   WHAT'S THE DIFFERENCE interpolates PARCE QUE TU CROIS by Charles Aznavour Editions Musicales Djanik (SACEM) for the World except Pompidou Music (ASCAP) for USA. All Rights Reserved . International Copyright Secured.

Index